LAND-USE/TRANSPORT PLANNING IN HONG KONG:
THE END OF AN ERA

Land-use/Transport Planning in Hong Kong: The End of an Era

A review of principles and practices

Edited by
HARRY T. DIMITRIOU
ALISON H.S. COOK

Routledge
Taylor & Francis Group
LONDON AND NEW YORK

First published 1998 by Ashgate Publishing

Reissued 2018 by Routledge
2 Park Square, Milton Park, Abingdon, Oxon, OX14 4RN
52 Vanderbilt Avenue, New York, NY 10017

Routledge is an imprint of the Taylor & Francis Group, an informa business

A Library of Congress record exists under LC control number: 98071406

ISBN 13: 978-1-138-36193-5 (hbk)
ISBN 13: 978-0-429-43237-8 (ebk)

Contents

List of Figures

Part One: Historical developments

Part Two: Current issues and problems

Editors' Overview

Chapter 4

Chapter 5

Chapter 6

Part Three: Policy and planning responses

Part Four: New directions

Chapter 11

Chapter 12

List of Tables

Part One: Historical developments

Part Two: Current issues and problems

Part Three: Policy and planning responses

Part Four: New directions

Foreword

This book is certain to be the definitive work about the practice of land-use and transport planning in Hong Kong for this century, and gains greatly from the balanced input of practitioners and interpreters, and some who are both. It shows very clearly where and how the sectoral interfaces have sometimes worked, sometimes broken down. But it also shows the steady determination of planners and others in Hong Kong to raise the aspirations of public policy even while contending with some very intractable current pressures. When Lord (then Sir Murray) Maclehose announced the Ten Year Housing Programme which in so many ways reoriented the process of development in Hong Kong in 1972 there was a target of one million people to house. Ten years later this target was in sight: but there were another one million people to house, and a knowledge that the easier options were long exhausted, or in some cases discredited.

There is a particular fascination in the sections of this book which look forward. Even in the first years of this decade Hong Kong civil servants had maps on their office walls which showed a broad expanse of white paper north of Sha Tau Kok, even though for the previous ten years we had all become increasingly conscious of the interaction between China and Hong Kong. Planning in Hong Kong must now contend with two primary imperatives: first, the need to achieve that greater integration of aspects of sectoral policy within a spatial context so as to tackle the domestic issues; and second, the need to see the strategic setting for that effort as the social and economic geography of southern China, with which it must mesh as the chief engine of growth in that very large region. The facets of those two tasks are described clearly in this book, and should strike chords for planners in many other parts of the world: we should have no doubt that the problems and the pressures will continue to come thicker and faster in Hong Kong than elsewhere, and that the lessons from Hong Kong will have an important degree of universal relevance.

Robert W. Upton
Secretary General, Royal Town Planning Institute, London
Former Director of Planning, Hong Kong Government (1989-1991)

Preface

Hong Kong has experienced phenomenal economic growth and physical transformation as the 'Gateway to China' over the years, particularly during the 1980s and 1990s. This has enabled the territory to finance numerous studies, real estate developments and mega transport infrastructure projects on such an extensive scale that it has become the envy of the region. So much so, that many of the experiences of transport infrastructure and land-use development in Hong Kong have now become models for the development of other cities in the region, particularly China.

A process of the 'Hong Kong-isation' of parts of the Asia Pacific Region has resulted whereby high-rise developments, land leasing systems and transport infrastructure projects, together with associated urban life styles and values, are very apparent in metropolitan areas. Such developments have been effectively disseminated by Hong Kong-based media networks, real estate interests and consultancy and construction activities and are strongly supported by Hong Kong-based investment capital and financial institutions. The successes and failures of Hong Kong's plans and projects, particularly with regard to their environmental and economic implications, have thus now taken on a significance that goes well beyond the territory's boundaries.

Because of the dominant role played by an efficient and financially viable public transport system, the low (although fast-rising) private motor vehicle ownership, and the compact and high density urban structure, many of Hong Kong's citizens enjoy high levels of accessibility, in marked contrast to those of other major cities in the region. These characteristics have led many international land-use and transport planning experts to suggest that Hong Kong has a great deal to offer in the global search for sustainable urban development and transport strategies.

In part, the purpose of this book is to establish the extent to which this is the case. The publication was embarked upon in 1994 to provide an informed account and critical review of the territory's major land-use/transport planning exercises and their environmental implications, up to mid 1996. The last chapter contributions were received before the hand-over of Hong Kong to the People's Republic of China and final editorial work on the manuscript was completed following the transition from British Colonial rule to Chinese sovereignty. The book thus documents an era of Hong Kong which has now drawn to an end.

Given the fast pace of on-going infrastructure development that has taken place since the completion of this book, the editors do not claim that the publication provides a completely up-to-date coverage of current developments. Articles in professional journals more appropriately do this. Rather, the book seeks to record the development of the land-use and transport planning system, and for the past two decades in particular, to offer a unique insight into how the planning exercises were prepared and how decisions emerged during this very economically active period of Hong Kong's history.

The book is expected to be of interest to students, academics, professionals and government officials concerned with the built environment, and to those in the private sector involved in transport, urban development and environmental protection. The fact that many of the contributors personally worked on the plans and projects cited in the publication also makes it potentially useful as an historical source.

Harry T. Dimitriou,
Aalborg, Denmark
and
Alison H.S. Cook,
Hong Kong, PRC

List of Contributors

Harry T. DIMITRIOU Dip.TP, MSc, PhD, MCIT, MRTPI

Dr. Dimitriou is Professor of Planning Studies at The Bartlett School, University College London. He was Research Professor at the Department of Development and Planning at Aalborg University in Denmark between 1996 and 1998 when he was Project Manager of the Research Programme on Large Scale Transport Infrastructure Investments (LASTIN) in Europe funded by the Danish Transport Council, and sometime consultant to the World Bank, UNCHS, UNDP and ESCAP. He was Reader in Transport Policy and Urban Development at the University of Hong Kong where he taught between 1986 and 1996, and a past planning consultant to the Hong Kong Land Development Corporation (LDC) and the Hong Kong Government. Prior to taking up his position at Hong Kong University, Dr. Dimitriou was Director of Transport and Development Consultants (TDC) S.A. (Lausanne, Switzerland) working in Indonesia; lecturer at the Development Planning Unit, University College London; consultant to Doxiadis Associates (Athens, Greece) working on projects in West Africa and the Middle East; lecturer in the Department of Town Planning at Sheffield University and Planning Officer in Coventry Corporation, England. He is the author/editor of several books and of numerous journal articles and international conference papers.

Alison H.S. COOK BA, MURP, MHKIP, MRAPI, MCIT

Ms Cook is an Assistant Professor in the Centre of Urban Planning and Environmental Management at The University of Hong Kong, specialising in applied planning and research on metropolitan structure. She is also a Director of Urban Connections Limited, a Hong Kong based urban planning consultancy. In the latter capacity she has participated in planning studies in Hong Kong and in Fujian, China. In 1992-93 she was involved in a research project in the Department of Geography and Geology of The University of Hong Kong on transport and land use in Hong Kong and Asia-Pacific rim cities. Prior to coming to Hong Kong Ms Cook was employed by Dwyer Leslie Pty Ltd, an Australian economics and urban and regional planning consultancy. In the 1970s she worked for the Australian government in the Departments of Urban and Regional Development and of Housing. Ms Cook has also undertaken a short-term consultancy for the Urban Affairs Programme of the Organisation for Economic Cooperation and Development (OECD), Paris.

Colin J. ARNOTT BA, M.Phil
Mr Arnott is a partner of Roger Tym and Partners and directs their Hong Kong office. He is an economist and strategic planner with extensive international experience in regional planning and economic development strategies, institutional organisation, and financial and economic appraisal. In Hong Kong he has acted as institutional adviser on the preparation of Development Statements for the metropolitan area and on a series of studies for the industrial, office, retail and tourism sectors of the territorial economy. Mr Arnott recently completed an Urban Sector Review of China for the Asian Development Bank and the Ministry of Construction in Beijing. He has also worked on other projects in China and in Malaysia, Singapore, Macau, Japan and Thailand.

William F. BARRON BA, MA, PhD
Dr Barron is an Associate Professor in the Centre of Urban Planning and Environmental Management, the University of Hong Kong. He specialises in energy and environmental work and is Editor of the Asian Journal of Environmental Management. Since 1985 he has served as a resource person for the Energy Programme of the Asian and Pacific Development Centre, and since 1996 as a resource person for the Economy and Environment Programme for Southeast Asia. This work has involved training sessions in Vietnam, China, Thailand and Malaysia. Recently he has also worked in Hainan and Yunnan in China on an energy-environmental planning assessment. Before coming to Hong Kong, Dr Barron worked for the Oak Ridge National Laboratory (ORNL) on energy planning in the United States and in developing countries; as an energy advisor to the Pakistan and Liberian governments; and as a natural resources advisor to the U.S. Agency for International Development in Washington.

Alexander R. CUTHBERT DipArch, DipTT, PhD, MSc, RIBA, MRTPI
Professor Cuthbert is currently Head of the School of Town Planning at the University of New South Wales, Australia. He has over 25 years teaching, professional, and research experience, gained from working in Britain, Greece, the United States, Hong Kong and Australia. He undertook his doctoral studies at the London School of Economics and has academic qualifications in architecture, urban planning, and urban design. During the ten years to 1992 he worked in Hong Kong where he maintained both research and consultancy activity as well as teaching at the University of Hong Kong, enjoying a joint appointment in the Faculty of Architecture and the then Centre of Urban Studies and Urban Planning.

Alan W.Y. KAM BSc, MSc, C.Eng, MICE, MHKIE, MCIT
Mr Kam is currently the Assistant Director heading the Major Works Project Management Office of the Highways Department. From 1991-1995 he was the Chief

Town Planner in charge of Transport Studies and Central Data Section of the Planning Department. Before joining the Planning Department he was Chief Engineer of the Territory Transport Planning Division of the Transport Department and directed the Second Comprehensive Transport Study (1986-1989). The Second White Paper on Transport Policy in Hong Kong (1990) was based on the findings of this study. From 1982 to 1985 Mr Kam was the Project Engineer of the Land-use/Transport Optimisation (LUTO) Study and was responsible for the development and application of the LUTO model which was used in the formulation of a long-term territorial development strategy for Hong Kong. He was also a member of the study team which undertook the First Comprehensive Transport Study (1973-1976). Mr Kam has presented papers on transport planning and integrated land-use/transport planning to various seminars and conferences.

Ernest S.W. LEE Dip.URP, BA, MSc, PhD, FCIT, MRAPI, MHKIP
Dr Lee is Assistant Commissioner for Transport in the Transport Department. He is also an honorary lecturer at the Centre of Urban Planning and Environmental Management, and in the Department of Geography and Geology, at the University of Hong Kong. Trained as a town planner and transport planner, Dr Lee is a registered professional planner in Hong Kong. His professional career has focused on transport planning and administration in which he has been closely involved in public transport planning, licensing of vehicles and drivers, and management of road tunnels and of parking facilities.

Peter F. LEEDS ISO, ACIS, FCIT
Mr Leeds was employed by the Hong Kong government from 1951 until his retirement in 1987. He was Commissioner for Transport from 1982 to 1987 and in this capacity was a government representative on the Transport Advisory Committee and on the Boards of Directors of the major bus and ferry companies. Mr Leeds was involved in the re-organisation of the Transport Department in 1981 and in major revisions of the legislation affecting public transport and road traffic. Prior to joining the Transport Department in 1971 he served in a number of government departments including Resettlement and Housing.

Kelvin C.W. LEUNG MSc, C.Eng, MICE, MHKIE, DMS
Mr Leung is a Director of MVA Hong Kong Ltd., transportation, planning and management consultants. He has extensive experience in transportation planning, traffic engineering and environmental assessment studies. Mr Leung has been responsible for studies for many Hong Kong infrastructure projects including North Lantau Expressway, Shing Mun Tunnel, Tate's Cairn Tunnel, Route 3, Route 7, Yuen Long Southern By-pass, Green Island Reclamation and Tin Shui Wai new town development. Throughout the last 16 years he has contributed to many projects ranging from planning, feasibility studies, design and implementation, both in consultancy firms and in the Hong Kong government. Mr Leung is currently the representative of the Hong Kong Institution of Engineers on the Asia Pacific Intelligent Transport Systems Co-ordination Committee.

Richard T. MEAKIN LLB, MSc, MSocSc, FCIT

Since 1993 Mr Meakin has been an urban transport consultant and is currently working on a transport study in Bangkok. He has worked in ten developing countries with a variety of clients including the World Bank. He was previously Assistant Commissioner for Transport with the Hong Kong government and had been closely involved with the planning and development of the territory's public transport system since moving to Hong Kong in 1974. The early years of his career were spent in the United Kingdom with the National Bus Company. Mr Meakin is a past Chairman of the Chartered Institute of Transport in Hong Kong.

Edward G. PRYOR MBE, PhD, BA, Dip.TP, MNZPI, MHKIP

Dr Pryor's association with planning and development in Hong Kong dates back to 1965. He has been employed by the Hong Kong government in various planning positions related to housing, economic, environmental and transport issues. Dr Pryor was Principal Government Town Planner in the Planning Department until his retirement in early 1998. Since 1984 he has been responsible for co-ordinating the production of a number of studies of major long-term significance to the Territory. These include Hong Kong's Port and Airport Development Strategy, Metroplan, and the Territorial Development Strategy Review. From 1980 to 1984, as project manager for the new town of Tsuen Wan - Kwai Tsing, he co-ordinated a number of major development projects. Dr Pryor also led the team which produced Hong Kong's first Territorial Development Strategy in the early 1980s. One of Dr Pryor's main endeavours has been in the formulation and implementation of major housing development programmes, which provided the basis for a book 'Housing in Hong Kong' published by Oxford University Press in 1983. Dr Pryor has wide-ranging interests in various academic fields and is a regular contributor to both local and overseas planning journals, conferences and exhibitions, from which a new series of books on urban design in Hong Kong have been produced.

Barry D. ROBB BSc, C.Eng, MICE

Mr Robb worked for the Hong Kong government from 1978 to 1997. He was Senior Engineer of the Transport Studies Unit in the Planning Department with responsibility for all transport planning aspects of the Territorial Development Strategy and Metroplan, both of which provide the structural context for planning and programming of development in Hong Kong. Earlier in his career Mr Robb specialised in physical development planning for the Hong Kong New Town Development Programme.

Timothy V. RUNNACLES MA, MSc, Dip.TP, DIC, FCIT, FRGS, MTA, FPWI

Mr Runnacles is an independent transport consultant with over twenty five years experience in transport planning and operation. In recent years he has advised the Mass Transit Railway Corporation on the operational design of the intermodal station interchanges of the Airport Railway, as well as undertaking assignments in South Africa, Kazakstan and Malaysia. For fifteen years he worked in the Hong Kong government's Transport Department and Transport Branch, in both operational and planning capacities, including a long involvement with the Tuen Mun Light Rail

Scheme as well as with other major projects. Before coming to Hong Kong in 1977 Mr Runnacles worked for London Transport, and as a consultant in the United Kingdom.

Edward C.T. SO MCIT
Mr So has been working for the Mass Transit Railway Corporation for twenty years on the transport planning of the operating network and new extensions to it. His work has included market research, passenger revenue forecasts, evaluation of pricing and ticketing policies, and recommendations for service improvements. Prior to joining the Corporation he was employed by the Public Works Department and where he undertook traffic analyses. He was involved with the development, calibration and application of traffic forecasting models for the first Comprehensive Transport Study.

John SPIERS BSc (Hons), MSc, MIHT, MCIT
Mr Spiers, a director of MVA Asia and MVA Malaysia, is a transportation engineer with a wide range of experience acquired both in the U.K. and Asia. He specialises in design and implementation studies and has been involved in traffic and transport studies in Hong Kong, Singapore, Malaysia, Korea, Thailand, China, and India. Mr Spiers also has substantial network planning and operations experience and was project manager of the World Bank funded Comprehensive Bus Study in Madras. In Hong Kong he has managed the traffic and transportation aspects of the Western District Traffic Study, the Engineering Feasibility Investigation for the Mid-Levels, the Wanchai Traffic Study, the Western Harbour Crossing Study and the Green Island Reclamation Study. He is currently responsible for all major traffic studies in Malaysia/ Singapore on behalf of MVA.

Nigel G. WILSON MA, FICE, FHKIE, FCIT
Mr Wilson is a Chartered Engineer who from 1987 to 1996 held the position of Railway Extension Manager with the Mass Transit Railway Corporation. He was responsible for preliminary planning of all extensions to the MTR system, in particular the Island Line and the Lantau and Airport Rail Projects. Prior to joining the MTRC in 1975 he worked for Brian Colquhon and Partners in the U.K. undertaking transport priority studies and road and rail planning schemes.

Acknowledgements

While all contributions to this edited book have not been published elsewhere, many rely on already published reports, official government documents and other material in the public domain that warrant acknowledgement.

The editors wish to express their gratitude to the Hong Kong Tourist Association for their permission to reproduce numerous photographs from its archive collection for use in the Editors' Overviews.

Thanks are due to the Census and Statistics Department of the Hong Kong Administration for the permission to reproduce the chart on passengers by mode of transport in Chapter 1. Acknowledgements are also due to the Hong Kong Museum of History of the Urban Council for the photographs used from their Museum collection for this same chapter.

The authors of Chapter 5 are grateful to Robert H. Noble, sometime Marketing and Planning Director of the Mass Transit Railway Corporation, for his assistance in the preparation of this chapter. Any errors of fact or expressions of opinion stated, however, are solely the responsibility of the authors.

The authors of Chapter 6 wish to express their gratitude to the Commissioner of Transport of the Hong Kong Administration for the permission to publish the chapter. The views expressed, therein, however, do not necessarily represent those of the Transport Department of the Hong Kong Administration.

Thanks are also due to the Planning Department of the Hong Kong Administration for permission to reproduce the figures attributed to it in Chapter 8.

The author of Chapter 11 wishes to thank Peter F. Leeds for the use of his unpublished notes on the institutional aspects of Hong Kong's transport sector.

The editors are particularly indebted to Oliver Trueb and Debbie Chan, who were responsible for the entire desk-top publishing work for this book including the preparation of many maps, diagrams and photographs.

Introduction

Harry T. Dimitriou and Alison H.S. Cook

Aims and scope of the book

The Hong Kong government's White Paper on Transport Policy, published in 1990, presented a daunting agenda of transport problems which have since become more critical. While there have been numerous other reports since then published which highlight many of the territory's urban land-use, transport and related environmental concerns, the transport policy agenda (albeit it, in some respects, rather narrowly defined) offers a good starting point for this book. Among the problems raised by the White Paper were:

- The fact that Hong Kong's road and rail-based public transport systems, supporting 9.8 million trips each day in 1989 (this having since increased to 10.5 million), operated at full capacity during peak periods.
- A forecast that transport demand, in terms of the total number of passenger trips, would grow by 40 per cent by the end of the century and freight transport movement, in terms of number of trips, by 130 per cent (since revised to 145 per cent) for the same horizon date. Freight vehicles, which had emerged as the dominant component of traffic, were forecast to increase from 42 per cent of road traffic in 1986 to 51 per cent in 2001.
- Options for meeting this growth, in terms of infrastructure and supporting resources, were likely to be costly in economic, environmental and social terms.

With the above and other related problems in mind, the aims of this book are fourfold: firstly, to explain the fast growing nature and magnitude (as well as origins) of transport, land-use and related environmental problems in Hong Kong; secondly, to give an account of the government's approach to tackling these problems and the role of the private sector in this endeavour; thirdly, to critically review these approaches and trace the implications of their adoption; and fourthly, to offer some new directions in land-use/transport planning and environmental protection that could more sustainably tackle the kind of problems cited, as well as offer lessons for cities elsewhere in the region.

The book is comprised of twelve chapters contributed by persons with Hong Kong government or consultancy experience in land-use, transport and/or environmental planning in the territory. These chapters are complemented by editorial comments that seek to provide a contextual account and broader international and regional context. The book is edited from a standpoint of a strong awareness that land use and transport are 'two sides of the same coin' and that they, therefore, need to be planned together, and that the pre-requisites of an effective integrated set of efforts are suitable institutional and regulative frameworks. The publication also reflects a concern that environmental considerations will increasingly and inevitably govern future urban land-use and transport policy responses.

As already indicated, the book is anticipated to be of interest not only to those involved in the planning and development of Hong Kong but also to those who have internationally followed the rapid and impressive changes in the territory over the years and the implications they generate for the region. Students of urban planning, real estate development, transport planning and engineering, and geography, will find this a useful text offering an account and critique of the principles and practices of land-use/transport planning and its environmental impacts in Hong Kong, with particular emphasis on the last twenty years.

The interfaces of land use and transport, and of transport and the environment (two of the main underlying themes of the book), are viewed by the editors as the 'Achilles heel' of much traditional urban transport planning practice. This conclusion echoes the findings of international research sponsored by the Transport and Road Research Laboratory in the UK in the 1980s which, among other things, noted:

> *... transport policies are enacted with only a vague and unreliable knowledge of what the effects might be on land use in the longer term, and how they in turn might react on transport itself.*

A report of this work (Webster et al 1988) commented that while government in the UK typically imposes planning regulations which dictate the siting of much new urban development and its related infrastructure, the success of such developments greatly depends on the efficient provision, operation and management of good transport links. Accordingly, the practice of land-use and transport planning by city governments as separate activities, subject to the control of different agencies (increasingly outside the public sector), is now seen to pose major shortcomings (DOE 1996). Since cities are a function of complex interactions between land-use patterns, transport systems and the fast changing wider environment, mechanisms for interaction and integration of the different policy fields are increasingly essential.

Both the aforementioned report and Webster and his colleagues observe that planners and policy makers in one field make decisions affecting other fields, the impacts of which may not be immediately apparent, despite the fact that in the long-run they may prove to be of critical importance:

> *.... planners tackle one problem only to find another growing from the solution to the first. In this way, new roads intended to cure congestion in one*

> *place may attract development away from areas which other policies are trying to revitalize, and transfer congestion to places which are even less able to cope with it.* (Webster et al 1988)

The above example epitomises the type of policy conflicts which occur in a system where urban planning and development decisions are sectorally based. As some of the following chapters will testify, the results of conflicting objectives of different arms of government in Hong Kong are particularly apparent because of the city's compact high density urban form.

However, recourse to land-use/transport modelling as an integrative mechanism is not an 'effective' antidote to the problems of co-ordination and integration as one of the co-editors of this book (Dimitriou 1992) argues. Many present day urban land-use, transport and related environmental problems in Hong Kong to a large extent have their origins in the misconceived premises and policies employed by past planning exercises. In similar vein, Goodwin (1991) in a recent UK study, demonstrated what many land-use planners long-time knew, namely that more roads generate more traffic in major urban areas.

Another underlying theme is the impact of planned land-use and transport changes on Hong Kong's natural, physical and human environments. This book not only provides an insight into particular aspects of this concern but in more general terms seeks to relate these changes to environmentally sustainable perspectives of urban development.

Principles for discussing sustainability are drawn from a number of international sources, including The Brundtland Report (1987), Blueprint 3 (Pearce 1993), The Eighteenth Report on Transport and the Environment (Royal Commission on Environmental Pollution 1995) and Sustainable Transport: Priorities for Policy Reform (World Bank 1996). The observations and recommendations of these (and others) raise critically important questions for the future of Hong Kong, particularly if the territory continues with its currently charted course of economic growth at almost any cost.

Structure of the book

This book comprises twelve chapters plus an *Introduction*, *Conclusion* and *Editors' Overviews*. It is organised into four parts, each comprised of three chapters with its own overview.

The *Introduction* sets out the framework for the book. It highlights the main conceptual and empirical relationships and issues concerned with land-use/transport developments and with the state of the environment in Hong Kong. Another major structuring element is the extent to which the current and growing urban transport problems in Hong Kong can be attributed to the lack of integrated land-use/transport planning in the past, and whether the present 'push' for further transport infrastructure development is environmentally sustainable.

Part One is concerned with historical developments of land use and transport in Hong Kong and the evolution of their planning up to the territory's political transition. The three chapters together take the reader through the territory's history from its colonial beginnings of transport and urban development, to the current plans for the future. The first two chapters highlight the separate development of land-use and transport planning. The third chapter traces the efforts to reconcile this mis-match on a territorial basis.

Part Two addresses key issues of land-use and transport planning that presently confront the territory, with particular emphasis on public transport. Two of the chapters are modal specific, while the third covers the role of public transport in Hong Kong more broadly. The experience recounted here is particularly significant given the major new public transport initiatives and investments embarked upon in the territory and the levels of financial viability achieved.

Part Three outlines and reviews selected Hong Kong policy and planning responses to the land-use/transport problems of the territory, particularly within the metropolitan area. One chapter critically examines the contribution of modelling in land-use/transport planning studies. Another reviews the opportunities (missed and taken) of transport planning within the territory's metropolitan development strategy (Metroplan). The third chapter discusses how environmental protection considerations are increasingly being taken into account in recent traffic planning exercises.

Part Four highlights those aspects of land-use/transport planning in Hong Kong that would benefit from additional concerted attention and new directions of thought. Each chapter emphasises different aspects: the important emerging environmental agenda in land-use/transport planning (especially with regard to air pollution); the potential role of railways in a more sustainable approach to urban transport planning; and the need for institutional development efforts to help better integrate land-use planning, transport planning and environmental protection efforts.

Conclusions are drawn from the numerous analyses and proposals offered by the different chapters and editorial comments. A final editorial position is presented on the successes and failures of integrating land-use and transport planning efforts in Hong Kong and their environmental impacts. The question of whether the present territorial transport infrastructure development endeavours are environmentally sustainable is discussed. The conclusions are intended to offer constructive thoughts and directions for the future, both locally and globally, in what are very new, dynamic and complex areas of study and professional practice for both public and private arenas of operation.

References

Brundtland Commission (1987) *Our Common Future* The World Commission on Environment and Development.

Department of the Environment (1996) *The UK Round Table's First Annual Report on Sustainable Development* London: HMSO.

Dimitriou, H.T. (1992) *Urban Transport Planning: A Developmental Approach*, London: Routledge.

Goodwin, P.B. (1991) *Transport, the New Realism* Transport Studies Unit, Oxford University, Transport and Society Final Report.

Pearce, D. (1993) 'Defining Sustainable Development' in *Measuring Sustainable Development*, London: Earthscan Publishers.

Royal Commission on Environmental Pollution (1994) *Eighteenth Report on Transport and the Environment*, Oxford: Oxford University Press.

Transport and Road Research Laboratory (1980) *The Demand for Public Transport* Report of the International Collaborative Study of the Factors Affecting Public Transport Usage (Webster, F.V. and P.H. Bly eds) Berkshire, England: Crowthorne.

Transport Branch (1990) *Moving into the 21st Century: The White Paper on Transport Policy in Hong Kong* Government Secretariat, Hong Kong: Government Printer.

——————— (1994) *Report of the Working Party on Measures to Address Traffic Congestion* Government Secretariat, Hong Kong: Government Printer.

Webster, F.V. et al (eds.) (1988) *Urban Land-use and Transport Interaction: Policies and Models*, Aldershot, England: Avebury Press.

World Bank (1996) *Sustainable Transport: Priorities for Policy Reform*, Environmentally Sustainable Development Transportation, Water and Urban Development Department, IBRD, Washington D.C.

Part One
HISTORICAL DEVELOPMENTS

Historical developments: Editors' Overview

Harry T. Dimitriou and Alison H.S. Cook

The British occupied Hong Kong Island from 1841 to 1997 on the basis of somewhat dubious trading activities in the nineteenth century and their victory over the Chinese in the First Opium War. China ceded parts of the Kowloon peninsula and Stonecutters Island in 1860 at the end of the Second Opium Wars. Whereas Lord Castlereagh (Britain's foreign minister at the time of Hong Kong's acquisition) is reported to have dismissed Hong Kong Island as 'worthless' and Kowloon as a 'mere duck-farm and rice-paddy adjunct', today, the territory is light-years away from its former self. It is one of the ten strongest trading economies in the world with a gross domestic product of US$142,538 million in 1995 (up from US$3,463 million in 1970) and a per capita income of US$22,990 for the same year (World Bank 1997). This latter figure has since increased and surpassed that of the UK.

Figure 1 *Hong Kong cityscape* *Source: A.H.S. Cook*

Perhaps influenced by this stark contrast with its past, an internationally famous west-coast US academic, once proclaimed while visiting Hong Kong that '.... as a city planner I am not concerned with history but only the future'. Looking back on how Hong Kong has developed, this sentiment must have been strongly shared by many of the 'makers-and-shakers' of the territory actively involved in shaping today's urban Hong Kong (see Figures 1 and 2). While the territory has a rich and very interesting heritage dating back to the nineteenth century, not much of this visibly remains. Investigation of the territory's past, however, suggests that its very history, and its location as one of the world's great deep-water ports[1], as well as its long-standing function as the southern 'Gateway to China', have in fact laid down the foundations of Hong Kong's present day immense success (Cameron 1991).

On the premise, therefore, that the history of Hong Kong is indeed an important starting point to understanding its present day culture, values and growth - from its humble beginnings as a settlement of 12,000 in 1841 and 23,817 inhabitants in 1845[2] to its population of six million in 1996 - the first part to this book traces three sets of critically important developments that have molded Hong Kong over its 134 year old colonial history. These include:

- The evolution of Hong Kong's urban transport system (see Chapter 1), now transformed into one of the world's most efficient and economically viable systems and a model to many other cities in the region.
- The emergence of land-use planning and urban redevelopment activities (see Chapter 2) which through the territory's land lease system produces revenues for the Hong Kong Administration in excess of HK$19 billion (Information Services Department 1996)[3]. This system has proven so successful as a generator of public revenues that, again, it is increasingly looked upon as a model for other Asian cities.
- The parallel development of land-use and transport planning exercises (see Chapter 3) that have in recent years sought to guide the territory's rapid urban growth, providing a rational framework for the largest public works programme in the world costing an estimated US$24 billion[4] for the replacement of the territory's airport at Kai Tak, up-grading its sea port capacity and providing support infrastructure, including roads, railways and a new town (see Figure 2).

Figure 2 New airport at Chek Lap Kok
Source: Hong Kong Airport Authority

Robert Cottrell, in his introduction to his fascinating book, The End of Hong Kong (1993), cites the Peak Tram (see Figure 3) as '....one of the few vestiges of nineteenth century history which Hong Kong has yet to dismantle ....'. Cottrell goes on to claim that '...the tram was, like most things in Hong Kong, conceived as a speculation....'. This speculative quality - in the transport industry of Hong Kong - is also referred to by Peter Leeds (sometime Commissioner of Transport for the Hong Kong Administration between 1982 and 1987) in Chapter 1 of this book.

Figure 3
Peak Tram,
Hong Kong Island
Source: Courtesy of Hong Kong Tourist Association

Leeds provides an insight into the entrepreneurial spirit that underlaid (and still underlies) the development of the public transport industry in Hong Kong. He traces the evolution of urban public transport in the territory from the the early years of British colonial settlement to the present. He points out that many of the oldest transport routes remain key transport arteries today, despite the numerous reclamation works that have taken place almost annually (see Figure 4) and the construction of complex elevated road transport infrastructure (see Figure 5). In his contribution, Leeds highlights:

- the early introduction of vehicular licensing and mechanised transport;
- the critical role of the tram system as an agent of urbanisation on the northern shore of Hong Kong Island;
- the extensive reliance on good ferry services between Hong Kong Island and Kowloon (at least until recently)[5];
- the belated introduction of railways, now more than ever, poised to play a strategic role in the development of the territory;
- the very successful development of bus services, which in the case of minibuses, have set a level of regional and world-wide acceptability and respectability[6].

The above developments are described against a backcloth of historical events, migration flows and population growth, rising demands for housing, and a growth

Figure 4 Construction of West Kowloon Expressway
Source: Courtesy of Hong Kong Tourist Association

(and decline) in industry and commerce - culminating in a whole array of urban development initiatives, including reclamation schemes, redevelopments and new towns construction. With economic growth of the territory averaging 5.5 per cent per annum between 1980 and 1993, these developments have generated a huge transformation in the socio-economic profile of the travelling public, both in creating a more affluent market for public transport travel, as well as a steep rise in private car usage. The growth in personal trip-making has been accompanied by a dramatic rise in freight traffic (especially by road)[7], and the resurgence of the railway as a potentially very significant transport mode for Hong Kong's future.

Chapter 2, by Alexander Cuthbert (a past faculty member of the University of Hong Kong), offers a very different perspective to the preceding chapter. Apart from focusing much more on land-use development and the socio-economic and political pressures that have transformed the 'commodity' of land in Hong Kong, the reader is reminded, that for all its economic successes, Hong Kong is void of natural resources. Cuthbert argues that the territory owes its success almost exclusively to an atypical colonial experience that defies the normal chapters of history. Today Hong Kong continues to follow an atypical course in that it is the first capitalist economy in a decolonisation process to be returned by a colonial power to a communist government; except that there is a very valid debate as to whether Britain still qualifies as a world power and whether the current government of the Peoples Republic of China (PRC) is, by traditional definitions, still a communist administration given its heavy reliance upon the market system as a vehicle for its development.

Perhaps above all, Cuthbert's chapter highlights the longtime importance of two major sources of revenue for the Hong Kong Government: revenues from the sale of monopolistic franchises, and from land leases. While the former (especially revenues from utility and transport franchises) became more significant later in the territory's development, revenues from real estate go back to the very beginnings of Hong Kong, when in 1841, Captain Elliot declared Hong Kong a Free Port. Soon after the annexation, developments mushroomed uncontrollably along the northern coast of Hong Kong Island,whereupon, Elliot introduced the first sale of land lots in

June 1841 in an attempt to control the rampant development which was occurring (Cameron 1991). This planted the seed for revenue which has resulted in an excellent harvest over the years, and a yield of US$2.5 billion for 1994-95,[8] representing 11 per cent of the total coffers of the Hong Kong Administration from land based revenues which the British Government is obliged to share with the Hong Kong Special Administrative Region (SAR) upon transition.

Cuthbert sequentially outlines the establishment of the territory's town planning legislation and briefly describes the intermittent property booms that have since taken place and which have ultimately raised residential and commercial property prices to among the most expensive in the world. He firmly discounts the simplistic notion that Hong Kong has been managed by a non-interventionist administration. He points out the government's support for restricted market access to various aspects of its economy and the massive government subsidy provided to the housing sector by accommodating in public housing, much of it in the new towns, approximately 50 per cent of the territory's population; albeit within apartments with very confined

Figure 5
Elevated roadway on Hong Kong Island
Source: A.H.S. Cook and H.T. Dimitriou

space standards[9]. Arguing that land-use developments are merely an expression of economic and political forces, Cuthbert gives an account of:

- the strategically growing importance of Hong Kong as an international trans-shipment centre;
- the rise and decline of its manufacturing sector;
- the rise of its Central Area as a world-class financial centre;
- the dramatic increase (especially in the 1980s) of the middle class;
- the development of tourism;
- and the associated urban development/redevelopment and infrastructure these changes have spawned.

The latter part of Chapter 2 outlines some of the government's responses to the above developments in the form of programmes, plans and projects introduced on the one hand, to facilitate further economic growth and, on the other hand, to control the direction of this growth. The combined set of 'enabling' and 'control' measures ('carrot and stick approach') has largely been developed and adapted by government in close alliance with the captains of industry and trade. For the most part they have not been proactive efforts but rather responses (often belated) to bloated economic and physical growth with little regard to democratic rights, community participation and environmental concerns.

Chapter 3 continues the theme of Chapter 2, but from an entirely different perspective. It takes a much more technical and detailed approach to discussing the plans and programmes over the last two decades for guiding the future urban and infrastructure developments of Hong Kong. Edward Pryor, the Principal Government Town Planner responsible for strategic planning at the time of writing this book, provides a more officially oriented account of the main methodological steps and principles employed in strategic land-use and transport planning in the territory, particularly the Territorial Development Strategy (TDS). Whereas the preceding chapters essentially focus on different sectors - the first on transport and the second on land use - this contribution, in Pryor's own words: '..investigates whether the respective roles of land-use and transport planners in Hong Kong have been complementary and what endeavours (if any) are needed to rationalise their roles'.

Pryor, in presenting the case for the need for territorial-wide land-use/transport planning, outlines:

- various models employed in this task;
- the principles employed;
- the inevitable inter-dependence of land-use and transport planning;
- the three levels of planning pursued;
- the role of transport models in the planning process; and
- the new way forward (in the form of a simplified TDS Review Process) with its constituent stages and sequential tasks.

Close scrutiny of this chapter suggests that a very strong technocratic tradition has been developed and utilised by the Hong Kong government to support and justify its investment proposals for road, port and airport infrastructure in particular, and urban development in general, in response to specifically targetted needs and problems. The Hong Kong administration's high regard for, and dependency on, international consultancy expertise in responding to such perceived problems is made very apparent[10]. Pryor acknowledges that extensive use of this expertise and the values it has imported from the post-industrialised countries have generated a number of problems. In particular, these include problems associated with the use of algorithmic-based models and quantitative data of past trends for projecting into the future, and suggest that the territory's planners have relied too much on extrapolated growth trends as a basis for planning and designing the future, rather than considering any alternative vision(s) that might have been more environmentally sustainable.

Pryor's articulate account of the steps of territorial land-use and transport plan-making, describes a process with a technical precision much to be admired. However, the commitment (even determination) of government to follow this up with action, especially in relation to the core projects of the Port and Airport Development Strategy and reclamation, has been at a scale and speed of execution that has left very little time for the community to ponder on the implications of such major changes.

Facilitated by first class project management capabilities, abundant public funds, often a collaborative private sector, a simple (some would argue weak) land-use development control system and very limited public consultation and participation, the Hong Kong government has produced a collection of bold blueprints for its future under extremely uncertain circumstances. Ironically, at a time when planning studies have not been in vogue, an administration that is often presented as the champion of free enterprise has, in effect, used these plans, programmes and projects as a means of trying to reduce uncertainty[11]. It is, however, too early to be sure of the exact merits and demerits of these various efforts. In any event, one's conclusion will inevitably depend upon on the criteria employed to evaluate their success[12]. More ominous perhaps, given the likely departure from Hong Kong of many of the key plan-makers and project implementors with the transition to Chinese rule, many key personnel will no longer be present to answer for, and deal with, the consequences of their recommendations.

It is clear from the Territorial Development Strategy and other land-use/transport planning exercises outlined by Pryor (especially the numerous reclamation schemes and transportation planning exercises, although, perhaps not Metroplan) that every effort has been made to 'optimise' the economic growth benefits they offer; especially to the territory's transportation industries, real estate and construction interests, trading houses and finance community (i.e, the backbone of Hong Kong's economy). Such action presumes that the 'trickle down' benefits of these policies and schemes to the territory's inhabitants are sufficiently positive to warrant their aggressive pursual. If the economic growth rates experienced in recent decades by Hong Kong are a guide to the future, then this presumption may indeed seemingly be justified. Closer scrutiny of the aims and impacts of these plans, utilising more sustainable concepts and criteria, however, places in question the level of confidence one should have in this presumption.

Notes

1 In 1842 only 381 vessels called into Hong Kong harbour. This has increased to the phenomenal figure of 42,000 ocean going vessels and 175,000 other vessels in 1995 (river steamers, hydrofoils, hoverferries, catamarans and river trading vessels).
2 This excludes the garrison soldiers.
3 Land sales generated 11 per cent of total revenue in 1994-95 and a further HK$26 billion was generated through property related sources - property tax, stamp duty and rates. In total about 26 per cent of the Government's total annual revenue was generated directly from these sources (Information Services Department 1996). If in addition profits tax from companies involved in various aspects of the property development industry is included the contribution to revenue is estimated to be about 40 per cent (K.W Chau, Department of Surveying, University of Hong Kong, personal communication May 1996).
4 Airport core projects and Lantau Port terminals 10 and 11 at MOD prices.
5 Ferry patronage has suffered a steep decline from a high of 681,000 per day in 1971 to 215,000 in 1996, while cross harbour movements by motor vehicle (private vehicles, buses and taxis) and by rail have increased. Cross harbour passenger trips by bus and rail are around 1,327,000 per day and daily cross harbour vehicle trips average 217,600 (Transport Department 1997).
6 The economic efficiencies of minibus public transport were hailed by Professor Alan Walters in the 1980s, drawing heavily on Hong Kong's experience as the basis for his support. Whether by formalising vehicles previously part of the informal sector, or by introducing a new type of public transport service, minibuses have become an important component of the public transport scene in the 1990s in cities as different and geographically separated as Kuala Lumpur and London.
7 Cross border road freight traffic has increased very dramatically from 71,300 tonnes in 1978 (when movement of freight by road between China and Hong Kong was in its early days) to 14,802,400 tonnes in 1996. Sea freight (by ocean vessels) has increased from 27,832,400 tonnes in 1970 to 125,838,000 tonnes in 1996. In the same period air freight growth has increased from 13,430 tonnes in 1970 to present day estimates of 1,563,500 tonnes in 1996. Rail freight has increased to a much lesser degree, from 858,500 tonnes in 1970 to a high of 2,216,718 tonnes in 1990 and has since fallen to 938,961 in 1996 (Census and Statistics Department 1997 and Information Services Department 1997).
8 Revenue from land sales (Information Services Department 1996).
9 It has been earlier argued (see Henderson 1991) that this housing subsidy acts as a knock-on subsidy to Hong Kong's labour costs, becoming more significant as housing costs rise.
10 A regard not always shared by the growing number of environmental and community groups both in Hong Kong and the countries from which many of the more conventional of these consultants originate.
11 The number of planning and feasibility studies cited by Pryor suggests that in

recent years more have been commissioned per acre of land in Hong Kong than in any other part of the globe.

12 There are indeed signs that exploration of new evaluation criteria (even new planning goals) may be on the horizon. Towards the end of 1995 the Hong Kong government issued an invitation to selected consultants to tender for a HK$40 million study on the sustainability of the territory, seemingly responding to the Rio Agenda of June 1992 which (incidentally) the Hong Kong Government refused to sign.

References

Cameron, N. (1991) *An Illustrated History of Hong Kong*, Oxford: Oxford University Press.

Census and Statistics Department (1997) *Hong Kong Annual Digest of Statistics* (and previous years) Hong Kong Government

Cottrell, R. (1993) *The End of Hong Kong: The Secret Diplomacy of Imperial Retreat* London: John Murray.

Henderson, J.F. (1991) 'Urbanization in the Hong Kong-South China Region: An Introduction to Dynamics and Dilemmas' *International Journal of Urban and Regional Research* Vol.15 (2).

Information Services Department (1997) *Hong Kong 1997* Year Book (and previous years) Hong Kong: Government Printer.

Leeds, P. (1984) *The Development of Public Transport in Hong Kong: An Historical Review 1841-1974* Hong Kong: Government Printer.

Transport Department (1997) *Annual Traffic Census* (and previous years) Hong Kong: Government Printer.

World Bank (1995) *World Development Report 1995: Workers in an Integrating World* Oxford: Oxford University Press.

World Bank (1997) *World Development Report 1997: The State in a Changing World* Oxford: Oxford University Press.

1 Evolution of urban transport

Peter F. Leeds

Introduction

When Hong Kong became a British settlement in 1841, sea transport, both international and local, sustained the commercial life of the territory. The then new township of Victoria, on the north shore of the island facing the harbour, covered a very small area and internal transport on land was of little importance (see Figure 1.1).

Figure 1.1 Hong Kong harbour, 1856
Source: Courtesy of Mr and Mrs Barry Yates

For many years the main communication line was the harbour with launches providing the easiest way of moving around between the public and private piers that were built along the waterfront. One of the first area development schemes (to drain and build in Happy Valley) envisaged a system of canals for transport. The settlement's streets were mostly narrow and, apart from the shoreline road (Queen's Road), were steep. Outside the settlement the roads on the island were little more than tracks constructed primarily to facilitate the movement of troops and police. Piracy on the sea and highway robbery on land were prevalent and real and imagined dangers of invasion by rival European powers resulted in the early communication links being largely related to military and law enforcement requirements.

The climate was unkind to horses and most conveyances were thus pushed, pulled or carried by manpower. Sedan chairs, simple carts and a few horse-drawn carriages were the alternatives to walking. As the settlement expanded along the north shore of the island and up the slopes (towards Mid-levels) the need for passenger transport for all was largely met by two conveyances - the sedan chair and the jinricksha (see Figure 1.2).

Figure 1.2 Sedan chair and jinricksha
Source: Courtesy of the Hong Kong Museum of History, Urban Council, Hong Kong

Sedan chairs had been common in China and Europe. Hong Kong sedan chairs varied considerably. Some were no more than two long bamboo poles with a seat slung between, while others were heavy and ornate requiring a team of eight or more bearers. Jinrickshas, or rickshaws as we know them now, are more recent and were Japanese in origin. Their introduction into China has been attributed to an American Baptist missionary (Jonathan Goble) in response to a request by his wife that something be done to relieve the distress suffered by the sedan chair carriers.

In Hong Kong most wealthy families had their own private sedan chairs or rickshaws and the bearers and pullers were part of the domestic establishment. Public chairs and rickshaws were also available for hire and these constituted Hong Kong's first means of public transport. The vehicles were usually hired to the coolies by contractors, a particularly rapacious group who kept the hiring rates at levels which barely allowed the coolies to survive. The coolies suffered grinding poverty, poor health and had a short life expectancy.

In 1863 a licensing system was introduced. Such was the state of law and order at the time that ordinary licences only authorised vehicles and chairs to be used between 5 a.m. and 8 p.m. The Registrar General was the licensing authority and had powers to establish rickshaw and sedan chair stands. The 1863 Ordinance contained many do's and don'ts - keep to the left, overtake on the right, give lost property to the police - plus the admonitions of not to drive carelessly, nor to use bad language, and the offences of refusing to accept hire and of demanding more than the authorised fare.

When this licensing system was introduced the population of Hong Kong was estimated to be 124,000 and urban development was still restricted to the Central and Western segments of the north shore of Hong Kong Island. By the end of the nineteenth century, when the population reached almost 300,000, most journeys were still made on foot. The only public transport modes on the road were rickshaws (500 in Hong Kong, 70 in Kowloon and 25 in Quarry Bay) and sedan chairs (570 in Hong Kong and 30 on the Peak). There were also just over 1000 horse drawn carriages. Even with the introduction of mechanised transport, rickshaws and sedan chairs continued in use for many years. Rickshaws reached their peak of 2,790 in 1926 and sedan chairs of 1,212 in 1920. By 1939 the numbers had dropped to 1,168 and 218 respectively (see Table 1.1). After the Second World War only a few tourist 'photo-opportunity' rickshaws have survived.

Table 1.1 Passenger-carrying on-road vehicles, 1896-1938

	Rickshaws	Sedan chairs	Horse carriages	Buses	Trams	Motor cars
1896	585	468	714			
1898	670	600	1,074			
1900	840	598	1,263			
1902	1,340	588	1,313			
1904	1,245	678	41		26	
1906	1,425	704	29		36	
1908	1,495	665	24		-	3
1910	1,495	734	30		-	20
1912	1,495	733	13		46	21
1914	1,535	719	9		-	54
1916	1,580	869	6		-	112
1918	2,940	1,200	7		80	258
1920	3,258	1,215			80	508
1922	3,379	1,139			80	796
1924	3,411	1,087		-	-	1,256
1926	2,790	819		-	86	1,270
1928	2,490	660		-	-	1,348
1930	2,196	516		-	-	2,013
1932	1,887	499		174	-	2,774
1934	1,600	453		174	-	3,384
1936	1,390	318		185	97	3,760
1938	1,168	218		231	103	4,490

Source: Transport Department, Hong Kong Government

Mechanised transport

Streetcars and tramways

The limitations of man-powered transport prompted interest in the use of steam-power and electricity for greater speed which opened up the possibilities of developing land that hitherto had had insufficient access.

In the late nineteenth century urban streetcars and tramways flourished in North America and Europe and promoters sought to follow this example in colonial territories. In 1881 a variety of tramway schemes for Hong Kong was announced and, although a detailed enabling ordinance was passed in 1882, the initial burst of enthusiasm subsided when the question of raising finance had to be faced. Three years later, the rights to what is now the Peak Tramway were sold and a new company, the High Level Tram Company, opened the cable car system, initially steam-powered, in May 1888. Although the Peak District was sparsely occupied at that time the new tramway encouraged the development of the Peak and Mid-level districts as residential areas for the wealthy. Its cooler climate provided relief from the heat in the days before air-conditioning. The catchment area of the tramway was extended by the provision of sedan chair, and later rickshaw, services, from the stations along the contour line roads. These transport developments represented examples of transport investment propagating property development and, on a small scale, of feeder services complementing a main line railway.

The tramway system along the northern shore of Hong Kong Island was taken over by new promoters in 1902. Track laying began in 1903. The first sections of the new electric tramway were opened in July 1904. The maximum fare of 20 cents was expensive in those days but 70 years later the fare was still 20 cents. Originally there were also charges for carrying goods and minerals, including 20 cents for a chest of opium! Initially the company was not a financial success and shareholders had to wait until 1913 for their first dividend. Nevertheless, the tramway stimulated development in Wan Chai, Causeway Bay and through to North Point and Shau Kei Wan (see Chapter 2). These developments have provided the passenger demand that has kept the system working profitably ever since.

Ferry services

The harbour as a transport link became even more important after 1860 when, under the Convention of Peking, the southern part of Kowloon came under British administration. It had been expected that Kowloon would provide a valuable and less hilly residential area with easy communications to Central. However, development was slow. In 1876 there were nine steam launches operating passenger services in the harbour and the first regular cross harbour ferry service, which later became the 'Star' Ferry Company, began in 1880. By 1901 the population of Kowloon was still only 44,000, about 15 per cent of the total population, and it was not until after the First World War, that the development of Kowloon began to gain momentum (see Figure 1.3).

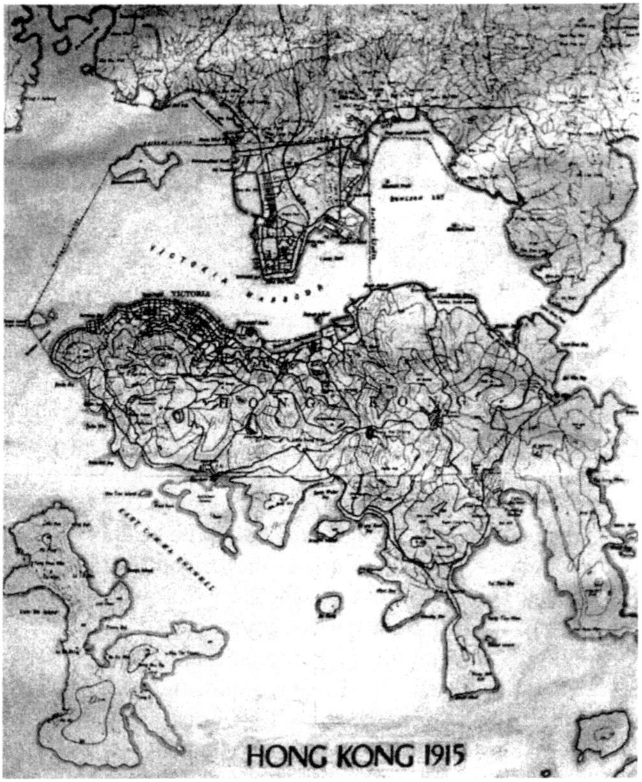

Figure 1.3 Urban development of Hong Kong, 1915
Source: Courtesy of the Hong Kong Museum of History, Urban Council, Hong Kong

With the development of Kowloon came a need for improved ferry services. Apart from the 'Star' Ferry, some sixteen individually owned vessels plied the harbour. However, the services were erratic and fights between boatmen competing for passengers were not uncommon. A licensing system for ferries was introduced in 1918. After tenders were invited for the operation of additional regular cross harbour services in 1923 an award was made to the newly formed Hong Kong & Yaumati Ferry Company (now Hong Kong Ferries). More cross harbour services were introduced in 1928 and a franchise for the first vehicular ferry service was awarded in 1933. In 1938, the company was granted a franchise to operate regular services to Cheung Chau, Silvermine Bay, Peng Chau, Tsuen Wan, Tsing Yi, Castle Peak (Tuen Mun) and Tai O. The introduction of these services was a major contributor to the development of the outlying islands of Hong Kong and the expansion of recreational activities.

By 1905 the City of Victoria (the urban area of Hong Kong Island) had a public transport system that was well suited to the prevailing conditions. Trams provided what was, in those days, a fast service along the north shore of the Island, while the Peak tram had opened up the Peak district and improved accessibility to the Mid-levels. Supplementing the trams were rickshaws and sedan chairs but soon afterwards came the beginnings of a major change - the arrival of the motor car.

There were only six motor cars in 1907. By 1918 there were still only 178 private cars and 102 'livery' or hire cars, quite a small number in contrast to the 3,700 rickshaws and 1,200 sedan chairs (see Table 1.1). Traffic congestion was never a major problem until well after the Second World War.

Railways

Railway development came relatively late to Hong Kong. The end of the nineteenth century and beginning of the twentieth saw commercial rivalry among foreign powers to secure trading rights and concessions in China (see Chapter 2). This began with a scramble for the major ports and then for railways to open up the interior beyond the confines of coastal and river traffic. The railway project that most concerned Hong Kong was a proposed link from Hangzhou (Hankow) to Guangzhou (Canton) which, it was thought, might divert trade from Hong Kong. A railway from Hong Kong to Canton was the obvious solution and in 1898 the Hong Kong & Shanghai Banking Corporation with Jardine Matheson formed the British & Chinese Corporation which was granted rights to build the railway. After lengthy negotiations between the foreign concessionaires and provincial governors the Corporation retained rights in China but the British Section from Kowloon to Shumchun (Shenzhen) was taken over by the Hong Kong Government. This was the first and, for a long time, the only direct government investment in public transport. Apart from its importance for international trade, locally it was the first major transport system to break through the collar of hills (the 'Nine Dragons' of Kowloon) and link the Colony with the leased 'New Territories' acquired in 1898. The British section of the Kowloon-Canton Railway (now operated by KCRC) was completed in 1910 and the annual local passenger traffic exceeded one million for the first time in 1924.

Buses

The most important change between the two world wars was on the roads, particularly the development of bus services. The first motor bus probably came to Hong Kong around 1921, the year in which the Vehicles and Traffic Regulation Ordinance was amended to give powers to grant licences for the operation of bus routes. Small operators proliferated. Among the early bus operators on Hong Kong Island were Hongkong Tramways, Hong Kong Hotels, the Aberdeen Kai Fong Association and, in Kowloon and the New Territories, the original Kowloon Motor Bus Company, Kai Tak Bus Company, China Motor Bus Company (see Figure 1.4), the Nam Hing Bus Company and the Chung Mei Bus Company.

The bus services were subject of many complaints. The government stated its intention to improve the situation by granting monopoly rights, based on the post-First World War experience in the United Kingdom, which was considered to be better than free competition, However, intention to action was a slow process. It was not until 1932 that tenders were invited to operate services for 15 years, starting in June 1933. The successful tenders were, for Hong Kong Island, 18 routes to be operated by China Motor Bus company (which had previously run only in Kowloon) and for

Figure 1.4
An early CMB bus
Source: Courtesy of the Hong Kong Museum of History, Urban Council, Hong Kong

Kowloon and the New Territories a further 18 routes to be operated by the reformed Kowloon Motor Bus Company (1933) Ltd. In a series of takeovers and mergers the other companies were absorbed and the new services began with 59 buses on Hong Kong Island and 115 in Kowloon and the New Territories. In the mid-1930s because of the economic depression the number of buses in operation did not grow very rapidly until 1938. In that year the population increased as a result of an inflow of refugees from China as Japanese military operations moved into South China. In 1939, the last year in which figures were published prior to the war, Hong Kong Island had 109 buses and Kowloon and the New Territories had 136.

Transport and urban development

Between 1841 and 1941 the rate of expansion was modest. Nonetheless, a relationship between the growth of public transport services and urban development was established: the Peak tram for the Peak and Mid-levels, Hongkong Tramways for the north shore of Hong Kong Island, and buses for the south side of Hong Kong and for Kowloon. The construction of the railway and the new roads through the hills bounding the urban area provided the infrastructure that subsequently allowed for the development of the New Territories. Road construction was, however, sometimes more related to providing access to new building lots, than to providing an efficient transport network. As all land in Hong Kong is held by the government the auctioning of leases has always been a major source of revenue (see Chapter 2).

Apart from the Kowloon-Canton Railway (KCR), which was run as a government department, the Hong Kong administration had little interest in public transport. It took the view that it was sufficient to provide reasonable facilities for private enterprise to undertake the job of providing public transport services. In the enabling legislation for public transport, government had powers to grant franchises, levy royalties, specify routes and control fares. However transport planning, such as there was, was left largely to the transport operating companies. So as long as development took place at a steady pace this worked well enough but, in the late 1930s, when the population increased rapidly, signs that policies of benign indifference did not work well in times of unusual stress emerged. It was to be in the post-war period, however, that the greatest changes would take place and the interdependence between urban development and transport services become all too evident.

Post-War urban reconstruction and planning

Recovery efforts

The Japanese occupation of Hong Kong ended in August 1945 by which time the population was reduced to around 500,000, compared with 1,600,000 when the war began in 1941. Public transport had almost ceased operation due to lack of fuel and maintenance, the confiscation of equipment, looting and war damage. The ferry companies suffered heavy losses. None of the six 'Star' ferries could be found and only 21 of the Hong Kong & Yaumati Ferry Company vessels were afloat, most of which were unserviceable. Despite the damage, recovery was rapid and by 1946 most of the services were back in operation. By 1947 services were again operating at pre-war levels although the increase in demand for the vehicular ferries resulted in waiting times of up to three hours.

The Kowloon-Canton Railway suffered from lack of maintenance as well as damage from air-raids and sabotage. By the end of 1946, the numbers of through-trains had reached their 1938 level, although local and goods services still suffered from shortages of rolling stock. Hongkong Tramways had only 15 serviceable trams at the end of the Japanese occupation by which time much of the overhead wiring and workshop equipment had been dismantled and removed. Fortunately, most of the track was intact and, with improvisation and the use of material hidden during the occupation by company staff, most of the tram routes were put back in operation on a restricted scale with 45 trams. By 1947 the fleet was restored to its pre-war level of 112 trams. The Peak tram, which had also suffered from looting, recommenced its services on Christmas Day 1945. Most of the buses had been confiscated by the Japanese. The Kowloon Motor Bus Company could find only six vehicles and these were put into operation within ten days. New buses were ordered, but for the immediate needs, converted war surplus lorries were put into service. By 1947, 68 buses and 75 'lorry' buses were operating on all the pre-war routes together with some new ones in the New Territories. On Hong Kong Island, although the China Motor Bus Company had restored its routes, the shortage of vehicles was severe.

Before new ferries and buses were delivered, various make-shift arrangements were employed, including the use of tandem bicycles to carry passengers for hire. The objective in the first four years after liberation was to restore services to their pre-war levels. Considering the cost and the world-wide shortage of equipment, as well as the limited shipping space, this was done with commendable speed.

Population increase

Having recovered the pre-war position, the public transport undertakings were then faced with the problems resulting from the almost explosive increase in population from 1949 onwards (see Table 1.2). Sudden inflows of people, as a result of unsettled conditions in China, had occurred often in the past but the post-Second World War migration exceeded anything that had happened before. Some of the refugees from the civil war in China had financial resources, technical skills and entrepre-

Table 1.2 Population of Hong Kong, 1851-1996

Year	Population	Year	Population
1851	34,000	1951	2,070,000
1861	119,300	1956	2,677,000
1870	124,200	1961	3,195,300
1881	160,400	1966	3,679,400
1891	224,800	1971	4,064,400
1901	300,700	1976	4,478,000
1911	456,700	1981	5,154,100
1916	528,000	1986	5,699,000
1921	686,700	1991	5,822,000
1926	874,400	1992*	5,811,500
1931	879,000	1993*	5,919,000
1936	988,200	1994*	6,035,400
1941	1,639,000	1995*	6,156,100
1943	500,000	1996	6,311,000
1947	1,800,000		

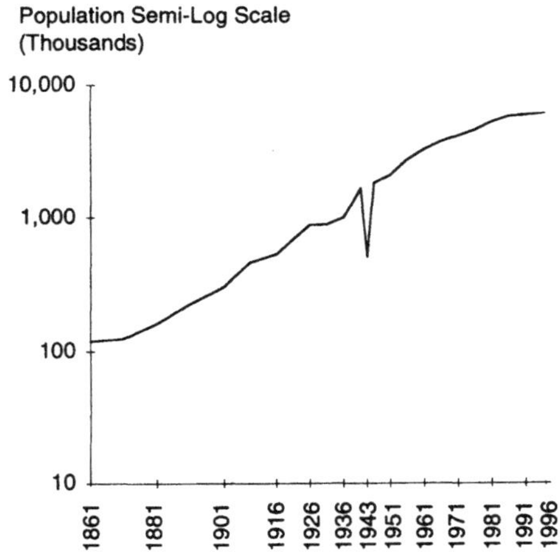

Source: Census and Statistics Department, Hong Kong Government
** estimated*

neurial judgement which stimulated expansion of the industrial activity in Hong Kong (see Chapter 2). The population increase, however, brought with it serious problems in housing, sanitation, water supply, fire prevention and social services.

On the outskirts of the urban area, and in parts of the New Territories, large shanty towns of wooden huts sprang up. In 1949, when the Nationalist (K.M.T.) government collapsed in mainland China, the population of Hong Kong reached 2 million. It was expected that, with the end of the war, many of the refugees would return home. Some did and in 1950-51 the population fell by 300,000, but it was not a continuing trend. Thereafter, with new immigrants and a high birth rate, the population continued to rise, reaching 3 million by 1959. This increase and the measures taken to deal with housing and the changes in the structure of industry were to have a great effect on public transport.

Housing, industry and transport

In 1953 a disastrous fire occurred in the largest of the territory's squatter areas at Shek Kip Mei in Kowloon when some 50,000 people were made homeless. Before this, some effort had been made to re-house squatters by providing prepared sites where they could build stone cottages. However, the Shek Kip Mei fire, and others which occurred soon afterwards, made it obvious that small scale efforts were insufficient. It was realised the only body capable of providing housing in quantity was the government. New multi-storey estates were subsequently built by the government and other agencies. Due to the shortage of land in the existing urban area, most of the new estates were on the fringes of the urban area, and later in the New Territories. While changes were taking place in the distribution of the residential population other important changes were taking place in the location of Hong Kong industry. The traditional workshop, where a family would both live and work, was replaced by large factories which were often located on the outskirts of the urban area or in the New Territories, particularly in the rapidly expanding industrial area of Tsuen Wan.

With many people changing their places of residence and employment, this period saw a greater dependency on public transport for going to work. A comparison of the total population and the number of passengers carried on public transport in the years between 1953 and 1965 illustrates the expanded demand. In 1953, when the population was 2.25 million, the number of public transport passengers was 471 million. In 1965, when the population had risen to 3.82 millions, the number carried was 1,162 million. In other words the population had increased by 53 per cent in twelve years but in the same period the number of passengers carried by public transport went up by 246 per cent - a phenomenal increase! It was to the credit of the transport companies that they were able to expand quickly from their own resources but the system was showing signs of strain with severe overcrowding and excessively long passenger queuing times.

The main burden of this growth fell on the two major bus companies (CMB and KMB). They both retained strong links with their respective founding families. Typically, expansion was financed from their operating profits in preference to loans or a broadened equity base. Although both companies remained profitable and were able to increase the numbers of buses, the quality of service deteriorated and, even with gross overcrowding, the capacity did not meet demand. Consequently, potential passengers were lost to other modes of transport - both legal and illegal. A number of factors contributed to the deterioration of bus services after 1965. Operating costs rose steadily (particularly staff costs) with the competition for skilled labour from industry. Increasing traffic congestion meanwhile constrained the number of passenger miles that each bus could operate.

Savings were made by reducing the manning of buses from four (driver, two conductors and gateman) to one, by using larger vehicles and by improving availability through better maintenance programmes. The operating costs were also eased by reductions in the amount of royalties paid to the government under the then terms of the franchises. These measures helped the companies to keep fares steady for many years. However, the public, aware of the companies' profits, had come to expect the low fares to continue.

The brief rioting incident that occurred in 1966 (after a five cent ferry fare increase) made both public transport operators and the government chary of increasing fares until public opinion was ready to accept that an increase was imperative. This meant that, for a time at least, increases were sanctioned later than they would have been if decisions had been based on prudent financial planning. Consequently necessary new investment was deferred, adding further to the deterioration in the standards of public transport service. The worst transpired in 1967, when public transport faced damaging political strikes, unrelated to any demands on the employers. For the bus companies this meant the loss of much of the operating staff for many months. The 1967 'troubles' were a local manifestation of the so-called 'Cultural Revolution' which was disrupting normal life throughout China at the time.

Improvement of services

From a low point in early 1968 bus services steadily improved. An agreement was made with the Hong Kong government which, subject to satisfactory performance, allowed fares to increase so that the operators could enjoy a reasonable, some would say generous, rate of return of 15 per cent of the net assets employed. The original franchises granted in 1933 had been continued under new enabling ordinances in 1960 but were replaced in 1975 (under the Public Bus Services Ordinance) by new rolling franchises for specified routes, thus ending in theory, the geographic monopolies. Financially the companies were more secure and, with ready loan finance and export credit for bus purchases, more able and willing to improve services both in quantity and quality: an improvement that was more obvious with KMB than CMB.

The ferry companies expanded their passenger services without difficulty but, even with two additional cross-harbour services, the vehicular ferries were inadequate until the first cross harbour road tunnel was opened in 1972. The bus services through the tunnel, operated jointly by the two bus companies, reduced patronage on some ferry services. A few were withdrawn but in the central harbour area additional services were introduced to serve developing areas on reclaimed land in East Tsim Sha Tsui and North Wan Chai. In 1971, before the opening of the tunnel, the two ferry companies carried 248 million passengers. In 1973, when 12.5 million vehicles passed through the tunnel, the ferry companies still carried 208 million passengers. The number of vehicles using the ferries dropped by 31 per cent. The Hong Kong & Yaumati Ferry Company built-up its services to the outlying islands for both commuters and recreational traffic. The quality of service was further improved with the introduction of air-conditioned accommodation and high speed hover ferries in 1971 and 1975 respectively.

By 1961 Hongkong Tramways had 143 trams and 20 trailers and was carrying around 400,000 passengers a day. Some suggestions had been made that trams should be removed to ease road traffic congestion. However, as cheap fares and high utilisation prevailed, these ideas were not pursued, added to which, in terms of passengers per vehicle, trams are recognised to be efficient users of road space (see Chapter 6).

Light bus services

Illegal public transport services, which eventually became an extensive 'para-transit' system, had their origins in the New Territories which by 1966 had a population of 645,000 residing in small towns and scattered villages. Bus services in these areas were mostly confined to the main roads. As the minor road network developed, agriculture changed from rice production to market gardening and generated a demand for better transport for passengers and their market produce. This was met by private cars and small vans often illegally plying for hire. In 1960 nine-seater vans, with some space for goods, were licensed as 'New Territories Taxis': a misnomer, as they were rarely hired for a single journey and were mostly operated as a regular route bus service. Dual purpose vehicles, intended for small business transport, were also misused in this way. With the shortcomings of the bus companies in the late 1960s and ineffective prosecutions of the illegal operators the trade flourished and spread into the urban area where public hire cars were also often used illegally as 'pirate' services, although the competitive style of driving was not appreciated by other motorists. A 1968 survey showed that some 2,150 mini-bus type vehicles were carrying about 528,000 passengers each day by which time it was recognised that they had become a useful part of the public transport system. The operations were, however, illegal and that situation could not be tolerated. As a result, in 1969 the owners of these vehicles were given the opportunity to take up licences for a new category of vehicle - the 'Light Bus' - either a public light bus (which could ply for hire) or a private light bus (which could not). Subsequently, under a revised 1982 Road Traffic Ordinance, some light buses were licensed to operate fixed fare / fixed route scheduled services to areas where franchised bus services did not or could not operate. The number of licences issued was determined by the government which in 1969 numbered 3,458 for public light buses. These increased to around 4,350 in the mid 1970s and remain around this level today (see Table 1.3). In addition there were 2,481 private light buses licensed in 1996. For 1996 passengers carried by light bus (public light bus and green minibus) are estimated at 640 million.

The beginning of transport planning

The need for long term planning in public transport took some time to be recognised. As in other areas of the economy the government adopted a policy that was de-

Table 1.3 Passenger-carrying on-road vehicles 1947-1996

	Private cars and motor cycles	Taxis/ public cars	Trams	Buses	Public light buses	Private light buses
1947	4,457	615	112	146		
1951	10,662	629	124	431		
1956	21,135	627	146	644		
1961	39,792	1,436	152	1,010	603*	
1966	67,460	3,728	162	1,920	2,065*	
1971	122,566	4,290	162	3,075	3,813	
1976	134,968	6,316	162	3,919	4,346	
1981	238,996	11,061	163	5,720	4,350	924
1986	179,049	16,538	163	6,253	4,350	2,265
1991	259,733	17,308	163	8,104	4,350	2,561
1992	290,645	17,537	163	8,447	4,350	2,525
1993	318,705	17,639	163	8,884	4,350	2,564
1994	340,325	18,111	163	9,292	4,350	2,589
1995	347,329	18,190	163	9,599	4,350	2,585
1996	355,297	18,126	163	10,265	4,348	2,481

Source: Transport Department, Hong Kong Government * *Dual-purpose vehicles*

scribed as 'positive non-interventionism' while the operators' forward planning did not extend much beyond what was required to satisfy shareholders in the company annual report. In 1947 a Hong Kong Preliminary Planning Report prepared by Sir Patrick Abercrombie proposed a cross harbour link for a re-aligned route for the Kowloon-Canton Railway but otherwise ignored public transport. The first serious study came in 1951 in a report by the Harbour Ferry Services Advisory Committee which, apart from making proposals on ferry services, took a somewhat broad interpretation of their terms of reference, and made recommendations on tram and bus services (some of which were accepted) and the setting up of a Transport Executive Board (which was not).

With a rising population, the onset of increasing traffic congestion and the deteriorating standards in public transport services the government commissioned the 'Hong Kong Passenger Transport Survey' which was carried out in 1964-66. It included an examination of the public transport system and the survey report noted:

> *If not entirely unique, Hong Kong's traffic problems are somewhat unusual and ... different from most urban areas of the world. Elsewhere traffic planners are attempting to arrest the transfer of as many passengers as possible from public to private transport. ... In Hong Kong such problems do not arise. Rather the situation is that each year the profit making public transport companies find it more difficult to put extra capacity on to already overcrowded routes.*

The main recommendations in the report had regard to the need for more efficient use of existing public transport facilities - including public transport priorities at intersections and reserve lanes - and the need for 'a new level of public mass transportation' which was received as a coded message for an underground railway.

The second age of the train

The establishment of the Mass Transit Railway Corporation

After the 1964-66 survey, a Traffic and Transport Survey Division was formed within the Hong Kong government and, with help from consultants, detailed studies were made on long-term road requirements and mass transport. The 1967 'Hong Kong Mass Transport Study' recommended an underground railway with four lines totalling 64 kilometres. However, with lower population forecasts government decided to concentrate attention on two lines in a 'Y' shape configuration to link central Hong Kong Island to Tsuen Wan and Kwun Tong on the mainland. After further studies a decision to go ahead was taken in 1972 and in 1973 it was announced that a corporation would be established to finance, build and operate the Mass Transit Railway - the MTRC.

Four international building consortia expressed an interest but one, headed by Mitsubishi of Japan, submitted a pre-emptive bid and a letter of intent was signed with the then Provisional Mass Transit Railway Authority in 1974. When the consortium's attempt to alter the terms was not accepted the offer was withdrawn, and a year wasted. The Authority abandoned the single contract concept and went ahead with the Kwun Tong line on a multi-contract basis within the original financial limit of $HK5 billion and a government equity holding of $HK800 million.

Construction of the new mass transit line began in 1975 in difficult working conditions, under congested streets in a mixture of decomposed rock, granite boulders and reclaimed land. It was a notable achievement that the line from Kwun Tong to Central opened in stages between October 1979 and February 1980 within budget and ahead of schedule (see Figure 1.5). The enabling legislation allowed the Corporation to develop its land-holdings as well as build the railway. The good financial returns from joint building projects over the depot in Kwun Tong and at the stations encouraged an early decision in 1977 to go ahead with the Tsuen Wan line and this was completed in May 1983, again ahead of schedule.

By 1983, the capital costs amounted to $HK10 billion for 26 kilometres of track, 52 train sets and 25 stations. With trains running at up to 2.5 minute frequencies and keeping within two minutes of schedule for 98 per cent of the journeys the railway earned a reputation for efficiency among the passengers: 1.2 million of them each day by the end of 1983. Peak hour travel was crowded but speed and air-conditioning pleased the commuters.

The next stage was the Island Line and here the cost advantages were not as clear cut. Hongkong Tramways had earlier put forward proposals to modernise its system and the government commissioned a study on an upgrading of the tramway route, using MTR tunnels through Central District, to create a fast light rail system which, it was suggested, could be operated by the MTRC. The light rail option was financially attractive but its passenger capacity was forecast to be inadequate by 1996. The Mass Transit Railway Corporation submitted its proposals for the Island underground line at an estimated cost of $HK7.095 billion. These were accepted by

Figure 1.5
The first MTR train
Source: Courtesy of the Information Services Department, Hong Kong Government

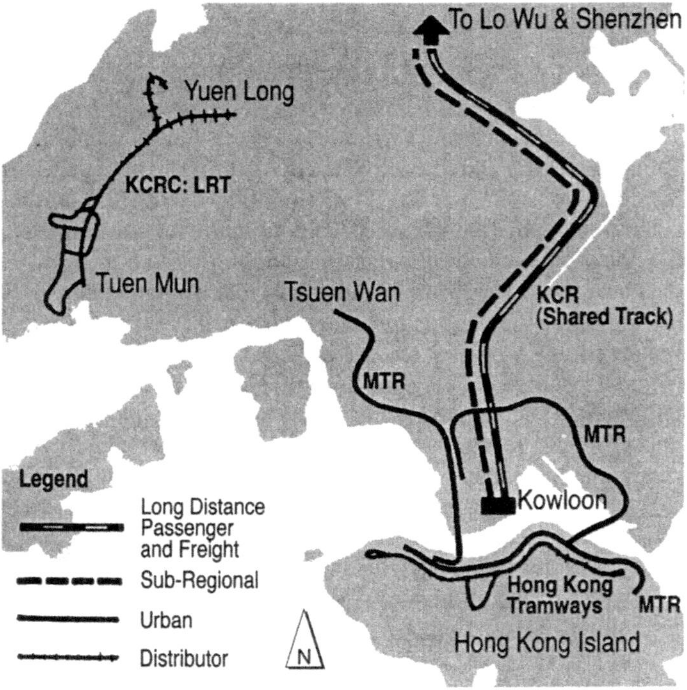

Figure 1.6 Hong Kong's rail transit system prior to new airport extension
Source: Transport Department, Hong Kong

the government in 1980. Construction was completed in May 1986 and by the end of 1987 the whole system was carrying 1.8 million passengers a day and by the mid 1990s over 2.2 million.

The railway was a success in its construction, in patronage and in the relief it gave to road transport. Although the financial return was better than in other metropolitan areas, the debt situation was a cause of concern. The Corporation, with the implicit support of the government, became a very efficient corporate borrower both locally and internationally. By 1985 the MTRC had outstanding loan debts of $HK18.7 billion. From 1986 onwards, increasing revenue earnings, lower interest rates and a rebounding property market improved prospects. A further section of railway was completed in 1989 joining the Kwun Tong Line to the Island Line at Quarry Bay. This was done at little capital cost to the Corporation as the harbour section was constructed by the Eastern Harbour Tunnel Company alongside its new tolled road tunnel and then leased to the Corporation (see Figure 1.6).

The establishment of the KCRC

It took some time to realise that the single track Kowloon-Canton Railway could have an important role in the development of Hong Kong's new towns. Although it served Sha Tin, Tai Po, Fan Ling and Sheung Shui, with its one passenger train every 54 minutes, it did not make any substantial contribution to the public transport services. Apart from the change from steam to diesel powered locomotives little change occurred in the first 60 years of the railway's history. However, when change came, it came fast.

Tsuen Wan was the first of Hong Kong's new towns, then Sha Tin and Tuen Mun and, following the establishment of the New Territories Development Department in 1963, the north east New Territories received greater attention with the new towns of Tai Po, Fan Ling and Sheung Shui, all on the route of the KCR. The double tracking from Kowloon as far as Tai Po was authorised in 1973. Further studies were undertaken to examine the feasibility of continuing this for the whole British section of the line to facilitate increasing cross border freight traffic and passenger services to the new towns. The recommendations, quickly approved in 1978, were for the completion of the double tracking and replacement of the antiquated diesel hauled passenger rolling stock by electric multiple units. The original 1910 Land Grant allowed for double tracking and the only major civil engineering was in a new Beacon Hill tunnel. The Tsim Sha Tsui terminus had been moved to Hung Hom in 1975 to allow for land development. Other stations were rebuilt as part of the modernisation programme, together with a new depot at Fo Tan. The modernisation took five years to complete at a modest cost of $HK3.5 billion. The first electric trains ran to Sha Tin in May 1982 and as far as Lo Wu by July 1983. A new station was built at Kowloon Tong where passengers could interchange with the MTRC network, a link which was further improved when the two railway corporations introduced a common stored value automatic ticketing system (see Chapter 6).

As a government department the KCR was hampered in its commercial operations. Following the successful example of the MTRC, the government changed its status to that of a public corporation in December 1982. In 1981, before the modernisation, the number of passengers carried was 17 million. With a rise in local commuter traffic and an increased number of trains on the re-introduced through service from Guangzhou, the number had risen to 80 million in 1984, 100 million in 1985, and had doubled to 200 million in 1992 (see Table 1.4).

Public transport and the New Towns

The transport requirements for the first new towns were erroneously based on the concept of local integration and demand for inter-urban services was underestimated. As a result the infrastructure provided for local public transport did not extend much beyond setting aside space for bus stops and termini.

However, by the 1970s, when the development of Tuen Mun began, the opportunity was taken to incorporate reserved public transport routes in the layout of the new town. A 1978 study concluded that a light rail system would, in the long term, have advantages in terms of fuel cost and environmental conditions over a bus network. After long negotiations (described in Chapter 6), the government decided ultimately to invite the KCRC to take over the light rail project. Enabling legislation was passed in 1986 and work started in the same year. The planned system was extended to include a link to the nearby town of Yuen Long and later to Tin Shui Wai. The 25 kilometre LRT network was opened in phases from September 1988 with five routes supplemented by ten feeder bus routes. Revenue was less than pre-

Table 1.4 Rail passengers by type of rail transport, 1981-1996

(thousands)

	KCR heavy	MTR	KCR light	Tram
1981	17,123	222,914		160,442
1982	22,215	351,082		143,371
1983	48,110	411,955		131,093
1984	79,973	410,556		122,502
1985	103,270	462,894		120,831
1986	115,866	532,074		122,245
1987	136,352	592,912		128,677
1988	154,615	629,857	16,112	131,917
1989	173,094	687,598	62,596	127,652
1990	181,077	719,111	73,149	127,643
1991	190,668	725,966	82,061	123,247
1992	199,905	751,005	92,273	124,087
1993	208,863	778,509	106,555	125,211
1994	222,290	804,062	116,980	124,636
1995	233,804	812,519	122,651	112,963
1996	248,331	816,572	125,124	107,816

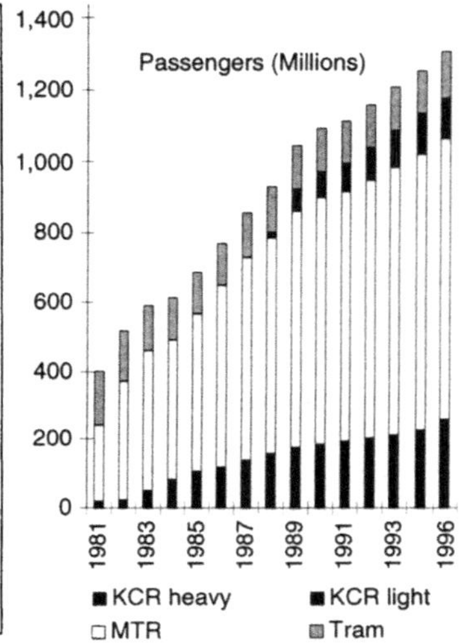

Source: Transport Department,Hong Kong Government

dicted but this is expected to improve as Tuen Mun expands and new lines are opened. Public transport in the other new towns was based on bus services plus, on the eastern side of the New Territories, the KCR heavy rail service.

The quality of public transport

The rapid development of urban rail transport in the ten years from 1979 to 1989 transformed public transport in Hong Kong. It meant that about 20 per cent of trips were taken off the congested road system and showed that many passengers, when given the choice, were prepared to pay for higher quality services. Other operators responded to the competition. When some light bus and taxi operators introduced air-conditioned vehicles in 1983 it was not long before most of the others followed their example. The Citybus Company, which started operating in 1979, introduced the first air-conditioned bus service from Sha Tin new town. This was a 'residential' coach service, outside the system of franchised bus services. The company later extended its range of upmarket services and was granted its first franchise in 1991 for a single route, took over 26 routes on Hong Island in 1993, and added a further 14 routes in 1995. Both the Kowloon Motor Bus Company and the China Motor Bus Company have followed the market trend towards quality with air-conditioned buses on selected routes.

In retrospect, the decade 1979-1989 may prove to be unique in the history of public transport in Hong Kong. In 1978, when the estimated population was 4.7 million, an average of 6.3 million passengers were carried each day on public transport (see Figure 1.7). In 1989 there were 9.8 million daily passengers when the population was 5.8 million; by 1996, 10.7 million passengers relied on public transport from a total population of 6.3 million. In 1978 those travelling by railway and tram comprised 7 per cent of total trip-makers; in 1989 railways, light rail and tramways carried 30 per cent of total public transport patronage, and in 1996, 33 per cent.

Transport planning, regulation, public opinion and politics

Introducing the ethos of planning and regulation

With Hong Kong's strong commitment to laissez faire economics any planning initiatives on the part of the government were viewed with some suspicion. Government therefore tended not to interfere until it really had to. A Transport Department in government was not formed until 1968 while a separate policy branch for transport did not come into being until 1981.

Since the first Hong Kong Passenger Transport Survey in 1964 there has been a continuing flow of planning studies on transport e.g. the Mass Transit Studies from 1967 onwards; the 1968 Long Term Road Study; and the 1976 and 1989 Comprehensive Transport Studies (see Chapter 3). These sought to estimate future requirements in the context of current transport policy and possible future policy options. However, as some 80 per cent of all passenger trips are made on public transport, public transport has been a major component of planned transport provision. Also of significance is the relationship between land use and transport. This has been the

Figure 1.7 Passengers by mode of transport, 1967 -1994

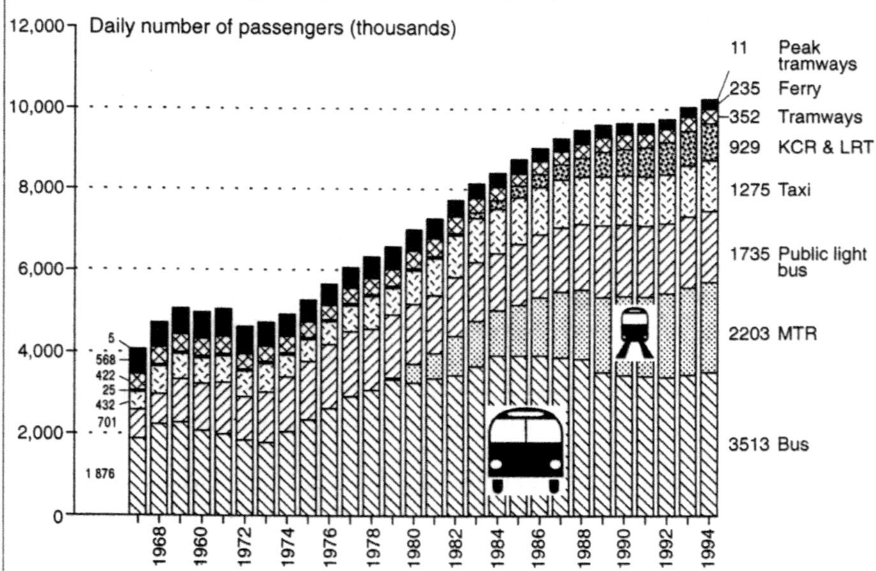

Note: KCR = Kowloon-Canton Railway; LRT = Light Rail Transit; MTR = Mass Transit Railway
Source: Census & Statistics Department, Hong Kong Government.

focus in the more recent studies in Hong Kong as the territory's economic and urban structure rapidly evolves and changes. It is an aspect especially dealt with in Chapter 3 as well as subsequent chapters of this book.

Public transport operators, who in the past did little long term planning, have been drawn into the government planning process. They have been consulted in the course of the government studies. In addition, in their enabling legislation, the ferry and bus companies are required to prepare rolling five year plans on changes and extensions of their services. Where the operators once kept to themselves and considered co-operation as potentially detrimental to their commercial interests, it became clear (with some coaxing from government) that there were advantages in working together and co-ordinating service provision. The introduction of cross harbour bus routes were a start and later, after some initial resistance from the bus companies, feeder bus services to new termini at railway stations enhanced inter-modal co-ordination. Some of the buses serving railway stations have now been equipped with ticket readers so that the railways' stored value tickets can be used.

Post-war regulatory legislation on public transport provided the government with considerable powers to control fares, determine routes and frequencies and the types of vehicles used. The powers were exercised to good effect in ensuring that adequate public transport was provided both to the new towns and to individual housing estates in the early stages of their development. Initially, these routes, like those in rural areas, had to be cross-subsidised from the profitable routes in the urban areas. Apart from these measures, government allowed operators fairly free rein to adjust their services to respond to public demand (as they saw it) except in some instances where the government decided to intervene to limit inter-modal competition.

This was most evident when the MTR was first opened and government policy makers wanted to remove or restrict bus services which ran parallel to the railway, particularly on cross-harbour routes. These routes commanded higher patronage and higher fare scales, and were therefore the most profitable to all the operators. The restrictions were, however, gradually whittled away as it became evident that passenger demand was growing at such a rate that none of the operators was going to be short of customers.

Some restrictions were also imposed by government for traffic control reasons, particularly on public light buses which, in the most congested areas, are limited to specific routes while those operating on scheduled routes are partly protected from competition by the 'free-range' public light buses.

Public opinion and politics

An interesting question to pose in retrospect, is whether the Hong Kong public have had any real say in the development of their public transport system? Given past developments, one would have to acknowledge that initially they did not. Although, in the absence of elected government, there was a system of advisory committees established in Hong Kong, and a Transport Advisory Committee (TAC) was formed in 1965. Even though the committee has no statutory powers any major decisions requiring approval by the Governor-in-Council, for example, fare changes or the granting of franchises, are made with the advice of this Committee. More recent constitutional changes have allowed for more public debate on transport matters. The elected District Boards (formed in 1982) each have a committee dealing with local traffic and transport matters. The Legislative Council, now with mainly directly elected members, brings to the planning process a political element which did not exist before, although no new transport policies have so far emerged from outside the administration (see Chapter 12).

Conclusions

This brief history shows that until recent times transport has had little influence on land-use planning. Public transport provision largely developed in response to demand although when services have been established they have influenced the speed and density of urban development. Government intervention was minimal until the post-World War II years when initiatives in public housing, the development of new towns and the disposition of industrial land made it obvious that the aims would not be achieved without concurrent forward planning for land use and the provision of transport services.

A combination of rapid economic growth, concentrated urban population, limitations on the use of private transport (lack of parking space and high vehicle taxation) has led government to create a planning framework and to develop policies which encourage public transport without too much direct involvement in the management of the operators. This has created an environment where public transport has been able to prosper financially and expand with a wide variety of modes and a choice of service quality, speed and fares which is the envy of many cities elsewhere in the world. The government has, however, been less successful in the related, and politically more difficult area, of sustaining an effective policy to deal with increasing traffic congestion or to grapple with what most transport planners regard as an oversized fleet of taxis. The following chapters expand on this experience and perhaps provide some useful lessons.

References

Arnold, J. (1974) *Hong Kong's Famous Funicular - The Peak Tram*. Peak Tram Company, Hong Kong.

Atkinson, R.L.P. & A.K. Williams (1970) *Hongkong Tramways*. London: The Light Rail Transport League.

Barnett, M. (1984) *Tramlines*. South China Morning Post, Hong Kong.

Chiu, T.N. (1973) *The Port of Hong Kong*. Hong Kong: Hong Kong University Press.

Endicott, G.B. (1958) *A History of Hong Kong*. London: Oxford University Press.

Hong Kong Government (1993) *Hong Kong - 25 Years' Development Presented in Statistical Data and Graphics (1967-92)* Hong Kong: Government Printer.

Hutcheon, R. (1986) *Wharf - The First Hundred Years*. Hong Kong: Wharf (Holdings) Ltd.

Leeds, P.F. (1975) *The Development of Public Transport in Hong Kong: an historical review 1841-1974*. Hong Kong: Government Printer.

Phillips, R.J. (1990) *Kowloon-Canton Railway (British Section) - A History.* Hong Kong: Urban Council.
Rimmer, P.J. (1986) *Rikisha to Rapid Transit.* Sydney: Pergamon Press (Australia) Pty, Ltd.
Transport Department (1979) *Keeping Hong Kong Moving: A White Paper on Internal Transport* Hong Kong: Government Printer.
——— (1990) *Moving into the 21st century: The White Paper on Transport Policy in Hong Kong.* Hong Kong: Government Printer.
——— *Annual Transport Digests* (various years) Hong Kong: Government Printer.

In addition to the above the following Hong Kong reports and surveys were consulted:
1908 Progress Report on Construction of the British Section of the Hong Kong-Canton Railway to 31st December 1907.
1919 Report on the Improvement of Jinricksha Services during the Year 1919.
1928 Report on the Vehicular Ferry, Hong Kong-Kowloon.
1948 Hong Kong Preliminary Planning Report, Sir Patrick Abercrombie.
1951 The Harbour Ferry Services Advisory Committee Report.
1960 Hong Kong Traffic Survey Road Research Laboratory.
1966 Hong Kong Passenger Transport Survey, Road Research Laboratory.

2 Genesis of land-use planning and urban development

Alexander R. Cuthbert

Introduction

In this chapter a rather broad-brush approach to the history of land-use planning and urban development in Hong Kong will be taken. Apart from setting the context for readers who may be unfamiliar with the territory, there are three other reasons why such an approach is necessary. Firstly, it is difficult to consider 'land' as an independent factor in urbanisation. Secondly, it is conceptually inappropriate to detach the concept of 'land use' from other economic, political and technical considerations within society. Finally, the term 'land use' is an analytically treacherous concept whose meaning is often best defined by the person using it. In this last sense, when referring to land use, an architect will think of buildings, a lawyer will conceive of leases, a surveyor of rateable value, a traffic engineer will focus on trip generation/attraction, a planner will think of zoning categories, and a farmer of crops. As early as 1959, Guttenberg had enunciated the six key dimensions of land use, namely:

- ownership;
- the site;
- the site facility;
- an activity or set of activities taking place on the site; an enterprise conducting this/these activities; and
- effects of activity(ies), such as pollution, energy consumption, traffic generation, density of development etc.

In providing the above breakdown, Guttenberg ignores, however, the relationships between the dimensions of land use and the social structure of the society the land use serves. While it can be appreciated that definitions of land use are clearly contextual, the inalienable fact is that the private ownership of land (or the right to utilise it through leases) is the cornerstone of successful economies such as Hong Kong, and as we shall see, revenue from land development has always played a vital role in the government's economic affairs. Current land-use distribution in Hong Kong is shown in Table 2.1.

Table 2.1 Land-use in Hong Kong, 1996

	Class of land	Area (km^2)	(%)
A.	**Developed lands**	**170**	**15.6%**
i.	Commercial	2	0.2%
ii.	Residential	42	3.8%
iii.	Public Rental Housing	12	1.1%
iv.	Industrial	11	1.0%
v.	Open Space	16	1.5%
vi.	Government, Institution and Community Facilities	18	1.6%
vii.	Vacant Development Land	41	3.8%
viii.	Roads/Railways	27	2.5%
ix.	Temporary Housing Areas	1	0.1%
B.	**Non-built-up lands***	**922**	**84.5%**
i.	Woodlands	220	20.2%
ii.	Grass and scrubs	519	47.5%
iii.	Badlands, swamps and mangroves	44	4.0%
iv.	Arable	63	5.8%
v.	Fish ponds	16	1.5%
vi.	Temporary structures/livestock farms	12	1.1%
vii.	Reservoirs	26	2.4%
viii.	Other uses	22	2.0%
C.	**Total lands**	**1092**	**100.0%**

Note: * *Within these are 413 km^2 of country parks and special areas designated under the Country Parks Ordinance for protection of vegetation and wild life*

Source: *Government Information Services, Hong Kong*

Hong Kong - origins

Today Hong Kong has a land area of 1,092 sq. kms. The territory is so small that it may be thought of as entirely 'urban', since even the land which is not built on services the urban area for agricultural, recreational and tourism purposes. As is well known, the forces which brought Hong Kong into being originate from the policies of the great world powers in the middle of the nineteenth century vying for dominance and supremacy through territorial expansion, as a means of acquiring raw materials and creating new markets. Today their influence comes more in the form of financial investment and trade associations.

In the beginning, none of the traditional concepts as to how colonies become established applied to Hong Kong. Firstly, at the cession of Hong Kong in 1843, the population of Hong Kong Island was just over 4,000 persons (see Table 2.2), equally engaged in subsistence agriculture and fishing. The fishing population were the 'boat people' many of whom did not even live on-shore. So, for all practical purposes, in the earliest days of Hong Kong's development, there was no significant existing population to exploit.

Secondly, Hong Kong Island had a land area of only 32 square miles (83 sq. km.). The agreement of 1869 only extended this by three and a half square miles. In

1898, the Convention of Peking which granted the lease of the New Territories for 99 years, established the final land area of the colony at some 390 square miles (1,070 sq. km.) much of which was mountainous and marshy or otherwise unproductive. The idea, therefore, of using Hong Kong as a territorial resource was not only unrealistic but was never the case. Thirdly, not only is the land area of Hong Kong minute and generally poor in agricultural quality, it has always been insufficient to support even a modestly expanding population. Much of the land is barren, and therefore has never been able to support a local supply of timber for shipbuilding or ship repairs. Local sources of energy have, furthermore, always been almost completely absent. Having no minerals to extract which could justifiably be hauled half-way round the world to fuel British industry, Hong Kong's value to Britain as a colony has historically never been typical.

Fourthly, it is only relatively recently that there have been internal markets to exploit, or wealth to be made from what grew on the land or what was stored beneath it. Exporting finance from Britain to 'capitalise' on this situation was therefore a futile exercise. As a treaty port, however, Hong Kong clearly served its purpose. But the rapid development of its own economy, particularly after 1890, conclusively demonstrated that its trajectory was entirely different to that of a simple trading

Table 2.2 Urban development of Hong Kong Island: 'Census' of 1841

			Population
Chek-chu	赤柱	The capital, a large town	2,000
Hongkong	香港	A large fishing village	200
Wong-nei-chung	黃泥涌	An agricultural village	300
Kung-lam	公岩	Stone-quarry-poor village	200
Shek-lup	石凹	Stone-quarry-poor village	150
Soo-ke-wan	掃箕灣	Stone-quarry-large village	1,200
Tai-shek-ha	大石下	Stone-quarry, a hamlet	20
Kwun-tai-loo	群大路	Fishing village	50
Soo-koon-poo	掃竿浦	A hamlet	10
Hung-heong-loo	紅香爐	Hamlet	50
Sai-wan	柴灣	Hamlet	30
Tai-long	大浪	Fishing hamlet	5
Too-te-wan	土地灣	Stone-quarry, a hamlet	60
Tai-tam	大潭	Hamlet, near Tytam bay	20
Soo-koo-wan	東鼓灣	Hamlet	30
Shek-tong-chuy	石塘嘴	Stone-quarry, a hamlet	25
Chun-hum	春坎	Deserted fishing hamlet	00
Tseen-suy-wan	淺水灣	Deserted fishing hamlet	00
Sum-suy-wan	深水灣	Deserted fishing hamlet	00
Shek-pae	石牌	Deserted fishing hamlet	00
			4,350
		In the Bazaar	800
		In the Boats	2,000
		Laborers from Kowlung	300
		Actual present population	7,450

Source: 1841 Census, Hong Kong

installation. Its later development into a full-blown urban economy denied all of the elementary assumptions by which colonies were typically established. More so, because at each stage of its development the British government must have received some kind of tacit support from China in order for it to exist at all.

The development of Hong Kong: 1842-1942

Government franchises and land development reserves

While in other parts of the world, colonialism has often had devastating effects on third world economies, Hong Kong has prospered in contradiction to this experience. Where nothing previously existed, a thriving economy was created in the face of remarkable adversity. The provisions for public housing, health, education, and limited democratic rights in Hong Kong are claimed to represent pinnacles of achievement. Because of Britain's need to confine the territory's function to that of an entrepot for China - for banking, finance, trade and as a military garrison - historically, investment funds exported from Britain went directly to China, and not to Hong Kong. As late as 1929, 77 per cent of Britain's investment in China went to Shanghai, and only 9 per cent to Hong Kong (Halliday 1974).

Hong Kong government revenue from the outset was tied to two major sources: the sale of monopolistic franchises and rent from land. These franchises have been, and still are, owned by a small group that enjoy extensive financial freedom and considerable influence on the economy of Hong Kong. Land leases have long been sold by auction to the highest bidders, most of whom belonged to this privileged small group. Historically, many such interests were established for the sale of opium, salt and the quarrying of stone. Subsequently, other markets were treated in the same way, and monopolistic licences were sold on this basis. Such practices were thus formative in the setting up of the colony, from as early as 1845. Hong Kong's reliance on the sale of monopolistic franchises was only exceeded in economic terms by its dependence upon revenues accrued from land leases and land development premiums.

The first British settlement in Hong Kong was at Possession Point, named 'Victoria' (see Figure 2.1). It centred on what is now Queen's Road, and terminated at the offices of Jardine Matheson & Co., the largest 'Hong' (Hong Kong trading company). Chek Chu (Stanley) was the biggest settlement in 1841 with a population of some 2,000 persons (see Table 2.2). By 1845, rent from land leases was the main source of government revenue, and land speculation was already widespread. Nonetheless, the Royal Asiatic Society managed to piece sufficient land together to open the Royal Botanical Gardens in 1854.

The combined effect of migration to Hong Kong resulting from the Taiping Rebellion in 1854 and the cession of Kowloon in 1861 placed increasing pressure on the physical resources of the city. The Surgeon General's report on the sanitary conditions of the colony at the time was such an indictment of the government that it was suppressed by the governor. Land speculation increased in the face of extreme

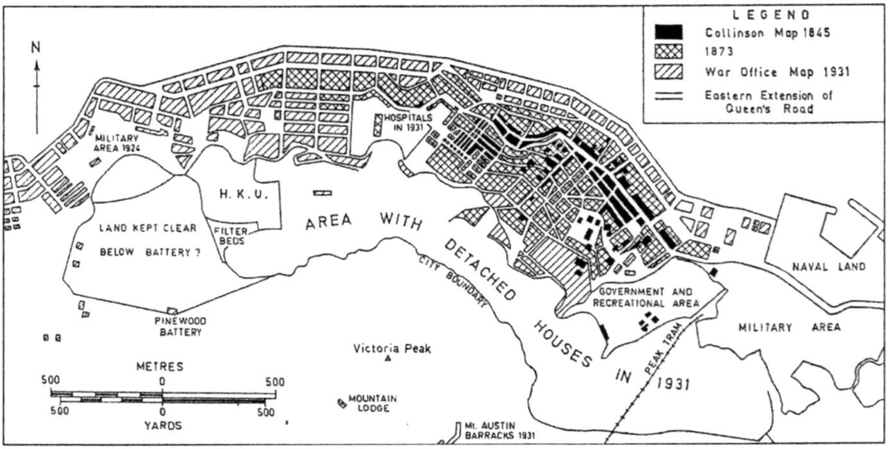

Figure 2.1 The changing face of Hong Kong
Source: Talbot, H.D. in Dwyer, D.J. (ed) 1971

overcrowding, while insufficient provision of public works had the effect by 1877 of government no longer maintaining the previous policy of segregating Europeans from Chinese.

Due to the widespread take-over of European property by Chinese speculators, a concession was granted allowing Chinese ownership of property in Hong Kong's central district. The initiation of the first major water project at Pok Fu Lam in 1863, and a much larger one in 1871, followed closely by the construction of Tai Tam reservoir in 1883 and the opening of the Peak Tram in 1884 were all partial responses to this development in that they allowed westerners to escape to yet another enclave of their own. By 1885, overcrowding had become so severe in central Hong Kong that a land commission was set up to advise the government on urban planning matters. Four years later, two important pieces of legislation were introduced.

The first was the Buildings Ordinance (1889), and the second, the Crown Lands Resumption Ordinance (1889) which was introduced to deal with the squalid housing conditions of the time. Although land reclamation in Hong Kong had been ongoing since the colony was founded, the first major reclamation schemes initiated specifically to alleviate problems of overcrowding took place in Causeway Bay between 1833-34, and in Kennedy Town from 1886. Thereafter reclamation became the major strategy used to resolve land-use problems in the city. However, it was not until 1894 when the bubonic plague ravaged Hong Kong that any real action was taken to improve sanitary conditions within the urban area (Bristow 1984).

Hong Kong's entry into world markets

In the 1890s the increasing intervention by Russia, France and Germany in the Far East, along with corresponding concessions made to them by China, created demands in Hong Kong to extend the colony's northern boundaries as a defensive measure against potential aggression by a competing European power. Hence, the area called the 'New Territories' and a small island called Lamma were leased to Britain from China on 11 June 1898. The same year also marked Hong Kong's entry into the world financial market, using merchant finance to underwrite the cost of constructing railways in southern China. By the turn of the century, the population of Hong Kong had grown to 300,000 persons of whom only 10,000 were non-Chinese civilians. This proportion, approximately two per cent of the total population, remains much the same today.

From this period until the end of the Great War in 1918, Hong Kong experienced steady economic growth and administrative development, surviving on the basis of increasing commerce and trade with all parts of the world. Hong Kong as yet had no industry to speak of, although some of the original industries related to shipbuilding and repair, dried fish products etc. continued to grow. The first cotton spinning company had been introduced in 1898, but quickly went out of business and moved to Shanghai. During this same period, two major transportation projects commenced: the electric tram service (inaugurated in 1905) and the Kowloon to Canton railway (opened in 1906) (see Chapter 1). The latter allowed the integration of the then urban area with Shau Kei Wan, an old established fishing and shipbuilding community on the north-eastern shores of Hong Kong Island.

The initial major development of Hong Kong's industrial growth came in two stages. The first was as a consequence of the Great War which eliminated many European produced commodities from the world markets. The second was due to the Ottawa Agreement of 1932 which gave Hong Kong preference and favourable trading opportunities within the British Commonwealth. Hong Kong, as a result, managed to survive economically. This was in spite of the fact that much industrial capital from Britain went primarily to Shanghai (rather than Hong Kong) right up to the beginning of the Second World War in 1939 (the year, incidentally, in which town planning legislation was introduced to the territory). The textile industry was re-introduced to Hong Kong during this period. Its revival, however, was short-lived since Hong Kong was overrun by the Japanese in December 1941. By the end of the war, Hong Kong was described as 'a run-down, war-damaged, pre-industrial society with no very evident future' (Youngson 1982).

Urban development: 1945-1975

The emergence of an industrial economy

The Japanese surrender on 16 September 1945, not only symbolised a transition from war to peace for Hong Kong, but also from a mercantile to an industrial economy. Up until 1945 industrial development in Hong Kong had barely begun. This of course was all soon to change. After the Japanese occupation in 1945, the population of Hong Kong shrank from 1.6 million in 1941 to 0.6 million in 1945 (see Figure 2.2); this was largely through emigration, deportation and death from Japanese aggression. Hong Kong's economic system was in a state of total collapse, the built envi-

Figure 2.2 Hong Kong population change, 1846-1996

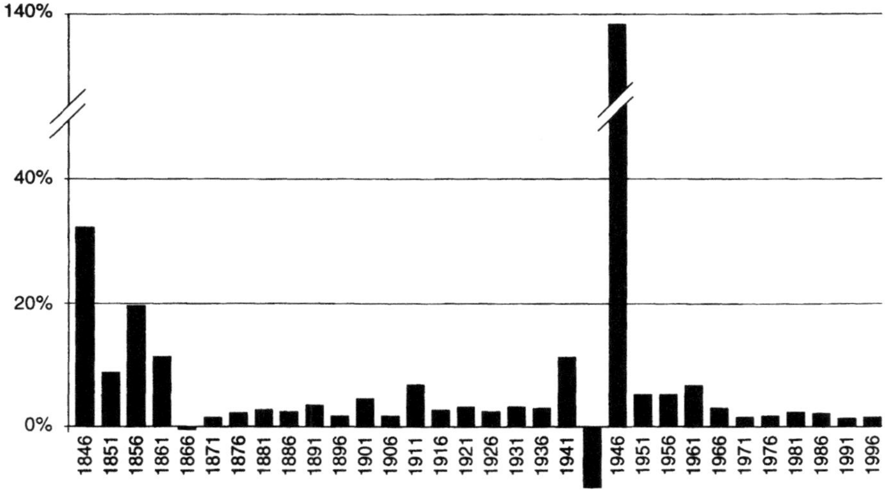

ronment was close to dereliction, but most important of all, the sense of confidence which characterised and sustained Hong Kong's historical evolution had vanished.

Immediately after the war, government revenues from taxation, the sale of franchises, land leases and land improvements, together with other revenue-producing services, such as stamp duties, coinage etc., all went largely into reclamation works and other public works projects. These included dock schemes, road engineering schemes, government building projects, and dam and reservoir schemes, the last of which provided a barely adequate supply of fresh water for a burgeoning population. To implement these projects, construction workers were imported, largely from China.

Establishment of the Town Planning Board

The first meeting of the Town Planning Board (established in 1939), took place in 1951. Its remit was to produce a workable master plan for the colony, and to tackle and accommodate a combination of economic and political problems, many of which were related to high rates of immigration, poor infrastructure provision and a much deteriorated built environment. The meeting of this board was, in some ways, merely a symbolic event. It marked a tacit recognition by the Hong Kong government that a burgeoning economy was now operational which required planned infrastructure. By embarking upon public utility improvements, the government furthermore accepted that outbreaks of disease could no longer be tolerated, and that adequate housing had to be provided to eliminate or (if necessary) control outbreaks of civil unrest. Enormous pressure was also brought to bear on government to provide land for commercial and industrial development. The Hong Kong government appeared to vacillate in coping with the realities forced upon it by the transformation of an economic system over which it appeared to have little control. This was reflected in the government's response to the city plan prepared by Sir Patrick Abercrombie in 1948. By and large, government ignored the plan, and disbanded the Town Planning Department in 1951. It decided that all future planning would be carried out within the Public Works Department, despite the complication that the department's control over land development matters did not extend over most of the New Territories; a problem that was to remain until 1982 (Bristow 1984). Government complacency about housing its vast labour force, and its limited attempts to come to grips with a rapidly industrialising society resulted in it being forced into positive action. On Christmas day 1953, 50,000 squatters were left homeless when an enormous fire destroyed their settlement at Shek Kip Mei. In 1954 both a Commission for resettlement and a Housing Authority were established, which were charged with the tasks of dealing with the complexity of Hong Kong's housing problem (Castells et al 1990).

First property boom

Prior to 1955, Hong Kong's building ordinance limited buildings to a plot ratio of three, or to a maximum height of 70-80 feet (21-42 metres). In 1955, new regulations permitted this to be doubled. However, because building heights are also related to street width, plot ratios of up to 20:1 could occur, with the result that very high population densities of up to 10,000 persons per acre (about 25,000 per hectare) were possible (Bristow 1984). This change heralded a new era in which redevelopment became a highly profitable undertaking. The workforce could be accommodated into smaller areas, developers could profit from immense increases in floorspace from the same land area, and government could increase land prices and rateable values. Compensation paid to those whose properties underwent compulsory purchase, however, remained unaltered. Hence by 1954, the first post-war building boom took place, with government simultaneously realising that it had no choice but to provide a considerable amount of housing to accommodate its labour force. After Singapore, Hong Kong's public housing programme, accommodating at its peak more than 50 per cent of its population, was to become the second largest in the world in proportion to its population.

In 1961 the first post-war Census was taken. This allowed, for the first time, some realistic assessments of Hong Kong's growth and development. Of this census, Youngson (1982) notes two major deficiencies. Firstly, the Hong Kong government's statistics on changes in the distribution of income were conspicuously absent. Growth instead was presented only in terms of GDP statistics, which effectively masked changes in the distribution or redistribution of wealth. Secondly, because GDP statistics present what is bought and sold, an increase in GDP can also easily coincide with decreasing social benefits, such as decreasing leisure time, an increasingly degraded environment, and overcrowding. Factors which related to the general quality of life or changes in the conditions of existence of the general population were not assessed. The developments of the period were further commented on by Halliday (1974) who points out that between 1960-70, manufacturing output rose by

275 per cent and labour productivity by 207 per cent, while the manufacturing wage index rose by only 71.5 per cent.

The Hong Kong economy began to grow at an average GDP of 10 per cent per annum from 1961 over the following twenty years, peaking at 16 per cent in 1976 (see Table 2.3). By 1964, in spite of a rigorous public sector housing programme during the previous ten years, the number of squatters in Hong Kong had actually increased from 300,000 in 1953 to 600,000 in 1964 (Drakakis-Smith 1979). It was in part these disturbing social conditions, together with the birth of the Cultural Revolution in China in 1966, that led to the rioting which broke out during 1966-67

Table 2.3 Growth of expenditure-based GDP 1961-1996

(Volume indices 1990=100)

Year	GDP	Year	GDP	Year	GDP	Year	GDP
1961	10.6	1970	22.6	1979	48.4	1988	94.3
1962	12.1	1971	24.2	1980	53.3	1989	96.7
1963	14.0	1972	26.7	1981	58.2	1990	100.0
1964	15.2	1973	30.0	1982	59.8	1991	105.1
1965	17.4	1974	30.7	1983	63.2	1992	111.6
1966	17.7	1975	30.8	1984	69.5	1993	118.5
1967	18.0	1976	35.8	1985	69.8	1994	124.9
1968	18.6	1977	40.0	1986	77.3	1995	130.8
1969	20.7	1978	43.4	1987	87.3	1996	137.0

Annual growth rate of expenditure-based GDP 1961-1996

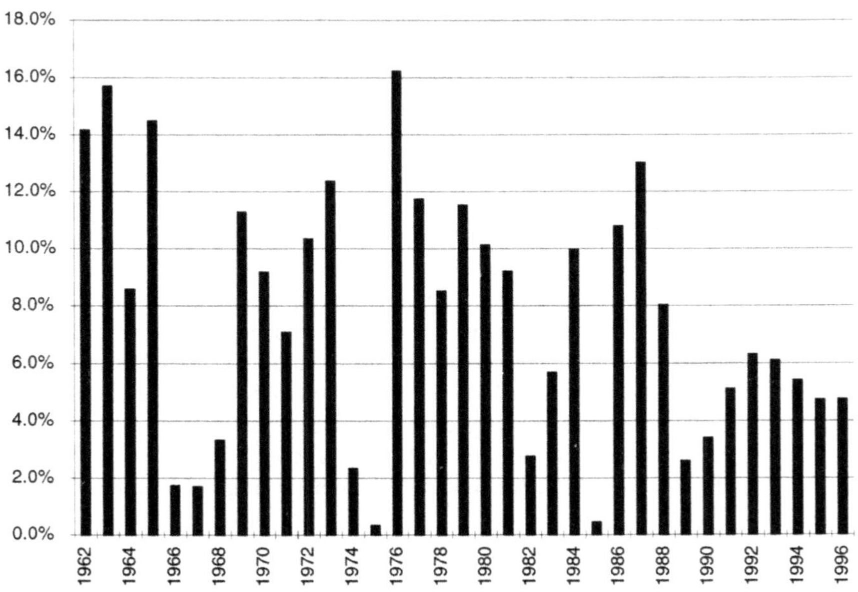

Source: Census and Statistics Department, 1997

in Hong Kong. One hundred people were to die in the territory as a result of the ensuing violence. One of the major economic effects of the riots was a crisis of confidence and a depression of the property market which was not restored until the early 1970s.

Hong Kong - an international finance centre

Changes in economic restructuring

The 1970s marked the start of a new period in global history. Beginning with the oil crisis of 1973-74, along with the economic and political consequences which flowed from it, the developed nations moved from modernism-industrialism into a new era of post-modernism/post-industrialism (Harvey 1989). The concomitant growth of service economies, increasing emphasis on the financial sector and the subsequent demand for central area sites was to transform Hong Kong in a matter of two decades. The physical space of the city was to become repackaged in line with developments in the world economy (Cuthbert 1994).

From 1973 onwards, Hong Kong began to emerge as a world-class financial centre, moving its economic centre of gravity in the process from one based on industrial production to one dependent upon the exchange of financial information and funds. Apart from the continuing importance of industrial development, Hong Kong was also to experience the increasing impact of its significance as a major international finance centre. The 1970s also marked the beginning of the internationalisation of Hong Kong in other ways, most importantly as a centre of tourism. This generated demands for new hotels, recreational activities and other tourist-related developments. However, the most important impact of Hong Kong's transition to finance capital and white collar jobs, later to have an enormous impact on residential land use, was the rise of a substantial middle class for the first time in the territory's history.

The general shift in Hong Kong's economic growth outlined above, the oil crisis of 1973-74, as well as the increasing protectionism of the world's core economies against the cheap exports of Hong Kong, Taiwan and South Korea, together had major detrimental impacts on Hong Kong's land use and economy throughout the 1970s. The first of these was a crisis in the territory's industrial production in the first half of the 1970s as a consequence of over-dependency on the manufacturing sector, which later led to extensive industrial decay in the territory. A similar crisis occurred from 1977-80, signs of which were reflected in the high rate of unemployment (9.1% in 1975 when the annual growth rate of GDP shrank to 0.3%), and in the drop in the real wage index (from 160 in 1973 to a low of 135 in 1975). The best lending rate also fell from 9-12 per cent in 1974 to an average of 5 per cent in 1977 (Leung et al 1986).

The second impact resulted from the stock market collapse of 1973, in part a product of the overall global economic recession of the time. This situation contributed to a consequent drop in property prices in Hong Kong in the latter part of the

1970s which obliged government to in effect subsidise the private sector by deferring the receipt of premium payments from land sales. This became the prelude to the property boom of 1976-81.

The third effect was the rapid rise in interest rates as capital flowed into the colony from overseas. This was accompanied by the movement to Hong Kong of a host of new international banking corporations, and extensive capital switching which took place from the manufacturing sector into the built environment (Leung et al 1986). The government's fiscal policy throughout this period was to maintain an environment which would encourage foreign investment in order to compensate for the balance of payments problems stemming from consistent trade deficits.

Housing, urban renewal and infrastructure programmes

The year 1970 marked an important point for Hong Kong's development, with the final draft of the Colony Outline Plan being submitted to the Land Development Planning Commission. While the document was termed a 'plan', it remained so in name only, since it was the government's first attempt at assembling a manual of basic standards to guide land-use development. While transport problems were addressed in the preparation of the document, the idea of an integrated transport and land-use plan for the colony was still an alien concept. The first comprehensive transportation study which was carried out for the government in 1973, did however have the necessary effect of prompting demands for changes to the Colony Outline Plan to more closely integrate land use and transportation.

During the late 1970s urban spatial structures in Hong Kong reached levels of intensification never previously encountered elsewhere in the world. This was especially reflected in the provision of public housing which became increasingly problematic, as more and more residential accommodation had to be provided in a city where there was intense competition for alternative land uses from more profitable industrial and commercial purposes. The Hong Kong government thus launched an extensive programme of infrastructural improvements to further open-up the territory. This began with the opening of the Cross Harbour Tunnel in 1970, and the beginnings of an underground mass rapid transit system in 1973. Both these projects stimulated new opportunities and patterns in transport circulation, land-use development, and population decentralisation.

The administration also began its largest housing programme since 1954, when in 1972 it commenced a new ten year programme with the aim of providing the entire population with improved affordable housing. This programme planned to provide 35,000 new units per year up to 1983 and to raise floorspace standards to 3.25m^2 per person from 2.2m^2 in 1955. None of this housing, however, was to come from any significant programme of urban renewal. The government's first renewal pilot project (between 1961 and 1979 in Western District) contributed little toward this need, except for the demolition of old tenements. None of those evicted by the clearances resulting from this project received any special consideration from the Housing Department (Drakakis-Smith 1979). Due to the lack of progress in urban renewal, there was little land left that could be developed within the existing city, particularly for social housing. This situation resulted in the construction of the new towns on the periphery of the metropolitan area, and the further urbanisation of the New Territories (see Figure 3.1 in Chapter 3).

These events were formalised by the formation of a New Territories Development Department (NTDD) in 1973. The task of NTDD was to coordinate the development of three new towns at Tsuen Wan, Sha Tin and Tuen Mun, as well as other projects in the New Territories, with two additional new towns (Junk Bay and Tin Shui Wai) being designated for the early 1980s. At the same time it must have become transparent to the government that the phenomenal residential densities being planned constituted an explosive political situation if the ordinary citizen had little contact with nature and leisure activities. Hence the Country Parks Ordinance (No. 10 of 1976) was passed, creating a country park system which now constitutes 40 per cent of the surface area of Hong Kong. Otherwise, land-use planning in Hong Kong was at the time, somewhat ad hoc and continued to be so throughout the 1970s.

Hong Kong - from 1980 to the present time

The housing question

September 1982 - when Prime Minister Thatcher visited Beijing to discuss the latent problem of Hong Kong's transition to rule by the PRC - marked the beginning of a period of economic and political uncertainty (Cuthbert 1987 and Henderson 1991). By this time the population of Hong Kong had exceeded five million, a problem in no way eased by the Joint Agreement between the British and Chinese governments reached in 1984 which limited the supply of new land to 55 hectares per year. Since then, an unprecedented acceleration has taken place in terms of redevelopment, urban development and housing tenure renewals, facilitated by new planning legislation. While this may be construed as purely coincidental, one can also speculate that these events represent an unstated part of the political agenda of the British government to capitalise, as far as is possible, on the economic opportunities offered locally prior to the takeover of Hong Kong by mainland China in 1997. This suspicion is somewhat reinforced in some people's eyes by the large number of British firms awarded contracts aligned to the current execution of the Port and Airport Development Strategy (PADS).

By 1982, at the end of the government's ten year housing programme, two and a quarter million people were housed in public housing. While this represents a considerable success by any terms, official estimates of the number of squatters at the end of 1981 were calculated to be a minimum of 720,000 (Castells 1986). This meant that squatters alone comprised 15 per cent of the entire population of the territory, without considering overcrowding, substandard, and other inadequate forms of housing. The government's inability to overcome the squatter problem was tacitly recognised in 1982 when a Squatter Area Improvements Division was established within the Housing Department, whose function was to facilitate the supply of services to certain long-standing squatter settlements which were likely to remain.

In the process of creating its housing programme the Hong Kong government has in effect created two cities within one, independent of each other both spatially and in terms of legislation. One accommodates, in public sector housing, half the population who cannot afford to purchase property from the private sector. The other relies on the private sector for its housing, and thus is subject to the speculative pressures of the real estate industry. Hong Kong has one of the few modern planning authorities which separates planning from housing, in that the original Town Planning Ordinance of 1939 only relates to private sector development. Public housing is instead covered by a separate housing act (1973). Although originally a government-run body, the Housing Authority is now an independent organisation which builds and manages not only housing but all associated land development functions within government estates. Because of the government's entrenched problems with urban redevelopment, one of the main avenues of housing provision had always been left out of its repertoire.

Territorial development and its outcomes

By 1984 existing problems in the urban areas had become so demanding that the government formulated the Territorial Development Strategy (TDS) as its foremost strategic planning process. Emerging from TDS came a series of sub-regional studies whose purpose was to identify areas of future growth. Transport data from each of these was required as a primary input to the government's Land Use and Transportation Optimisation model (LUTO). As is normal with this highly technocratic type of modelling, only certain quantifiable data were included (see Chapters 3 and 7). Larger political and economic considerations, social and community development as well as environmental and other important issues were all ignored, and it remains to be seen how accurate its outcomes are likely to be. The Study of Harbour Reclamations and Urban Growth (SHRUG) was the most complex of these and took place between 1982 and 1983, although the government had already suggested the site at Chek Lap Kok as the location for the new airport prior to the start of the study. SHRUG also included the other large economic generators such as the port, the railway terminus and the Central Business District. The development scenario was projected for the period 1991-2001, and was based upon development and transportation data using 1981 as a baseline. As main outcomes, the study proposed certain developments which are now taking place, in particular, the West Kowloon, Hung Hom Bay, Central-Wanchai and Western District reclamations, the expansion of the container port, and the relocation of Kai Tak airport to Lantau Island.

As LUTO gave form to TDS, so TDS gave form to 'Metroplan'. In conjunction with PADS, from 1991 the government adopted Metroplan, a new metropolitan development strategy for Hong Kong Island, Kowloon and Tsuen Wan, Kwai Chung and Tsing Yi areas. As PADS sketched out a new skeleton of communications for the territory, so Metroplan tries to address the substance of the urban fabric (see Chapter 8). Metroplan has the unenviable task of suggesting how the uncoordinated developments of the last one and a half centuries may be consolidated into a well organised and efficient metropolis of 6.5 million persons by 2001. To a certain extent it tries to compensate for the deficiencies of LUTO, its creator, by addressing issues of housing, recreation, tourism, landscape, built form, and other important considerations. Possibly the most controversial issue in Metroplan is the overall reduction of urban densities, in order to ameliorate some of the worst aspects of Hong Kong's highly concentrated population.

The next major phase of the process which is currently reshaping the territory's land use and infrastructure, constituting a further refinement of the most critical elements of SHRUG, was PADS, based on studies which took place between 1987 and 1989. These studies centred on the provisions required by the key elements of trade, namely the circulation of individuals, raw materials and commodities. PADS envisages massive infrastructure upgrading which will impact on the entire territory. The most important elements are the completion of the Lantau fixed crossing by 1996, the relocation of Kai Tak airport to Chek Lap Kok by 1997 (now due for completion in 1998), and the consolidation of port development around Kwai Chung-Northeast Lantau (see Figure 2.3).

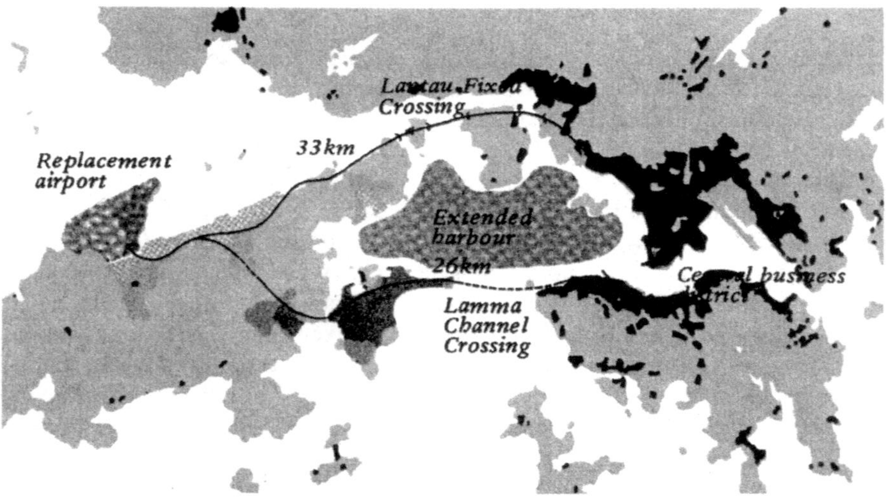

Figure 2.3 PADS proposals
Source: Hong Kong Government

With the adoption of PADS, the Hong Kong government set in train its strategy for the relocation of the airport from Kai Tak to North Lantau, expansion of the world's largest container terminal and a number of associated large infrastructure projects. Estimates of the cost of the airport have escalated with the end of 1995 estimated cost being $HK158.2 billion (at money of the day prices, i.e. allowing for inflation during the construction of the project). The cost of the first two terminals at Lantau Port, C.T. 10 and C.T. 11, is estimated at $HK27.6 billion. Implementation of both Metroplan and PADS has fallen victim to the political tensions in Sino-British relations concerned with the return of Hong Kong to Chinese sovereignty in 1997, particularly the impasse between the Chinese and British governments over electoral reforms for Hong Kong proposed by its Governor, which resulted in uncertainties and delays.

A review of TDS, commenced in 1990, is directed at incorporating the implications of the evolving relationship of Hong Kong with southern China, particularly the Pearl River Delta. The time horizon is 2011. In its wake follows a review of Metroplan which commenced in 1996.

Manufacturing industry

As an integral part of the TDS, the Industrial Land Development Strategy Report was produced (ILDS 1991) as well as a report on the Restructuring of Obsolete Industrial Areas (ROBINA 1993). Both these studies, along with the Survey on the Future Development of Industry in Hong Kong (SFDI 1991) and the earlier Survey of Land and Facility Requirements of Industrial Establishments (SLFRIE 1989) reflect the extreme concern of the Hong Kong government as to the economic future of its manufacturing industries and their spatial restructuring. Macro-economic trends towards outward processing in China have resulted in Hong Kong's manufacturing sector now being almost entirely in the production of light consumer goods. In 1996 the manufacturing sector accounted for only 14 per cent of employment in contrast to 42 per cent in 1984; similarly, estimated GDP contribution in 1995 was 8 per cent in sharp contrast to 84 per cent by the services sector. Despite these massive changes, the manufacturing sector will continue to play a role in the future and exert demands for industrial floorspace. The central problem is that the requirements of future provision will be different from before - in location, building form, intensity of use and associated facilities and services.

At the strategic level the problem has been exacerbated by the fact that the decentralisation policies of the seventies and eighties were largely successful in terms of population but unsuccessful as far as industrial development was concerned. Cheaper land and somewhat better servicing on the periphery were inadequate in persuading industrialists to move from inner urban locations, and 83 per cent of all industrial floorspace is currently located within the metropolitan area. The increasing obsolescence of these industries which permeates the entire urban fabric, is generating enormous problems in the spatial restructuring of the territory as a whole, not at least in part due to their small size. The average number of persons engaged in each manufacturing establishment has dropped from 55 persons in 1950 to 12 in 1996 (see Table 2.4). Hong Kong's land intensive approach to manufacturing in providing high-rise flatted factories is now proving unsuited to the increased automation and capital intensive industries which now characterise manufacturing industry. Part of the problem of controlling industrial and related land uses, and to a certain extent the cause of the existing problem, is the notational system adopted by the planning system. The actual technology of land-use control is as obsolete as the old planning regulations which were written on three pages of paper. Until recently only three broad industrial land use categories were in use, which did not even reflect the complexity of historical circumstances let alone current and future possibilities. A new I/O zoning, introduced in 1987, for dual purpose industrial and/or

Table 2.4 Number of establishments and persons in manufacturing, 1950-1996

Year	No. of establishments	No. of persons engaged	No. of pers. engaged per establishment (average)
1950	1,478	81,718	55
1955	2,437	110,574	45
1960	5,346	218,405	41
1965	8,646	341,094	39
1970	16,507	549,178	33
1975	31,034	678,857	22
1980	45,409	892,140	20
1985	48,065	848,900	18
1986	48,623	869,753	18
1987	50,409	875,250	17
1988	50,606	844,575	17
1989	49,926	802,983	16
1990	49,087	730,217	15
1991	46,276	654,662	14
1992	42,411	565,137	13
1993	36,847	483,628	13
1994	33,863	423,015	12
1995	31,114	386,106	12
1996	27,316	327,464	12

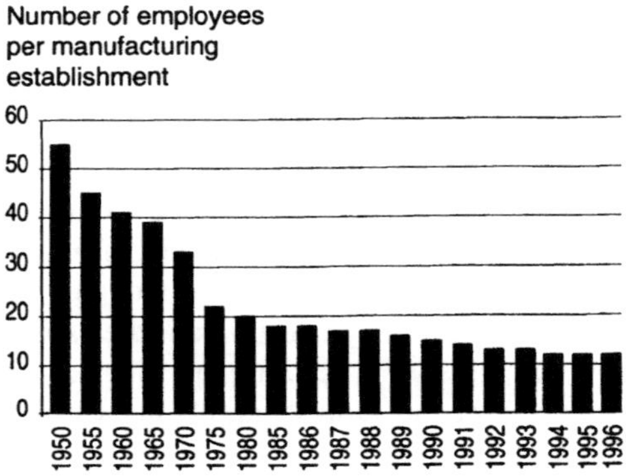

Sources: Figures for 1950 to 1970 from employment statistics, Labour Department; Figures for 1975 onwards from Information Services Department

office buildings, will merely extend the problem rather than solve it. A concurrent consideration of business parks and science parks is yet another extension of the old style 'zoning' mentality which, while appropriate to cleaning up the excesses of the industrial revolution, is completely inappropriate to the technical and social developments of post-modern societies. What is required is a highly sophisticated fine grained system of control which ignores crude land-use categories and instead looks at the properties of land-use activities and their effects. Ideally this should also consider the form of buildings as well as their economic over use, and integrate all environmental considerations with land-use controls. Such a system would also accommodate a greater complexity of architectural form than the 'podium-plus-tower environment' planned for the Hong Kong of the twenty-first century.

Apart from the changing nature of the industries themselves, which is creating massive obsolescence, improved planning standards, pressure on non-conforming uses, as well as demands for modernisation by industrialists, are accelerating the process of decay. In addition, there is now a general movement of industrial land use towards the west of Hong Kong due to the relocation of the new airport, placing even greater pressure on Kowloon, Tsuen Wan and Kwai Tsing to accommodate the new demands. At the same time, improved environmental legislation creates new standards for the form and design of industrial areas in terms of waste treatment, storage and servicing requirements, the most important of which is a demand for lower densities and less intensive land use. These factors effectively increase the demand on land supply. With the provision of new land limited by the Joint Agreement, reclamation and redevelopment become the most obvious ways to meet available needs. It is quite clear however that reclamation in proximity to the urban areas has reached its limit if the port is to continue to function and any semblance of the natural harbour is to remain. Hence urban redevelopment, by far the weakest element in the government's planning strategies to date, must represent the single greatest challenge to the efficient land-use planning of the city.

Urban redevelopment

In order to make some impact on the seemingly intractable problems of redevelopment, the Land Development Corporation (LDC) was established under the Land Development Corporation Ordinance of 1987. The establishment of this new organisation was a tacit recognition by government of its own shortcomings in tackling redevelopment when the adopted policy of 'positive non-interventionism' failed to cope with massive social problems. Using seed money from the Hong Kong government, the LDC attempts to marry private sector funds with the government's legal powers to resume land and thereby make joint LDC/private sector venture projects more viable as commercial concerns. The LDC's existence is predicated upon its fee for assembling development packages. In addition the LDC is considerably constrained by its inability to compulsorily acquire land for redevelopment on its own behalf, since these powers are embedded in the Town Planning Ordinance.

Through a number of large scale redevelopment studies of old urban districts throughout the metropolitan area, the LDC has identified a redevelopment and renewal need of proportions that are both monumental and expanding (Cuthbert and Dimitriou 1992). What also emerges from the LDC studies is that the massive areas most in need of redevelopment are likely to deteriorate even further since their location does not coincide with private sector interests for business opportunities, and the government has no redevelopment strategy of its own. This is truly a recipe for disaster. According to the ordinance establishing the LDC, the agency can only redevelop sites which are financially and commercially viable. There remains, therefore, a quantum leap between what the LDC is enacted to do, and the redevelopment needs of the city and its citizens. In terms of implementation, by 1995 the LDC had initiated 34 projects. Five major redevelopment projects have been gazetted, one of which was dropped when objections from local residents were upheld by the Town Planning Board, and several smaller schemes have been undertaken.

In addition to areas of deterioration in the private sector, the government's public housing stock is in a serious state of disrepair in many areas. Here there is an analogy between the situation the LDC finds itself in, and that of the housing authority. The government recently tried to carry out its own novel process of redevelopment, launched in 1991, when the first attempt to privatise public rental housing took place by selling-off 4,000 units to the existing tenants. The exercise, however, failed miserably as the overwhelming response indicated the units were far too highly priced. Had the exercise succeeded, experience from elsewhere has shown that it is the most desirable public housing in the best locations with the wealthiest tenants which sell first, with the net effect of downgrading what is left of the housing stock. Since this effort no further initiatives have been taken in this respect.

The proposed New Town Planning Ordinance

The Town Planning Ordinance which had guided development in Hong Kong, and remained in its original form for half a century, was finally determined obsolete in 1991 and a comprehensive review of the statutory planning system was undertaken. The key recommendations concentrate on enforcement of plans within statutory areas, the control of non-conforming uses and an appeal system. Along with this comes a greater awareness of the need for public consultation at all levels of the development process, and a greater need for areas of special control, for example in line with contemporary practice in conservation.

Despite these proposed changes, the basic structure of land-use planning in the territory will remain largely unaltered: principally because public housing still remains outside the scope of the ordinance; secondly, because redevelopment still remains mainly a private sector function independent of planning, except for its need to acquire land through the Town Planning Ordinance; and thirdly, because strategic level planning is somewhat insulated from the preparation of Outline Zoning Plans (OZPs) at the local level. The likely net outcome of this legislation, therefore, is for redevelopment, housing and planning to maintain the same kind of relationships as in the past but with relatively more power given to the Planning Department.

To date the recommendations of the review have only been partially adopted, by means of amendments to the Ordinance. The Town Planning (Amendment) Ordinance 1991 covered two problems that were considered in need of immediate attention. One of these was the introduction of direct enforcement provisions against unauthorised development in Development Permission Areas (DPA plans have now largely been replaced by OZPs) in the New Territories. This was particularly aimed at the container storage problem. The other was the setting up of the Town Planning Appeal Board to hear appeals against decisions made by the Town Planning Board.

Whatever the disadvantages of the planning system based on the previous Town Planning Ordinance, it was extremely elastic and responsive to certain development opportunities. A reasonable balance of power existed among the various departments in government concerned with the planning and control of urban development. With the proposed changes to the new ordinance, however, this balance will change in favour of an enormously expanded Planning Department whose authority in the broad fields of urban development will far exceed both its own capacity and those of other government departments and agencies. With this development, it is feared in some quarters that there will be an inevitable lack of accountability and reduced development inertia resulting from increased fractional conflicts within government, as well as confusion of purpose for those agencies whose powers have been effectively reduced such as the LDC.

Finally it should be noted that one of the singular weaknesses of the old system was the complete absence of any system of democratic representation, compensation or appeal against development proposals. It is fitting that these are being discussed in the context of a much wider debate over democratic electoral reforms with China. Yet even if a new planning system is eventually installed, it should be noted that every significant decision affecting Hong Kong's physical development over the next ten years has already been taken in the absence of any democratic process at all.

Conclusions

It is not surprising that land-use planning in Hong Kong has to a degree reflected the minimalist political philosophy of 'laissez faire' adopted by the Hong Kong government over the last century and a half. But as this philosophy denies the reality of massive government intervention in, for example public housing, education and infrastructure, so the minimalist nature of statutory planning practice denies the reality that the construction of social space has always been the responsibility of the state. A commitment to the illusion of non-interventionism has however had its effects. History is now claiming its due, and the real costs of Hong Kong's exceptional economic growth of the last quarter of a century are not merely surfacing, they are threatening its future success.

The government's strategy for the democratisation of Hong Kong's political system - too little, too late, - has its parallel in urbanisation. The flurry of activity in terms of studies, strategies, reviews and amendments of the development process since 1984 is on the one hand commendable, but on the other can be seen as an indictment of everything which came before it. The new town system has failed to produce the balanced development projected by the planners. Industrial decentralisation did not take place as predicted, and the existing urban areas became a morass of traffic jams, derelict residential and industrial areas, and environmental degradation, all of which now stand in the way of progress. The government's reluctance to use its compulsory powers of land acquisition for the purposes of redevelopment guaranteed that while development flourished in new areas and in the upgrading of older more residential areas to high quality housing such as in the mid-levels, redevelopment only occurred at the whim of the private sector, and usually only within central areas on sites for commercial purposes. Since the development potential of the urban area is near saturation, and is in fact planned to decrease in overall density in accordance with Metroplan, massive government intervention in the redevelopment process appears to be necessary. But the essential ingredient in the future development of Hong Kong lies in its transportation system.

With the takeover by the PRC in 1997, it is quite likely that considerable land-use integration of Hong Kong and China will take place, most notably between Hong Kong and Shenzhen special economic zone. If the political border were to be removed so that China could more directly benefit from Hong Kong's skilled labour supply, development pressure could be reduced in some areas of Hong Kong by the release of land across the border, and it is possible that considerable population movement could also take place. Beyond this lies Guangdong province, the hinterland for Hong Kong's economic growth. Clearly an integrated land-use and transportation plan is required not merely to alleviate local problems, but to fully integrate Hong Kong with China's rapidly expanding economy. The TDS formulated in 1984 marked the recognition of the problems noted above. Not only this, but for the first time, the concept of an integrated land-use and transportation strategy was seen as a point of origin for strategic action rather than an outcome of it. Such a plan is yet to be fully developed. The complexity of this task, and the problems surrounding its successful completion will be fully enunciated in the following chapters.

References

Bristow, R. (1984) *Land-use Planning in Hong Kong*, Hong Kong: Oxford University Press.

—————— (1989) *Hong Kong's New Towns: A Selective Review*, Hong Kong: Oxford University Press.

Castells, M. (1986) 'The Shek Kip Mei Syndrome: Public Housing and Economic Development in Hong Kong', Working Paper No.15. Centre of Urban Studies and Urban Planning, University of Hong Kong.

Castells, M., L. Goh and R. Kwok (1990) *The Shek Kip Mei Syndrome: Economic Development and Public Housing in Hong Kong and Singapore*, London: Pion.

Census and Statistics Department (1997) *Estimates of Gross Domestic Product 1961 to 1996*, Hong Kong: Government Printer.

Cuthbert A. R. (1987) 'Hong Kong 1997: the transition to socialism - ideology, discourse and urban spatial structure', *Society and Space: Environment and Planning D*, Vol. 5.

—————— (1991) 'For a few dollars more: urban planning and the legitimation process in Hong Kong' *The International Journal of Urban and Regional Research*, Vol. 15, No. 4.

—————— (1995) 'Under the volcano - post-modern development in Hong Kong' Chapter 4 in Watson, S. and K. Gibson (eds) (1995) *Post-modern Cities and Space*, Cambridge Mass.: Blackwell.

Cuthbert, A. and H. Dimitriou (1992) 'Redeveloping the fifth quarter: practices and prospects in Hong Kong' *Cities Journal*, Vol. 9 No. 3.

Drakakis-Smith, D. (1979) 'High society: housing provision in metropolitan Hong Kong, 1954-79. A jubilee critique', Centre of Asian Studies, University of Hong Kong, Hong Kong

Dwyer, D.J. (ed) (1971) *The Changing Face of Hong Kong*, The Royal Asiatic Society, Hong Kong.

Guttenberg, A. Z. (1959) 'A Multiple Land-Use Classification System', *The American Institute of Planners Journal*, August.

Halliday, J. (1974) 'Hong Kong: Britain's Chinese Colony', *New Left Review*, No. 87/88, London.

Harvey, D. (1989) *The Condition of Post-Modernity*. Oxford: Blackwell.

Henderson, J. (1991) 'Urbanisation in the Hong Kong - South China Region: an introduction to dynamics and dilemmas', *The International Journal of Urban and Regional Research*, Vol. 15, No. 2., London.

Information Services Department (1997) *Hong Kong 1997* Year Book (and previous years) Hong Kong: Government Printer.

Labour Department *Employment Statistics* (various years) Hong Kong: Government Printer.

Leung, S., J. Ng, P. So, and T. Wong (1986) 'Complexity and contradiction in the capitalist mode of production: an historical analysis of the urban process in Hong Kong', Master of Urban Studies thesis, Centre of Urban Studies and Urban Planning, University of Hong Kong, Hong Kong.

Talbot, H.D. (1971) 'An Outline of the Urban Development of Hong Kong Island During the Nineteenth Century' in Dwyer, D.J. (ed) 1971.

Tregear, T. R. and L. Berry (1959) *The Development of Hong Kong and Kowloon as Told in Maps*, Hong Kong: Hong Kong University Press.

Youngson, A. J. (1982) *Hong Kong: Economic Growth and Policy*, Hong Kong: Oxford University Press.

3 Parallel development of strategic land-use and transport planning: the case of the Territorial Development Strategy

Edward G. Pryor

This chapter outlines the main methodological steps and principles of strategic land-use/transport planning in Hong Kong employed by a number of studies, the most recent of which is the Review of the Territorial Development Strategy (TDSR) undertaken by the Planning Department. This chapter investigates the question of whether the respective roles of land-use and transport planners are complementary and what endeavours (if any) are needed to rationalise their roles.

Over the past two decades, numerous land-use/transport studies have been undertaken in Hong Kong. Those covering both long term and broad-brush aspects include: Metroplan (1991) (see Chapter 8), the Port and Airport Development Strategy (PADS) (1989), and the Territorial Development Strategy (TDS) (1984, 1986, 1988 and 1993). There have also been more detailed studies with a shorter time horizon, such as the Central-Wanchai Reclamation Study (1988 and 1993). In parallel with these studies, work has been undertaken on solely transport-led areas of concern, such as the Mass Transport Study (1967), the Long Term Road Study (1968), the Comprehensive Transport Studies (CTS I in 1976 and CTS II in 1989) (see Chapter 7), the Freight Transport Study (1993), and the Railway Development Study (1993).

A common element of many such studies has been their employment of algorithmic-based transport models based on quantitative data of past trends projected into the future. In the formulation of such studies, questions have often arisen in Hong Kong and elsewhere as to whether land-use planners, on the one hand, and transport planners, on the other, have gone along converging, parallel or diverging paths in their planning efforts (see Dimitriou 1992). Further points of concern have been those of taking fuller account of various environmental and institutional issues.

The need for land-use/transport planning in Hong Kong

Initial attempts to produce long-range strategic development plans for Hong Kong were initiated in 1965 when steps were taken to produce a Colony Outline Plan (COP). The broad aim of this plan was 'to provide a framework for a balanced approach to the planning of individual areas,' and for which purpose it was designed to 'include broad proposals for future land-use and land development programmes.' The plan incorporated the outputs of the Mass Transport Study (1967) and the Long Term Road Study (1968), with generally little account being taken of interactive relationships between alternative land-use patterns and transport systems. The Colony Outline Plan, which also set out various planning standards and guidelines for the provision of a range of community facilities and other land uses, was completed in 1970. It established a foundation for the development of three new towns at Tsuen Wan - Kwai Chung, Sha Tin and Tuen Mun to accommodate about 1.3 million people by 1986, leaving a balance of almost 4 million people for the main urban areas of Hong Kong Island, Kowloon and New Kowloon, and 0.4 million people in other rural and marine areas.

However, in 1972, it was recognised that in order to tackle the immense scale of housing problems in the territory a bolder, sectoral approach was required. Accordingly, a 10 year housing programme was quickly formulated with the aim of producing 1.8 million units of accommodation, principally through the provision of low-rental, public housing. A new Housing Authority was set up along with the New Territories Development Department responsible for the development of six, first-generation new towns comprising Tsuen Wan-Kwai Chung, Sha Tin, Tuen Mun, Tai Po, Yuen Long and Fanling-Sheung Shui with a collective capacity of about 2.3 million people (see Figures 3.1 and 3.2). The new towns and long-term housing programme were given a high level of priority in terms of financial resource alloca-

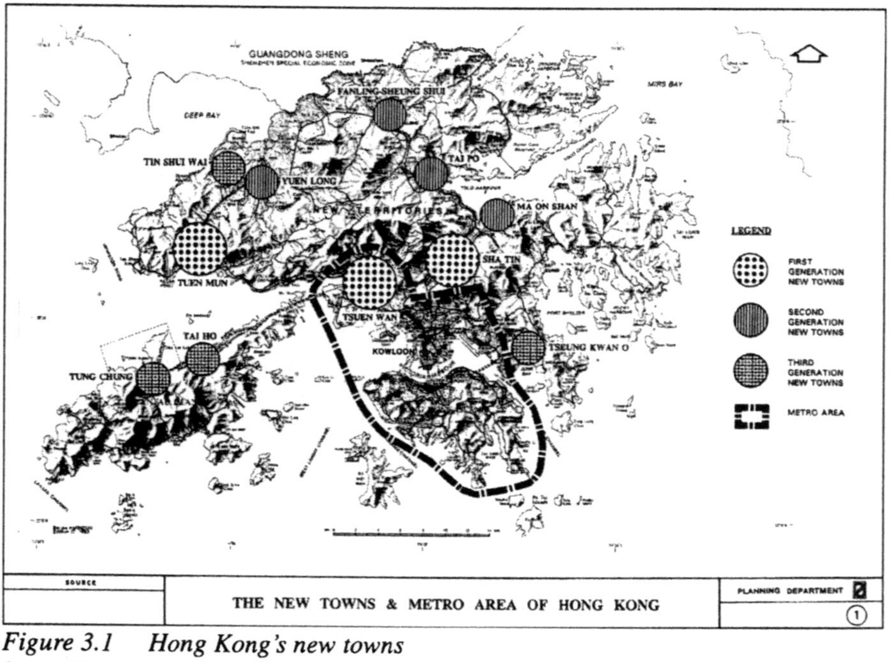

Figure 3.1 Hong Kong's new towns
Source: Planning Department

Figure 3.2
New town development: Tai Po
Source: Planning Department

tions, the preparation of master plans and the application of a comprehensive set of planning standards and guidelines intended to create 'balanced' and 'self-contained' communities with a high dependence on manufacturing industry. A parallel course of action was also taken to produce the first Comprehensive Transport Study (CTS I) (1976) to establish a more rational basis for the phased development of a hierarchy of transport infrastructure and services.

Given such a focus of endeavours, the main urban areas of Hong Kong Island, Kowloon and New Kowloon were generally left to 'fend for themselves' in that little was done to achieve co-ordinated planning and development. However, a considerable amount of redevelopment occurred, largely through private sector investment, initially prompted by the relaxation of building density controls in 1956. Also, the new Housing Authority built a number of new estates on 'infill sites' and, in parallel, launched an ambitious programme for the redevelopment of old housing blocks built in the 1950s and 1960s. The resultant surges of development placed enormous strains on the road-based transport systems and this prompted the staged construction of a mass transit railway system comprising three lines of 43.2 kilometres and 38 stations. The first line from Central to Kwun Tong was opened in 1979; the entire system in its present form was completed in 1989.

By 1981, it was evident that there was a need to take a comprehensive, long-term view of the future scale, timing and pattern of urban development and related infrastructure systems in the territory. Accordingly, steps were taken to formulate a Territorial Development Strategy (TDS) with the year 2001 as the long-term time horizon. The methodology for the TDS involved the commissioning of a number of studies to assess the development potential of five sub-regions with regard to a range of criteria relating to environmental conditions, transport services, land-use patterns, and engineering factors. From these studies, various potential 'solution spaces' were identified as a basis for meeting forecast demands for housing, employment and other community needs. In parallel, assumptions were made as to a range of alternative scenarios for the development of new port and airport facilities. A Land-Use/Transport Optimisation (LUTO) model was developed whereby for a number of objective functions (e.g. to minimise public sector expenditures), long-term patterns of development and infrastructure systems were derived (Choi 1986). At that stage, it was seen that there remained considerable uncertainties over a number of long-term assumptions (e.g. the availability of financial resources for major infrastructure facilities and the scale of future population growth), and it was therefore decided that a medium term strategy should be derived based on the 'common elements' of the alternative, long-term options. This approach led to the endorsement in 1984 of a strategy of major harbour reclamations along with further new town development at Tin Shui Wai and Tseung Kwan O (see Figures 3.3, 3.4 and 3.5).

Figure 3.3 Major harbour reclamation: Victoria Harbour
Source: Planning Department

Figure 3.4 Tin Shui Wai new town
Source: Planning Department

Figure 3.5 Tseung Kwan O new town
Source: Planning Department

The TDS was updated in 1986 and 1988 to take account of changed demographic forecasts. The use of a 'black box' model was regarded with some scepticism in certain quarters, one concern being that the LUTO technique assumed that each sub-regional development area was of equal environmental value.

The next step in the field of strategic planning in Hong Kong occurred in 1987. It was decided that more attention should be given to the development of a strategy that took greater account of the need to develop long-range plans and programmes for the provision of port and airport facilities to serve the domestic needs of Hong Kong and also the rapidly escalating demands arising from industrial-led urbanisation in the Pearl River Delta.

To that end, a Port and Airport Development Strategy (PADS) was formulated, involving the initial formulation and evaluation of 51 options that were eventually reduced to three preferred strategies from which a final decision was made in October 1989 to build a new airport at Chek Lap Kok and to provide for the major expansion of container terminals, anchorages, typhoon shelters and deep-waterfront sites for port-related industries (see Figure 3.6). The PADS proposals also included provision for new trunk roads and freight railway links and further new town development in close proximity to the new airport at North Lantau. Throughout the strategy formulation and evaluation process, there was a close interaction between various sectoral interests in an endeavour to ensure that the 'good' and 'modified' elements from a lower level of evaluation were taken forward to the next tier of option formulation. Whilst this process was in motion the Comprehensive Transport Study model

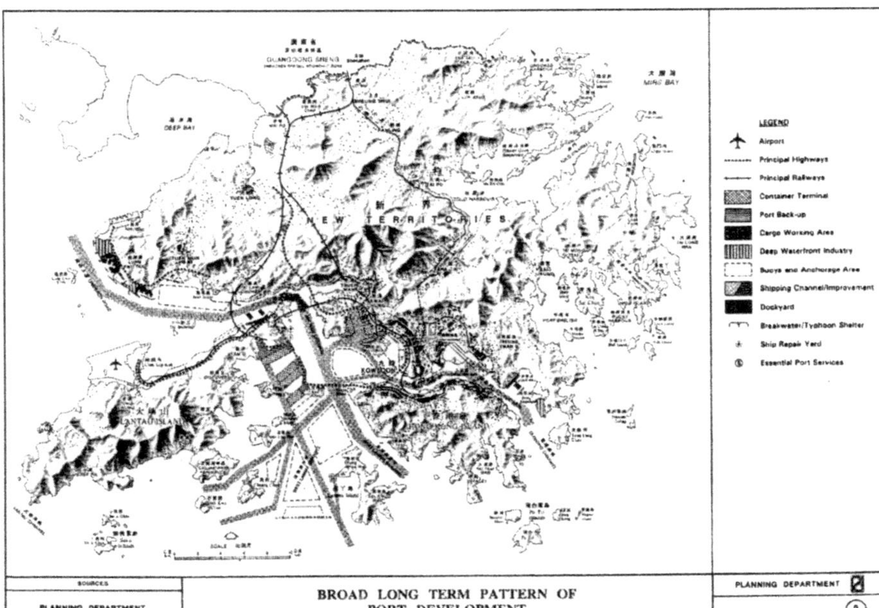

Figure 3.6 The Port Development Strategy
Source: Planning Department

was reactivated to provide a basis for the more precise definition of major transport development priorities and programmes.

Shortly after the commencement of the studies for PADS, it was decided that a strategic plan should be prepared for the restructuring of the metropolitan area comprising Hong Kong Island, Kowloon, New Kowloon and Tsuen Wan - Kwai Tsing (see Figure 3.7). The broad goal of the plan (known as 'Metroplan') is to 'bring about a better organised, more efficient and more desirable place in which to live and work'. An important principle followed in the plan formulation process was that the planning and development of new areas of reclamation and the replanning and redevelopment of obsolete, existing urban areas should be dealt with as a whole, rather than as separate entities. In that process, it was recognised that patterns and intensities of land use and transport infrastructure need to be planned and developed on an interactive, mutually self-sustaining basis taking close account of environmental objectives relating to noise, air quality, aesthetics etc. In this connection, a significant component of Metroplan is the provision made for relatively high density nodes of development around high-capacity, passenger rail/bus/ferry interchanges as in Kowloon (see Figure 3.8). Metroplan was endorsed by the Executive Council in September 1991 and is being carried forward to detailed planning through the preparation of a series of Development Statements for the component districts of the metropolitan area.

The strategic planning process in Hong Kong has now reached a stage at which it has become necessary to prepare a further review of TDS. The current review endeavours to establish a land-use/transport/environmental framework for long-term urban development, conservation and infrastructure provision, having regard to the

Figure 3.7 High density part of inner Hong Kong urban area
Source: Planning Department

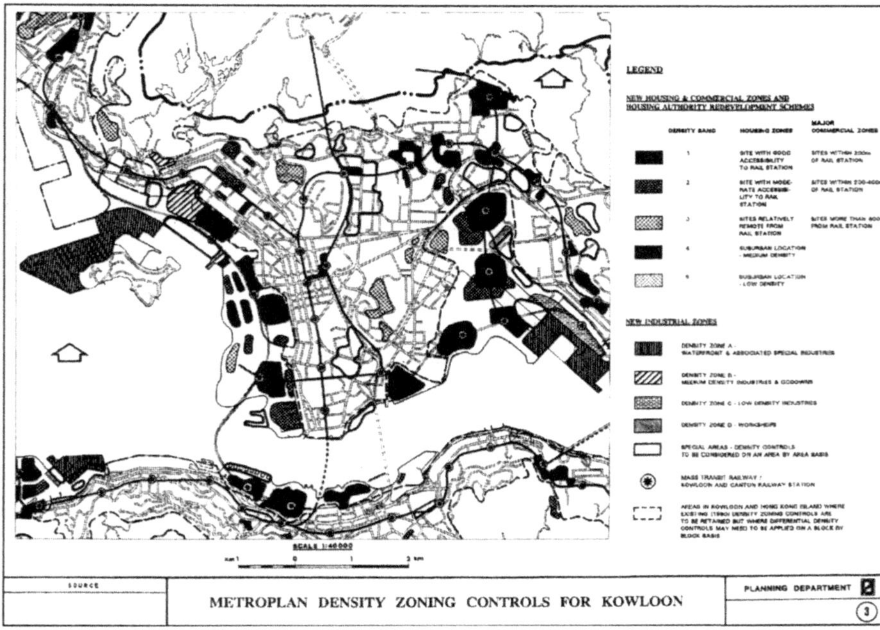

Figure 3.8 Metroplan zoning controls
Source: Planning Department

rapid economic growth of hinterland areas in South China, especially the Pearl River Delta. The plan-making process used for that purpose is dealt within the following sections with the issue of land-use/transport interaction being given a degree of emphasis as a major (but not the only) component of strategy formulation.

Some basic principles

In the context of the question as to whether land-use planners and transport planners are pursuing converging, parallel or diverging paths, it is first necessary to establish some basic points regarding the movement of people and goods relative to land-use activities.

In simple terms, the movement of people by any mode of transport arises principally from two major factors. On the one hand, movements occur from the generation of trips at places where people live, and the attraction of those trips to places where they work (or to places for some other purposeful activity). All such trips may be termed 'home-based' trips. On the other hand, traffic movement may also be derived from the generation/attraction of trips that are not home-based (e.g. between two places of business). A clear account of such home-based and non home-based trips is given in the seminal work of Mitchell and Rapkin (1952).

Equally, the movement of goods by any mode arises mainly between points of production and points of consumption, sometimes passing through points of interchange (e.g., container ports). Such movements may also occur between two or more points of production for activities involving mutually interdependent processes (as in the case of a fabric production plant supplying a garment-making factory).

The interdependence of land-use and transport planning

If the above are accepted as some of the most basic principles governing land-use/transport interaction, certain other aspects also warrant acknowledgement.

It is a basic premise of this chapter that land-use and transport planning are mutually interdependent to the extent that, in the context of a holistic planning process, the type, scale, location and timing of land-use developments govern the magnitude, nature and distribution of trips generated and attracted between them. Since transport movements and associated infrastructure per se often create substantial demands on resources, land uses need to be planned in a cost-effective and environmentally sensitive way that also minimises the need for travel (see Figure 3.9). This might entail, for example, the comprehensive development of high density, mixed-use urban forms around high capacity, multi-modal transport interchanges. An approach of this kind not only facilitates the balanced use of transport systems but also supports their financial viability. It is an approach that has been pursued in the formulation of Metroplan.

In the context of a planning process of this kind, the type, capacity, routing and phasing of urban transport systems need to be planned in an environmentally acceptable and cost effective way that facilitates the development of a coherent, efficient land-use pattern, minimises the need for land reservations for transport infrastructure and promotes the optimal use of land for productive and socially beneficial purposes (see Figure 3.10). A simple example would be the routing of high-capacity

Figure 3.9 Valuable space consumed by roads creating environmental problems
Source: Planning Department

Figure 3.10 Off-road mass transit system
Source: Planning Department

public transport systems within dedicated, well-insulated rights of way to serve a number of urban development zones within a linear corridor. Such a principle was recognised in the upgrading of the KCRC passenger rail line serving Kowloon, Sha Tin, Tai Po and Fanling-Sheung Shui.

If, with the above thoughts in mind, it should be found that for an initially assumed land-use pattern and associated transport system, certain links are congested to unacceptable levels then, four basic options for redressing the problem exist. One option might be to change land-use patterns and/or their associated attributes (e.g. building density) so that the utilisation of the transport network is better balanced. A second choice might be to provide additional transport links and capacities to relieve any congested corridors. A third possibility might be to introduce policies and related management schemes that suppress transport demand arising from those modes that consume the most road space and contribute least to productive activities. One can also undertake a combination of all the above.

Three levels of planning

The issue of land-use/transport interaction may be viewed at three different levels of planning in Hong Kong. One level is strategic planning which is usually 'broad-brush' in character and typically has a long-term time horizon of up to 20 years. Planning at this level indicates the overall future pattern of land use and related infrastructure for the territory as a whole. The TDS is an example of such a study, the most recent up-dating of which produced six 'hybrid' options.

Another level is sub-regional planning which is undertaken within the planning framework of a strategic study that uses the same time horizon but produces more detailed development proposals for broad areas in a given sub-region. These proposals are made in a particular sequence of time related to specific phases of identified development. An example of this kind of study is Metroplan (see Chapter 8), some outline proposals of which are shown in Figure 3.8.

The third level is district planning, which is undertaken within the planning framework of a sub-regional plan to provide a higher level of detail for specific sites and for project implementation, usually within a short to medium term time frame of between five to seven years. An example is the development plan for the Central-Wanchai reclamation, the configuration of part of which is illustrated by Figure 3.11. At each of these three levels, there may also be various sectoral plans, covering such matters as housing, commercial office development, business parks, water supplies, and solid waste disposal etc.

It has been the tradition in Hong Kong for the planning process at each of the three levels to be conceived and developed on a 'top-down' basis. In reality, however, this has not always been practical, particularly because of the fast-moving situation in the territory where a buoyant economy is driving investment along at an extremely rapid pace. For example, between 1987 and 1992, the index of gross value of construction work performed by main contractors rose from 124 to 258. Over the same period, the number of public and private sector residential flats/units constructed in the territory totalled over 441,000. Under such conditions, government agencies involved in planning and the provision of land and infrastructure are hard pressed to keep 'ahead of the game'.

The role of transport models in the planning process

Conventional land-use/transport planning exercises and their transport models have been in international use for many years. There are numerous, well written textbooks that explain their aims, objectives and techniques to which readers can be

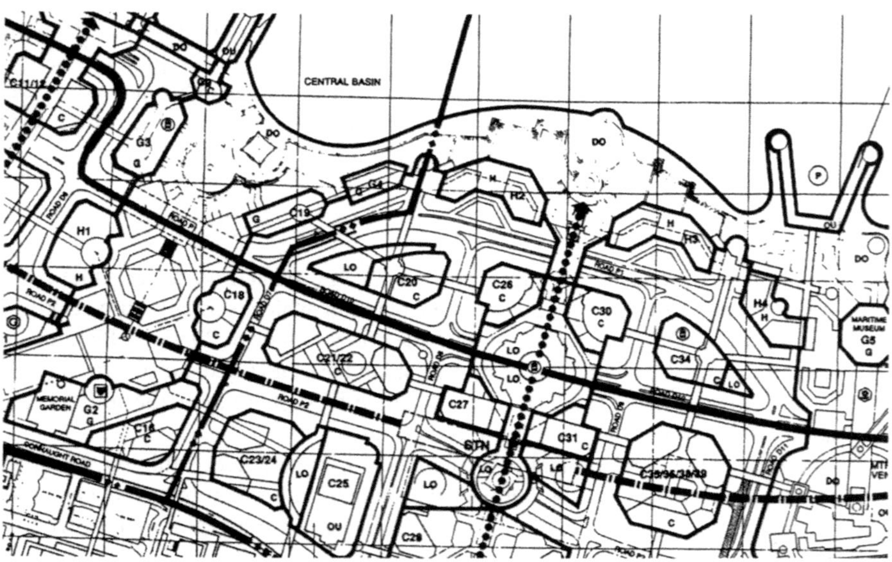

Figure 3.11 Extract from Outline Development Plan for Central-Wanchai reclamation

Source: Planning Department

directed (see for example, Black 1981, Bruton 1985, Creighton 1970, Morlok, 1978 and Ortuzer and Willumsen 1990). An extensive critique of their technology-transfer is given by Dimitriou (1992). Typically, such models entail a number of discrete steps involving, firstly, the estimation of volumes of trips generated by and attracted to various zones; secondly, the breakdown of trips by transport modes (public/private); thirdly, the assignment of traffic by mode on to assumed networks; and, fourthly, assessment of the resultant performance of the assumed networks and the derivation of modified networks/policies to produce a preferred transport plan (see Chapter 7).

A criticism of transport planning models is that they have frequently been used on a somewhat independent basis to urban planning studies, thus limiting their validity and use to a narrow field of interest. This situation, however, may be considered defensible for the purposes of designing and implementing transport projects in the short to medium term.

Furthermore, in a strategic and sub-regional context, transport planning studies have often been based on one or two predetermined, long-term end-state development patterns, as in the case of CTS I (1976) and CTS II (1989) for Hong Kong (see Chapter 7). Clearly, one of the weakest elements of this kind of planning process is the proposition that the assumed end-state development pattern from (and to which) trip ends are generated and attracted, is itself 'optimal'. In reality, this is seldom the case, except for a fairly limited time scale over which the distribution of population and jobs (the main trip generators/attractors) can be predicted with a reasonably high level of certainty. The LUTO model, used for the first TDS in Hong Kong, was created in an endeavour to deal with this issue.

Another weak element of transport models is that they depend very heavily on a projection of parameters established through past trend analysis, which can unjustifiably pre-cast the future too much in terms of past experiences. This has significant limitations for both urban development and transport. For example, growth and manufacturing jobs in Hong Kong have been closely linked to levels of ownership and usage of goods vehicles. However, with changes in the type, scale, location and modus operandi of manufacturing enterprises, such a correlation may become increasingly less tenable.

Urban transport models have also frequently implicitly assumed that the number of trips predicted can be accommodated at the trip-end destination. In Hong Kong, however, this is not so. There is very limited land available in the territory with which either to increase transport infrastructure capacity or accommodate more vehicles in already heavily congested areas. To take a simple example, container terminals Nos 1 to 6 at Kwai Chung were not originally planned with sufficient capacity to accommodate the container vehicles attracted at peak periods. As a result, tailbacks occur along the main external highway links (see Figure 3.12). Also, in the district of Mong Kok (the most densely developed area in the world), off-street parking is acutely in short supply. In both cases, however, current transport models (based on technology from more space-abundant societies) do not allow for the negative impacts which the lack of parking space can have on trip generation rates and/or on the capacities of road networks. The limited capacities of road links themselves can have a similar impact on trip generation rates.

Figure 3.12 Container terminal at Kwai Chung
Source: Planning Department

A further limitation of transport models is that by using GDP-related factors as a basis for deriving trip forecasts and vehicle ownership levels, the end result can be forecasts of traffic volumes that are far in excess of what a more 'normative' approach to transport planning would produce. Conventional transport models simply deduce that the wealthier the community, the more vehicles each household will own, and therefore, the more trips they will make. From this, a quick conclusion is then made that more roads are needed. An alternative approach to this logic offered by conventional land-use/transport planning is one that constrains the modelling process for passenger transport by adopting an assumed range of transport policy options, whereby (say) 75 per cent of all personal travel takes place on public transport. In the case of freight transport, an equivalent policy option would be to assume that a certain percentage and/or type of freight movement relies on railway modes of travel. By adopting more policy-orientated, directive approaches to land-use/transport planning, one could then argue that more mature judgements regarding the type of city we think we should have can be made.

A new way forward: the case of the TDS up-date

Background

Whilst there is clearly a specific role for conventional transport modelling exercises as a means of providing a basis for the formulation of specific transport projects, investment programmes and network management schemes, there is now a wider realisation, both world-wide and in Hong Kong, that medium and long-term planning have to be undertaken in a composite way. If this is done it will help ensure that all principal components of urban development are dealt with in an interactive way whereby transport model tests form one integral part of a much broader complex planning process. Transport models in this context are applied to establish the relative performance for option evaluation using agreed policy constraints, as well as mathematically derived parameters. An approach of this kind was employed by PADS (1989), Metroplan (1991) and the latest up-date of TDS (1996).

Although each of the above studies has been configured by a purpose-designed planning methodology, the general process of logic for each study follows a fairly consistent path, which is outlined in Figure 3.13 for the TDS. In very general terms, the process starts in a conventional way with the definition of a broad goal(s) and principal working assumptions. An analysis of key issues leads to the derivation of land-use budgets and the formulation of specific objectives and related evaluation criteria and performance measures. Various development themes or scenarios are postulated providing a foundation for the generation of a number of significantly different initial options that take account of committed patterns of development, potential solution spaces and environmental capacity constraints. Various land use, transport, environmental, financial, economic and community value assessments are then made against the specified objectives. The 'good' and 'modified' elements of strategy options are progressively taken forward to a point where, for each scenario, a preferred, long-term strategy is established. A choice may then be made either to select one such strategy or, should there be high levels of uncertainty over long-term assumptions, the common elements of each preferred strategy may be brought together to help create a medium-term plan and outline works programme. The final stage also incorporates a number of broad, sectoral development strategies.

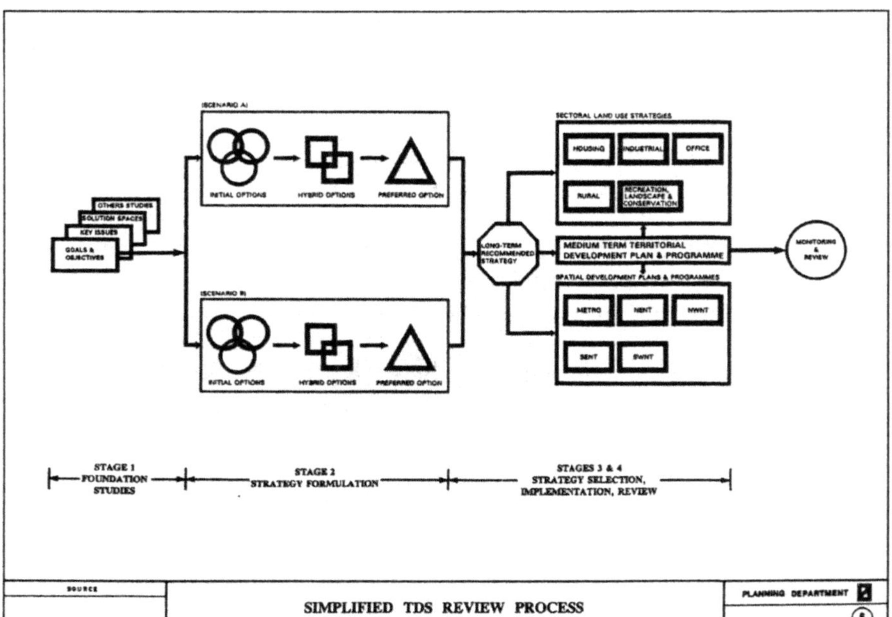

Figure 3.13 Simplified TDS review process
Source: Planning Department

It is this process of logic that is developed by the Hong Kong Planning Department in the belief that, allowing for geo-political and socio-economic differences in the territory and its rapidly developing surrounding region, such an approach can be adapted for application at both micro and macro levels. It is a planning process that relies on a mixed application of modern technology (through the use of low-cost, computer-based models) and high levels of corporate judgement in the making of decisions. Balancing the two depends on the level of expertise and resources available and the degree of sophistication required. While this planning process is one that is not without its weaknesses, it has led to the top level policy approval of several strategic plans of bold dimensions. These, in turn, have led to firm commitments in the substantial allocation of resources from the public purse, as well as from private sector developers. Together, these two sectors use the newly formulated planning framework as a basis upon which to rationalise their own investment decisions.

Task analysis for strategic land-use/transport planning

A strategic planning exercise, such as the up-date of the TDS, typically begins with the preparation of a carefully compiled 'Inception Report'. This sets out such matters as the general background and purposes of the proposed study; the principal tasks to be undertaken; the scope of the study; and its interface with other related studies that have been completed or are on-going. It also gives the study work programme, study manpower and other resource requirements, as well as the organisational structure for the control and management of the study.

It can be argued that the proposed task analysis is in fact the most important part of an inception report. It is somewhat akin to a software programme that has to be 'booted-up' to make a computer perform a particular job. If the programme is flawed, then the final output will also be deficient. In broad terms, the strategic planning process comprises four broad stages of work. These are described below.

1. Foundation Studies - such studies establish the strategic goals/objectives to be achieved; identify a set of principal working assumptions; assess a range of future needs for which a development framework is required; identify the scale and composition of existing and committed development; and map out various potential solution spaces for satisfying projected land use and infrastructure needs.
2. Strategy Formulation - this package of tasks first requires the formulation of a number of broad, long-term strategic planning scenarios from which, through an iterative evaluation process derived from the objectives agreed in Stage 1, a series of options for each scenario is produced. These, in due course, lead to a 'Preferred Strategy' for each scenario. At this stage, no cross-scenario comparisons are made.
3. Strategy Selection - this stage involves a comparison of the relative merits of the 'Preferred Strategies' produced for each scenario and leads to the selection of a 'Recommended Long-Term Strategy'; the derivation of a 'Medium-Term Implementation Plan'; and an 'Outline Programme of Works' formulated within the framework of a rationalised institutional structure.

4. Implementation, Monitoring and Review - clearly, for the strategic planning process to be meaningful, the various proposals need to be carried forward to detailed planning and project implementation through a process that involves the setting of development priorities and resource allocation. Another essential step involves the setting up of an effective system for recording development trends and monitoring progress on project implementation so as to provide indicators of the extent to which strategic development goals and objectives are being achieved. Results from this process will, at some future date, prompt a partial or comprehensive review of the adopted strategy.

Within these stages a number of tasks, which are described below, are undertaken in the plan-making process (see Table 3.1).

Table 3.1 Stages and tasks for the update of the Territorial Development Strategy

Stage 1 -	**Foundation Studies**
Task 1	Define broad study goal
Task 2	Establish scope of territorial development policies relevant to the TDS review
Task 3	Produce a suitable base plan
Task 4	Identify key spatial and sectoral issues
Task 5	Define objectives and evaluation criteria
Task 6	Formulate a set of working assumptions
Task 7	Develop models for transport options testing
Task 8	Establish potential demand and timing of infrastructure needs
Task 9	Identify existing and committed developments
Task 10	Identify potential 'solution spaces'
Task 11	Foundation Report
Stage 2 -	**Strategy Formulation**
Task 12	Postulate broad development scenarios
Task 13	Formulate initial development options
Task 14	Test performance of each development option
Task 15	Produce and evaluate hybrid options
Task 16	Strategy Formulation Report
Stage 3 -	**Strategy Selection**
Task 17	Select preferred strategy
Task 18	Formulate broad sectoral policies
Task 19	Formulate implementation programme
Task 20	Final Technical Report and Executive Summary
Stage 4 -	**Implementation, Monitoring and Review**

Stage 1: Foundation studies

The first of the eleven principal tasks of the foundation studies for the TDS up-date is to define a broad study goal, being a general statement of intent that identifies in broad terms the expected final outcome(s) of the study process. The principal goal assumed for the TDS is

> *to establish a long term, land-use - transport - environment planning framework within which the necessary land and infrastructure can be provided, having regard to resource availability, to enable Hong Kong to continue to grow as a regional and international city and to become a better place in which to live and work.*

In this context, it is necessary to set both the geographic limits of the study area and a time horizon (say to the year 2011) over which it is assumed the resultant planning proposals take shape in, say, five year tranches. Goal definition needs to take account, inter alia, of current and potential future development trends in the wider hinterland beyond the assumed study area boundaries.

The second task is to establish the scope of the territorial development policies relevant to the plan-formulation exercise. Policy areas covered include such matters as the role of manufacturing industry and the need for economic diversification; the future use of agricultural land; the achievement of certain levels of housing standards; the rate of expansion required for major port facilities, etc. The outcomes help to provide inputs for identifying various key issues at task 4 and also for deriving later working assumptions at task 6.

The third task of the first stage of the study is to produce a suitable base plan of the designated study area to be used for the mapping of relevant information and the formulation of development proposals. It is prudent at this juncture to install a computer-based mapping system to facilitate the production, amendment and analysis of plans. Potential sources of 'hard copy' information and information 'gaps' also need to be identified, and records assembled in a well-documented library. A computer-based information system for data storage, processing and retrieval is required to set up and tie-in with the computer-driven mapping system.

The fourth task is to identify key sectoral and spatial issues as the basis for the later formulation of action-oriented, investment programmes. The issues examined cover both sectoral fields of interest and spatial areas of concern, involving matters of some substance.

Some of these sectoral issues include estimating demands for new jobs and associated land requirements for various urban-based employment groups; establishing demands for agricultural production and resultant land and labour needs; forecasting demands for new housing and other associated uses, and land needs; producing estimates of future needs for tourist activities and facilities; reviewing the effectiveness of institutional organisations and mechanisms for strategy implementation, monitoring and review (see below); and establishing potential resource availability for strategy implementation.

Some of the spatial issues include the type, scale and causes of land-use problems, (squatter settlements, obsolete urban areas, non-productive/under-utilised farm land, poorly managed forestry areas, badly drained marshlands, over-utilised country parks/tourist attractions); the type, scale and causes of transport and other infrastructure problems (congested rail, highway and river-based transport networks and terminal facilities; under-utilised rail, highway and river-based transport networks and terminal facilities; the adequacy of water supply systems, drainage canals, telephone networks) and the type, scale and causes of major environmental problems, producing a baseline profile concerning such matters as areas subject to recurrent flooding; eroded uplands; blighted areas of potential landscape and/or recreational value; important ecological areas under threat; potable and non-potable water sources affected by or under threat from pollution; air sheds affected by or under threat from pollution; areas subject to or under threat from noise impacts from various sources (airport, trunk highway); areas where the quality of the living and working environment is affected by interface problems caused by the proximity of incompatible uses; areas of habitation/employment put at risk by hazardous installations.

From the analyses of the above key issues, a budget of needs is drawn up to be used as a basis for strategy plan formulation. Such a budget comprises an inventory of current problems to be resolved (e.g. the need to clean-up pollution black spots) and a 'shopping list' of new land-use needs for which additional site reservations and action programmes are required.

The fifth task is to define a set of specific objectives and related evaluative criteria. Usually there is a wide range of key sectoral and spatial issues, the identification and analysis of which need to be carefully undertaken so as to help formulate a set of specific land-use, transport and environmentally related targets to be accomplished so as to achieve the overall strategic goals. Such objectives, together with associated evaluative criteria and performance measures, provide a basis both for formulating and for progressively testing newly developed sets of options.

This stage of the planning process is a critical activity that needs to be undertaken with some care. The main requirements involve the statement of a number of principal objectives to provide 'stepping stones' for achieving the main study goals e.g., to enhance the role of Hong Kong as an international city and regional centre for business, finance, information, tourism and entrepot activities. For each principal objective, the statement of a number of related sub-objectives is needed. Examples might be to ensure the efficient use of resources; to enhance infrastructural linkages between Hong Kong and neighbouring territories; and to facilitate the development of new tourist attractions.

The relevance of each sub-objective to the key phases of the study planning process needs to be identified. Certain objectives are essential requirements at all stages. Some are relevant only to the evaluation of initial options. Others relate specifically to the final strategy selection. Under each sub-objective needs to be listed the associated performance targets e.g. minimise land development costs; maximise economic rate of return; reduce residential/industrial interface problems; achieve a better sub-regional distribution of white collar worker jobs. We also need an indication of the performance measures to be used e.g. amount of land zoned for commercial office development; p.c.u./km; linear metres of industrial/residential interface; volumes of sewage discharges etc.; and the sources of data e.g. professional judgement; outputs from transport models. Finally we need to specify the method of evaluation scoring to be employed, e.g. on a scale of say 1-10.

An issue often causing difficulties at this stage of a strategic land-use planning study is the question as to whether the main objectives and/or sub-objectives should be given any particular weighting. Should environmental issues, for example, be considered more or less important than, say, transport efficiency issues? Debates of this kind can (and do) cause considerable dispute both in government and the community, and often lead to unresolved conflicts. The approach adopted for the TDS (particularly for the 'coarse screening' of initial options) is to treat all contenders as equally important, and later when composite scores are produced, to apply sensitivity tests. Such tests give different objectives various weightings, the results of which establish the robustness of a strategy. For the evaluation process to be complete, before any substantive work is undertaken, it may be prudent to carry out a 'dummy run' of test options in order to see how the system works.

The sixth task of the foundation studies entails the formulation of a set of initial working assumptions. It is a stage of the study that is closely related to the second, fourth and fifth tasks, and, as such, has as its purpose the establishment of a number of working guidelines.

These guidelines cover such matters as the principles to be followed to achieve the integration of land-use/transport planning components, taking account of other important factors such as environmental objectives, financial and economic cost-benefit performance, and the interface with institutional mechanisms for plan implementation; assumed rates of GDP growth; basic standards for the provision of housing and other community needs; a range of average development densities for a range of broad land-use types, including light industry and housing; average productivity rates for various kinds of port facilities; assumed levels of dependence on various transport modes; average consumption rates per capita for potable and non-potable water supplies; and unit rates for land development and the provision of infrastructure.

These assumptions are required to help define the scale of future metropolitan needs and the potential costs of various land uses and infrastructure systems. During the course of the study, it is inevitably necessary to 'fine tune' the original assumptions or even define new ones.

The seventh task involves the development of sets of models for transport option testing, as well as the use of other simulation tests. Once a suitable base plan for land use and transport development has been produced, the study area is typically broken down into a hierarchy of numbered geographical units (zones) that provide a framework for the derivation of data for various kinds of computer-based transport model tests. The boundaries of these units, as far as possible, should tie in with Census boundaries, administrative units, patterns of land use and significant geographic characteristics. Ideally, geographic units in planning studies should be of such a size and configuration that they can be assembled into different functional polygons (e.g. traffic zones, water catchments etc.), with the understanding that the smaller they are the greater their potential use for the aggregation of data for different levels of planning.

The eighth task in the TDS foundation studies establishes the potential scale of demand and timing of key infrastructure needs. Once each aspect of the fourth task is identified, the next step is to estimate in broad terms the likely scale and timing of infrastructure demands for such requirements as power supplies; telecommunication links; waste disposal and treatment facilities; water supplies; and new/expanded port and airport facilities. For each facility a cumulative net shortfall is established by deducting from the cumulative forecast demands the estimated capacities of existing and committed facilities.

The ninth task involves the identification of existing and committed development in the study area. This is conducted as a parallel exercise to the above study tasks. In doing this, an early start can be made to identify the nature, scale and location of existing and committed development that is assumed to be retained over the study period. Such committed development includes rural settlements and urban land uses (providing the 'base growth' of the study area); special installations, such as military installations, water catchments, cemeteries, etc.; and trunk highways, railways, and river-based transport systems (providing what may be referred to as 'baseline networks').

From the base growth development, a matrix of data (derived from the seventh task) needs to be produced for each geographical unit of the study area relating to such matters as number of households; age profiles; job opportunities; school places, etc. Such data are used as a basis for calculating trip generation and trip attractions for each traffic zone. The baseline transport networks and services are subsequently tested relative to the demands generated by the base growth so as to ascertain which links have spare capacity and which are overloaded. The results from these provide an important input to the formulation of the first set of long-term development options. Model tests of base growth are also undertaken for air quality and water quality attributes.

The tenth task of the foundation studies entails the identification of potential 'solution spaces' for the planning study. This task focuses on a careful but broad-brush analysis of the potential of various parts of the study area to be designated for a range of uses that satisfy forecast long term needs. The general range of these land uses cover land that is potentially suitable for new general urban development and also for redevelopment; land/sites that are potentially appropriate for special/key infrastructure facilities such as ports, airports, power stations, waste disposal facilities etc.; land and coastal marine waters that are suitable for rural/marine based uses such as intensive farming, mariculture, afforestation, tourist resorts, etc.; and land and marine areas that are suitable for various conservation purposes, including water catchments, country parks and wild life refuges.

The identification of such areas for the TDS review required a number of inter-related studies to be undertaken using currently available information and local knowledge. The end result of these investigations is a series of large scale plans with associated sectoral reports that indicate broad areas suitable for various types of land uses and development. Some such areas inevitably overlap so that it is necessary in subsequent stages of the study to make decisions regarding the best, long-term use of such areas, relative to forecast needs that have to be met. The data so assembled are stored in a comprehensive, geo-coded information system.

The eleventh and final task of the foundation studies is to produce a summary report (incorporating all previous tasks) for submission to a high-level policy committee via the study steering group. The approved report is then translated into a form suitable for public consultation and comment.

Stage 2: Strategy formulation

This stage of the study commences with the postulation of broad development scenarios as the twelfth task. It is the first key step toward generating a sequence of possible, long-term development options. This stage of planning is a conceptual process which provides a broad idea of the range of possible, long-term regional patterns of development. It is important that this stage does not concern itself with detail or local development issues. The intention is simply to create a framework within which more specific planning proposals can be later made.

The TDS review assumes three broad scenarios. The first is Scenario A1, which assumes that the Pearl River Delta would be the primary economic hinterland of Hong Kong and within which growth would be maintained at present rates. The second is Scenario AII which assumes that the Pearl River Delta would be the primary economic hinterland of the territory but within which there would be accelerated growth. The third, Scenario B, assumes that the primary economic hinterland of Hong Kong would include the inner provinces of China.

For each scenario, various assumptions are made relating to key regional development components so as to provide 'pegs' around which to formulate sets of strategy options (see Table 3.2). Such scenarios may be regarded as 'driving concepts' so as to give a structured direction to the subsequent formulation of more detailed ideas.

The above three scenarios are not mutually exclusive. Rather, it is a question of timing as to the rate of economic growth within and geographic extent of Hong Kong's primary hinterland. The assumption of such scenarios is designed to enable long-term contingency plans for major land uses and infrastructure to be formulated leaving decisions on the implementation of works to be taken within a shorter time scale, having regard to feedback from recurrent reviews of development trends.

The thirteenth task involves the formulation of initial development options for each scenario based on the results of the preceding tasks (i.e., the fourth, fifth, sixth, eighth, ninth, tenth and eleventh tasks). For each scenario, a limited number of significantly different land-use/transport options is produced to satisfy the various, forecast land-use needs and also to help resolve identified spatially based problems. Each option represents a possible strategy for the assumed end design year of 2011. Various kinds of data need to be derived from the plans produced within the context of the zonal framework established in the seventh task.

Testing the performance of each initial option, the fourteenth task, is conducted by first taking previously defined objectives and evaluation criteria (developed in the fifth task), making use of the results of transport and other simulation models/

Table 3.2 TDS scenarios and strategy options

Main Component	Scenario A (PRD as Primary Hinterland)		Scenario B (PRD and Inner Provinces as Primary Hinterland)
	AI Steady Growth	AII High Growth	B Extra High Growth
Airports	International airports at Chek Lap Kok and Fuyong (Huangtian) Regional airports at Macau, Guangzhou, Huizhou	International hub airport at Chek Lap Kok Regional airports at Macau, Fuyong, Guangzhou and Huizhou	International airports at Chek Lap Kok, Fuyong (Huangtian) and Macau Regional airports at Guangzhou and Huizhou
Ports	Container hub ports at H.K. and Yantian Bulk cargo port at Yantian Feeder ports at Shekou, Macau, Zhuhai and Pearl River towns	Container hub ports at H.K. Bulk cargo port at Yantian Feeder ports as for AI	Container hub ports at H.K, Yantian and Gaolan Bulk cargo port at Yantian and Gaolan Feeder ports as for AI
General Land-Use Demands in H.K.	Relatively low	Relatively high	Extra high
Cross Border Transport Demands	Relatively low	Relatively high	Extra high
Industrial Land Demands in H.K.	A range of relatively low, medium and high land needs derived from different assumptions on densities of development and locational preferences.		
Commercial Office Land Demands in H.K.	A range of options based on the balance of land supply from new sites and redevelopment as well as on assumptions relating to densities of building development and locational preferences.		
Housing Land Demands in H.K.	A range of low to high land requirements based on the assumed mix of public/private sector housing and densities of development.		
'No Go' Areas	These will comprise areas of high landscape value and ecological significance which should be generally the same for all the scenarios.		
Common Components	These will comprise 'base growth' urban development and baseline transport networks which already exist, are committed or with policy approval.		
Development Themes	Public sector-led and private sector-led options for each Scenario.		

Source: Planning Department, 1993c

analyses using the data obtained in the thirteenth task. This is done with the assistance of a small selected panel of specialists and advisors in various fields of expertise. The testing requires each major land-use, transport and environmental element of the study to be examined and ranked according to a scale of performance values which, for each option, is brought together to produce an overall evaluation. In this process, certain elements of each option perform well and warrant being carried forward to the next stage of strategy formulation. On the other hand, certain elements that do not perform well, unless they can be modified to achieve an acceptable level of performance, are discarded. It should be noted that, at this stage of strategy formulation, it is not necessary to make cross comparisons between the initial options in each broad scenario. Each round of evaluation tests, however, needs to be well documented in the form of a technical report.

The production and evaluation of a series of hybrid options as a basis to derive 'preferred strategies' for each scenario constitute the fifteenth task of the study. This takes the results from the preceding task for each broad scenario to produce a hybrid strategy or suite of hybrid strategies. Another round of careful evaluations is then undertaken. For the TDS review, three tiers of evaluations are used. The first tier comprises a broad assessment of the overall performance of each hybrid using aggregated results related to land-use planning, environmental, transport, financial and economic objectives. The second tier focuses on the suitability for development of potential 'solution spaces' having regard to territorial, sub-regional and district considerations. The third tier evaluates from an environmental point of view the potential development capacity of each sub-region. From these analyses a set of broad development guidelines is produced for the derivation of a preferred strategy for each scenario and also for the formulation of a medium-term strategy and outline programme of works. From this process a 'preferred strategy' for each scenario emerges, representing what may be considered to be a 'reasonably derived optimum state'.

A summary report of Stage 2 of the TDS planning study is then prepared as the sixteenth task. This document highlights the principal findings of tasks 12-15 relating to strategy formulation for submission to a high level policy committee via the study steering group. Following this, a public consultation programme is launched, allowing up to three or four months for comment and debate.

Stage 3: Strategy selection

It is at this step in the study planning process that a decision may be made to select which of the 'preferred strategies' to put forward as the 'recommended strategy'. A decision of this kind needs to be made on the basis of a broad evaluation, task seventeen, which compares the relative merits of each 'preferred strategy' in terms of various factors.

These are financial performance i.e. potential return on investment; economic costs/benefits relating to the effective use of resources; robustness, in terms of the degree to which each strategy can retain its structure to cope with unexpected changes to key working assumptions; flexibility, in terms of the degree to which alterations could, if necessary, be made to cope with changes to key working assumptions; ease of implementation, in terms of making the best use of existing institutional mechanisms; and likely impact on the potential capacity of the construction industry and on the availability of human and other essential resources resulting from the implementation of the finally selected 'preferred strategy'.

A report on the results of the above, especially highlighting the institutional aspects of the exercise, is submitted to the policy committee for approval. Following this, the report is sent on to other appropriate (mainly government) bodies for consideration with a view to endorsement.

Clearly, the more distant the strategic planning time horizon, the greater is the degree of uncertainty surrounding the long term viability of the assumptions and forecasts made. Economic, political, environmental and technological factors can easily undermine the validity of a long term 'end-state' plan. Hence, it may be considered more appropriate to derive a 'medium-term strategy' for a period over which there can be a higher level of confidence. A medium term implementation plan and programme can be produced by incorporating the 'common components' from each of the 'preferred strategies' assuming a time period of, say, seven to ten years. The degree of commonality, of course, depends to some large degree on the extent to which the broad scenarios overlap and on the potential availability of resources.

Also at this stage, proposals (task eighteen) need to be formulated for a number of broad, sectoral strategies covering such matters as housing, employment centres, port development, conservation and recreation etc., with each strategy addressing the key issues examined in the fourth task.

Once a 'medium-term strategy' has been derived, a logical phasing plan and outline programme of works is produced, along with forecasts of expenditures and estimates of possible financial returns (task nineteen). This requires a clear definition of the roles and responsibilities of all the key agencies involved. These are the preparation of detailed plans; the acquisition and clearance of land; the formation of land; the provision of regional and district infrastructure; and the disposal of sites and the implementation of various construction projects.

The last step, the twentieth task, of the above outlined planning process for the up-date of TDS, has two sub-steps. Firstly, a Final Technical Report is produced to tie together the principal results from the strategy selection stage of the study (i.e., seventeenth, eighteenth and nineteenth tasks) for submission to the policy committee. Secondly, an Executive Summary is produced setting out the principal findings and recommendations.

Following the policy acceptance of both reports, a well structured and on-going public relations programme is embarked upon to keep constituent communities and participating agencies informed of how it is intended to carry the plan and its recommendations forward.

Stage 4: Implementation, monitoring and review of strategy

For the effective implementation of strategic development plans there need to be detailed, 'top tier', sectoral plans and programmes for major urban development/redevelopment schemes (e.g. for the development of new towns); for the provision of regional facilities (such as hospitals, port and airport facilities, crematoria, tertiary educational institutions, regional parks etc.); and also for key infrastructure services (e.g. for trunk highways, railways, power supplies, irrigation and flood control systems, as well as international/inter-regional telecommunication systems). Within the same framework, a 'second tier' of detailed plans and associated development programmes for each major geographic area within the study area need to be formulated. These, in turn, become vehicles for translating the more broad-brush proposals of the 'medium-term strategy' into more specific, district-based projects.

In Hong Kong, there is a relatively compact and efficient governmental planning system for the formulation of land-use/transport plans, and for the programming of resultant public works and the allocation of resources in support of their execution. Although not necessarily appropriate for application elsewhere, due to different political, social and economic circumstances, this planning system takes on the form of a sophisticated set of interlinked executive committees.

In light of the importance of the institutional aspects of plan implementation, it is highly desirable that aspects of institutional mechanisms for plan implementation are addressed at an early stage of strategy formulation. This is important because the effectiveness of the institutional framework will, to a substantive degree, govern the type of plans and programmes produced, and their performance.

It is also very important that a compact but highly competent core team of professionals is maintained to monitor changes that might affect TDS proposals of the land-use/transport plan and associated environmental protection. These might include new trends in economic development and growth, and changes in public policy regarding various aspects of city and socio-economic development. One is thereby able to formulate fairly quick adjustments to the 'medium-term strategy'.

A point will be reached, after, say five years, when it is necessary to undertake a comprehensive review of the strategic plan arrived at, following much the same process as described above. To do this, it is necessary to ensure that an effective, high level planning and development authority is maintained to provide a forum for guiding subsequent reviews and for taking executive decisions on matters of regional significance, especially regional infrastructure systems and facilities (an aspect further discussed in some depth in Chapter 12). In support of this, a well-structured computer-driven, geographic-based data information system needs to be established to help assess the progress made toward achieving strategic planning objectives.

Conclusions

The question was posed at the beginning of this chapter as to whether land-use planners and transport planners have taken convergent, parallel or divergent paths in their planning efforts in Hong Kong. The preceding account of the experience in strategic planning for the latest review of the TDS suggests that transport modelling in Hong Kong has a key but not predominant role to play.

As may be appreciated from the preceding discussion, results from the transport model tests are used as relative rather than absolute measures of plan performance. Such results, furthermore, are interpreted in an interactive way with outputs from other evaluation processes of the overall planning process arrived at through the application of a well-structured and consistent, multi-criteria evaluation system.

The inputs and outputs of such an evaluation system have both quantitative and qualitative attributes. In both cases arriving at these attributes entails high levels of professional judgement which need to be applied in the interpretation of the collectively produced results. Subsequent to TDS strategy selection, and the formulation of 'medium term' plans for which (as a result of the above outlined planning exercise) consistent sets of socio-economic data can be produced, there is a role for the more specific application of transport (and other) models as a basis for formulating sectoral development programmes.

References

Black, J. (1981) *Urban Transportation Planning: Theory and Practice*, London: Croom Helm.

Bruton, M.J. (1985) *Introduction to Transport Planning*. London: Hutchinson Press.

Census and Statistics Department (1993 and 1994) *Hong Kong Monthly Digest of Statistics* Hong Kong: Government Printer.

Choi, Y. (1986) 'Land Use Transport Optimisation (LUTO) Model for Strategic Planning in Hong Kong'. *Asian Geographer* Vol. 5, No.2, Hong Kong.

Creighton, R.L. (1970) *Urban Transport Planning*. Urbana: University of Illinois Press.

Dimitriou, H.T. (1992) *Urban Transport Planning: A Developmental Approach*, London: Routledge.

Freeman Fox, Wilbur Smith & Associates (1967) *Hong Kong Mass Transport Study*, Hong Kong: Government Printer.

——— (1968) *Hong Kong Long Term Road Study*, Hong Kong: Government Printer.

Lands and Works Branch (1985) *Planning for Growth*. Strategic Planning Unit, Hong Kong: Government Printer.

Mitchell, R. and C. Rapkin (1954) *Urban Traffic: A Function of Land Use*. New York: Colombia University Press.

Morlok, E.K. (1978) *Introduction to Transportation Engineering and Planning*. New York: McGraw-Hill.

Ortuzo, S.J. and L.G. Willumsen (1990) *Modelling Transport*. Chichester: John Wiley.

Planning Department (1991) *Hong Kong's Port and Airport Development Strategy: A Foundation for Growth*, Hong Kong: Government Printer.

——— (1992) *Metroplan: The Selected Strategy: Executive Summary*. Hong Kong: Government Printer.

——— (1993a) *Territorial Development Strategy Review: Foundation Report* Hong Kong: Government Printer.

——— (1993b) *Territorial Development Strategy Review: Environmental Baseline Conditions*. Planning Department Hong Kong: Government Printer.

——— (1993c) *Territorial Development Strategy Review: Consultation Digest*, Hong Kong: Government Printer.

Town Planning Office (1970) *The Colony Outline Plan: Vol.1, Concepts and Outline Proposals*. Hong Kong Government unpublished report, Hong Kong.

Transport Branch and MVA Asia (1993) *The Railway Development Study*, Highways Department, Hong Kong.

Transport Department (1989) *Hong Kong Second Comprehensive Transport Study*. Hong Kong: Government Printer.

Transport Department and Scott Wilson Kirkpatrick (1993) *The Hong Kong Freight Transport Study*, Hong Kong.

Wilbur Smith & Associates (1976) *Hong Kong Comprehensive Transport Study*. Hong Kong: Everbest Printing Co. Ltd.

Part Two
CURRENT ISSUES AND PROBLEMS

Current issues and problems: Editors' Overview

Harry T. Dimitriou and Alison H.S. Cook

Over the last decade Hong Kong has become a recognised world-class city and financial centre[1] that can boast of accommodating among the tallest, most impressive and expensive architectural artifacts in the world (see Figure 1). It has some of the most costly real estate, possesses the busiest container seaport (see Figure 2), the fourth-busiest airport, and owns one of the very few (some claim only) underground rail systems that makes a profit. Hong Kong also offers places of recreation and leisure that are the envy of other cities in the region, hosts the largest number of Rolls Royces, restaurants and banks per square kilometre and - for what it is worth - employs among the highest paid academics and civil servants.

While the territory is well known for the many stereotypes with which it has been associated in the past, beneath this surface is hidden one of the most dynamic and (in terms of growth) some would claim, successful, economies in South-east Asia (Cuthbert and Dimitriou 1992). Despite a recent slackening of consumer demand and an uncertain political future, Hong Kong has maintained it economic buoyancy principally because it has benefited from (after Information Services Department 1996):

- a consolidated property market;
- a stock market that rebounded in 1995;
- a sustained high growth in the export of goods and services;
- a robust spending in machinery and equipment; and
- an extremely active infrastructure construction sector.

Other statistics from the same source show that over the last two decades, economic growth in Hong Kong has more than quadrupled, attaining an average GDP growth rate of 7.5 in real terms with per capita incomes at a 1995 value of almost US$23,000. These income levels have tripled and not only surpassed the UK, Canada and Australia but are next only to Japan in Asia. Not having any natural resources, and now experiencing a declining manufacturing base, much of this growth can be directly attributed to the territory's performance in international trade. In terms of goods, this has increased more than 44-fold, and in services, about 26 times.

Figure 1
Contemporary office towers
Source: A.H.S. Cook

The decline in the manufacturing sector has led to the re-structuring of the economy much more toward servicing since the mid 1980s. This in turn has resulted in considerable industrial obsolescence and unemployment as the on-going relocation of manufacturing processes to China continues[2]. Such trends reflect the fast rising business opportunities that exist in the territory for a wide spectrum of service activities (including freight transport, telecommunications, banking, real estate development and professional services). These have, by and large, arisen from Hong Kong's role as a 'gateway' to China whereby Hong Kong's 6.3 million people generate a GDP equivalent to 18 per cent of that produced by China and its firms are the conduit for over 60 per cent of the total stock of foreign direct investment in China (The Economist, 1996). Many such opportunities have been spawned by the economic reforms and trade liberalisation measures that have taken place in mainland China and also (to a lesser extent) other economies in the Asia-Pacific Region.

The manifestations of these growth trends are numerous, diverse and often dramatic. In some cases, their root causes are indirect and complex, in other cases, less

Figure 2
Container trucks queuing at Kwai Chung Port
Source: A.H.S. Cook and H.T. Dimitriou

so. The most apparent of the manifestations are physical - i.e., land use changes, reclamation, and the construction of new building and transport infrastructure (see Figure 3). Industrial, commercial and residential obsolescence and new towns development have also been widespread (see Figure 4), while housing and commercial/retail redevelopments have made their mark too (see Figure 5). Less apparent are those trends that have led to greater expectations from working conditions and recreational opportunities, and rising expectations in the quality of public transport services and infrastructure provision. Perhaps the least apparent of the manifestations are the deteriorating social circumstances of the disadvantaged, the declining health of inhabitants arising from the increase in pollution[3] and the strain placed on the legal and institutional bodies as they attempt to cope with the rapid changes. Among the most worrying of the apparent manifestations, however, has been the rapid growth of motorised traffic and traffic congestion (see Transport Branch, 1994), with all that this implies in terms of rising costs, erosion of the economy's competitiveness and environmental deterioration.

Figure 3
Reclamation for transport infrastructure: Hong Kong Island West
Source: Courtesy of Hong Kong Tourist Association

Figure 4
New Town development: Tin Shui Wai
Source: A.H.S. Cook

These developments are by-products of the territory's growing affluence and need to be acknowledged as such. In line with this perspective, traffic congestion is thus a manifestation of other root problems which in Hong Kong's case is mostly associated with fast economic growth. In retrospect, while there are serious grounds for concern, it is surprising that Hong Kong has gained so much mileage out of its transport system to date and has not encountered more congestion before now. Adding more airport and seaport capacity, redeveloping commercial and residential areas at higher densities, along already congested infrastructure corridors, and reclaiming and then developing new areas in the heart of metropolitan Hong Kong, all generate additional traffic and related environmental degradation, and suggest an inevitable mis-match of land-use planning with traffic and transport planning endeavours.

The fundamental underlying question that arises from this interpretation of developments is how much more growth can and should Hong Kong absorb. The associated lessons to be learned are twofold: firstly, one cannot avoid increasing congestion if an economic strategy is pursued at almost any cost; secondly, the idea that the 'solution' to traffic congestion problems lies within the transport sector itself is a myth which if believed, can exacerbate rather than alleviate the situation. Given these observations, the editors are of the view that it is essential in the preparation of land-use/transport policies and plans that a clear distinction is made between critical 'root problems' and 'manifestation problems', while the extent that these problems pose obstacles to the achievement of these pre-stated goals should be simultaneously monitored to facilitate revisions as required and in light of on-going developments[4].

With the above in mind, Part Two of this book concentrates on selected transport and related land-use concerns that have accompanied Hong Kong's recent rapid growth. These include:

- The relationship between public transport networks and land use developments (see Chapter 4).
- The constraints and opportunities presented by mass transit planning in the metropolitan area (see Chapter 5).

Figure 5
Times Square shopping complex: Hong Kong Island
Source: A.H.S. Cook

- The principles and practices of planning road-based public transport services in the territory (see Chapter 6).

Timothy Runnacles (sometime of the Transport Department of the Hong Kong Government and consultant to the Mass Transit Railway Corporation) argues in Chapter 4 that the urban fabric of a settlement is rarely able to adjust itself to socio-economic changes as rapidly as transport technology, and on this basis, it is better to introduce 'transport solutions' which service specific land-use arrangements rather than the converse. With the very distinctive urban density and topographical conditions of the territory, Runnacles explains that Hong Kong has benefited from the absence of a 'motor car culture' of the kind that is beginning to sweep other parts of South-east Asia. Despite the fast rising incomes, he points out that the territory is void of the low density suburban developments typically generated by widespread motor car use elsewhere. He cites the geographical compactness of the city as a major asset to public transport operators and inhabitants alike, with walking playing a critical role within the wide spectrum of modes open to the public.

Runnacles laments the decline in the use of waterborne transport. He argues that this will have undesirable land-use implications as previous ferry points (attraction nodes of economic and social activities) are closed down. He applauds, however, the very significant role of the tramway system on Hong Kong Island, and its efficient operation, notwithstanding its old rolling stock and the recent decisions not to extend it into new reclamation schemes. He acknowledges the very significant role of buses and their rising quality of service, pointing out that almost 52 per cent of all public transport travel is undertaken by this mode. Without bus operations at their present scale, Runnacles claims Hong Kong's urban areas would simply become dysfunctional.

The strategic role of rail transport - both its mass transit system and its suburban network - is praised in Chapter 4. Particular attention is paid to the commercially profitable highrise developments this has created around MTR stations in the metropolitan area, and the decentralised corridor developments, both existing and planned, based on the KCR lines in the New Territories[5]. The property dealings of the two railway corporations are seen as offering useful lessons for other cities on how to finance their rail systems. Notwithstanding these achievements, however, Runnacles concludes that the Hong Kong government has been reactive rather than proactive in its transport policy formulation. He concludes that the administration has been over-cautious and obsessed with short-run transport costs to the point that too many important beneficial urban development opportunities have been lost.

An in-depth examination of railway planning is offered in Chapter 5 by Timothy Runnacles, Nigel Wilson and Edward So (all associated with the MTRC at the time of writing this book). In this chapter, the contributors consider policy, engineering and financial considerations of mass transit railway planning in Hong Kong. The authors commence by examining the overall passenger transport policy context of MTRC operations and within this, the MTRC policy regarding the development of its system and the main benefits the mass transit system offers, from lower on-road accident rates to reduced air pollution.

The very high carrying capacity of the network, together with the high densities it serves, are clearly among the key factors underlying the success of this well managed railway, although heavy loadings during peak periods and the rising quality of competing public transport services are detracting from the perceived overall quality of service offered by the MTRC. Apart from the many benefits a mass transit system offers to both the city and its inhabitants, in terms of less congestion and a better environment (benefits, incidentally, which are not typically included in the appraisal and comparative cost studies or rail projects), much is made of the profitability of the Corporation and the absence of its need for direct subsidies from government. It is pointed out, however, that despite the challenging topography of Hong Kong, the engineering constraints on mass transit planning are ultimately less problematic than other more loosely definable areas of concern, including the problem of how to pre-finance a rail link to a new town, a new airport or a redevelopment area, so that the potential of these areas can be realised.

Runnacles et al argue that the conflict between short-run costs and benefits, and long-run impacts need to be resolved by government and not by operational agencies. They claim that this is because only government both controls, and is responsible for, financing expenditure. Furthermore, they argue the role of the MTRC in the implementation of territorial development strategy should be clearly assigned by the government. The reality is, however, that as the MTR network is extended into less densely populated areas, and as decentralisation and de-densification take place (as part of Metroplan efforts to reduce densities in the metropolitan areas), the possibilities of operating a profitable and affordable service declines, and the need for government involvement and guidance thereby increases. While there are clearly new opportunities of land development and appreciation accruing from close co-operation with private sector real estate interests, many complicated implications arise. These are in part addressed by the authors and covered elsewhere in the book.

Chapter 6, written jointly by Ernest Lee and Richard Meakin (Assistant Commissioner for Transport of the Hong Kong Government, and past Assistant Commissioner and consultant, respectively) outlines the principles and processes employed in planning road-based public transport services in the territory. The authors claim that public opinion has played an increasingly influential role in bus route development. They explain the numerous types of transport plans and appraise the principles used in road-based public transport planning. In so doing, they highlight the passenger carrying capabilities and advantages of buses and this mode's role in the hierarchy of public transport services on offer.

Lee and Meakin place considerable value on the flexibility and relatively low operational costs of road-based public transport. They acknowledge, however, that these advantages are easily eroded by the presence of traffic congestion where no exclusive bus rights of way are provided. The dilemma regarding the differences of opinion concerning feeder and trunk bus routes, and competition and integration of services, is admitted by the authors to pose significant problems. The conflict between the desire by government to implement consultancy recommendations for enhanced integration of public transport services, and the pressures from the bus operators to free themselves of any restrictive operations that prevent competition with rail travel, echoes the debate of many cities throughout the world, with the difference that Hong Kong has a stronger tradition of allowing market forces to decide policy.

The authors also tell the story of the rise in use, popularity and acceptance of public light buses (PLBs), and their crucial role within the overall provision of public transport services in Hong Kong. This experience has not only provided a model for the introduction of similar bus services elsewhere in the region but also in the world. The PLB service perhaps constitutes the most flexible component of public transport provision in the territory, albeit a source of increased air pollution and Triad involvement. While other road-based public transport modes are covered by this chapter, including private operated buses and taxis, perhaps the most significant section of the chapter concerns the land-use and transport planning interface of such modes. Lee and Meakin advocate that land-use planning exercises should have due regard to the transport requirements before they are finalised, with every effort to tie-in transport development programmes with the target population intake timeframes of new developments. As in the case of the previous chapter, problems of reconciling the financial viability of public transport operations with broader urban planning and environmental interests shouts out for government guidance, better co-ordination of public and private sector interests, and innovative measures to tackle short-term problems in the interest of longer-term (and broader) benefits.

Notes

1 It is the world's fifth largest banking centre in terms of the volume of external banking transactions, and the fifth-largest foreign exchange market in terms of turnover. Its stock market is the second-largest in Asia in terms of market capitalisation. In 1995, the World Competitiveness Report published by the International Institute for Management Development and the World Economic Forum ranked Hong Kong as the third most competitive economy in the world (after the USA and Singapore).

2 The concern for this development is reflected by the fact that the Harvard Business School was commissioned by a group of concerned local leading businessmen to investigate the underlying implications of these developments and recommend what could be done to make Hong Kong a more competitive business environment (see Enright et al 1997). Another study reflecting similar concerns was commissioned by the Industry Department of the Hong Kong government. This study, undertaken by consultants from Massachusetts Institute of Technology, investigates how industry in Hong Kong can be revitalised (see Berger and Lester 1997).

3 The levels of air pollution in the summer of 1995 were the highest ever recorded. Given that much of this pollution is believed to be sourced from the transport sector, and given the further likely motor vehicle growth of both Hong Kong in particular and Southern China as a whole, the future in terms of air quality looks extremely bleak.

4 This view was first expressed by one of the editors in the South China Morning Post in 1995 and subsequently on RTHK in Viewpoint in October of the same year. The ideas presented were well received by a number of parties on the basis that it provided an insight of the 'big picture' whereas much other coverage of this area was focused on projects, project performance, cost over-runs etc.

5 The Western Rail which is proposed to serve the rapidly expanding population of the Northwest New Territories is long overdue.

References

Berger, S. and R.K. Lester (eds) (1997) *Made by Hong Kong*, Hong Kong: Oxford University Press.

Cuthbert, A. R. and H.T. Dimitriou (1992) 'Redeveloping the Fifth Quarter' *Cities Journal*, Vol.9 No.4 August 1992, Butterworth-Heinemann Ltd.

Dimitriou, H.T (1995) 'The Groaning Infrastructure of Hong Kong' *Viewpoint*, RTHK, Hong Kong.

———— (1995) 'No Easy Answer to a Growing Problem' South China Morning Post, Hong Kong.

Enright, M.J., E.E. Scott and D. Dodwell (1997) *The Hong Kong Advantage*, Hong Kong: Oxford University Press.

Information Services Department (1996) *Hong Kong 1996 Year Book* Government Printer, Hong Kong.

The Economist (1996) *Hong Kong: China's Gamble*, London, 22nd June.

Transport Branch (1994) *Report of the Working Party on Measures to Address Traffic Congestion*, Government Secretariat, Government Printer, Hong Kong.

4 Public transport network and land-use impacts

Timothy V. Runnacles

Introduction

As explained in the previous chapter, transport and land use are invariably linked in a bi-lateral cause and effect process. Movement patterns at any given time reflect the spatial distribution, intensity and nature of land uses. These movement patterns are clearly recognisable and stable in the short term, and relate to the routine journeys that people make to and from work, school, shopping, and social activities. As time passes the pattern of land use itself comes to reflect the changes in accessibility afforded by new routes and modes, together with the evolving preferences of the travelling public.

However, the urban fabric cannot generally change as rapidly as transport technology. The resulting lag-effect typically produces an imbalance between transport supply and demand. Giannopoulos and Curdes (1992) have shown how this process applies to European cities, but the principles are ubiquitous, and find acute expression in the burgeoning metropolises of newly industrialised countries (Hall 1984). This imbalance between urban fabric and movement makes new transport facilities imperative, and largely determines the form that they take. Thus, specific land-use arrangements demand particular transport solutions. This is readily apparent in Hong Kong.

Hong Kong's urban characteristics

Hong Kong is certainly amongst the most distinctive of the world's major cities. Topography, cultural attitudes, political and economic circumstances, and (latterly) strong planning intervention have all jointly determined current settlement patterns. Contemporary Hong Kong is characterised by the very dense urban agglomeration around Victoria Harbour and planned clusters of urbanisation elsewhere. Since this polycentric metropolis has been largely created since the 1950s, it would be inappropriate to explain urban form merely in terms of successive waves of transport innovation. Thus, whilst Hong Kong can certainly demonstrate how the built environment has related to the transport facilities available at particular stages of development, the resulting urban pattern is quite distinct from that found in cities elsewhere.

Despite enduring the world's highest traffic densities (271 licensed vehicles per kilometre of road at the end of 1995), and notwithstanding a spectacular network of high capacity urban roads, typical 'motor culture' land uses are notably absent from Hong Kong. There are virtually no low density suburbs, and no arterial roads flanked by supermarkets, filling stations, motels, 'family restaurants', car repair shops or cinemas. Hong Kong has neither sprawling suburban shopping malls nor green-field office parks. A combination of rugged topography, a characteristic Chinese preference for communal propinquity and the government's strict control on the disposal of land for development have collectively thwarted any tendency towards low-density sprawl. The territory's urban fabric has remained tightly knit, with high densities, high buildings, concentrated activities and an unusually strong dependence on public transport for passenger travel. The extent of this reliance is revealed by the evidence that public transport comprised 79.5 per cent of all trips in 1992 assuming that taxis are regarded as private transport (MVA Asia 1993). However, if taxis are considered to be public transport, the 1992 proportion was 90 per cent (see Table 4.1).

High-density development is the most obvious feature that distinguishes Hong Kong from most other cities. With a land area of 1,092 km^2 and an estimated mid-1995 population of 6,189,800, the overall density of the territory was 5,668 persons/km^2.

Table 4.1 Hong Kong's internal travel by mode

Mode	Daily boardings (thousands)		%
	1995	1992	1992
Metro (MTR)	2,226	2,052	17.0
Railway (KCR)	641	546	4.5
Light rail (KCR)	336	252	2.1
Tram (HKT & Peak)	324	348	2.9
Franchised/public bus	3,601	3,475	28.8
Green minibus	831	705	5.9
Red minibus	899	1,016	8.4
Ferry	222	242	2.0
Private bus	n.a.	947	7.9
Taxi	1,307	1,267	10.5
Private car	n.a.	1,210	10.0
Total	**n.a.**	**12,061**	**100.0**

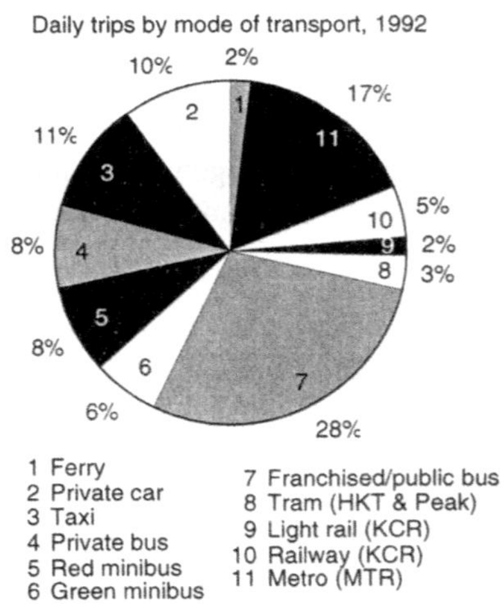

Note: *1995 figures for trips by private bus and private car are not available as these were obtained from the 1992 Travel Characteristics Survey*

Sources: *MVA Asia 1993 and Transport Department 1993 and 1996*

However, whereas Hong Kong's overall density may be of the same order as London and Tokyo, and significantly lower than Bombay, or even New York or Moscow, what distinguishes Hong Kong is the very limited extent of its developed area - only 15.6 per cent of the total territory. On developed land average residential densities sum to 36,335 persons/km^2. If this definition should appear rather artificial, the residential density of Hong Kong Island, Kowloon and New Kowloon in 1993 was as high as 26,180 persons/km^2, despite much unusable mountainous terrain on Hong Kong Island and a swathe of parks and recreation grounds in northern Kowloon. Indeed, in terms of its urbanised land, Hong Kong is probably the second most densely developed major city anywhere, the dubious accolade for first place being held by Shanghai, whose central district of 93 km^2 accommodated 4.81 million people at an astonishing density of 51,720 persons/km^2 in 1989 (Xu 1994).

Even in Hong Kong, urban densities are gradually declining as living standards improve and as new towns are built (Barden and Runnacles 1986). Nevertheless the overall impression is still that of a teeming metropolis with an almost overwhelming crush of humanity. In transport terms, such a concentration of people initially generated little need for mechanised transport, because many activities were within easy walking distance of each other. This characteristic was reinforced by the intermixing of industrial activities with living spaces, often in the same building, and even within the dwelling unit itself. However, this pattern has changed markedly in the last two decades in response to specific industrial zoning, the growth of the tertiary economy and the development of geographically independent new towns.

Mechanised transport is now absolutely essential in Hong Kong, and walking is little favoured in an urban environment where it is seldom a pleasure and where material expectations are swiftly rising (Dimitriou 1994). One consequence of this demand for mechanised mobility is that Hong Kong people now require a virtual door-to-door public transport service, even over the shortest distances. Other upwardly mobile societies might variously satisfy such movements by private motoring, walking, or cycling.

The evolution of urban form in relation to Hong Kong's transport systems

In the older industrialised societies of Europe, land-use development has plainly reflected the long term evolution of transport facilities through successive eras (although isolating these effects in a quantitative manner is usually supremely difficult). The locations of these cities have been largely determined by long-distance transport activities, whilst their size has been influenced by both long-distance and local transport facilities.

Local transport systems have invariably affected both urban form and structure (Vuchic 1981). In western cities the characteristic closely-spaced jumble of activities that had pre-dated mechanised travel was restructured by railways and tramways, whose radial networks both concentrated the growth of business districts in city centres and stimulated suburban housing and secondary commercial centres along their routes. Several decades later the widespread use of motor road vehicles encouraged low-density suburbanisation and reduced the importance of city centres.

Nothing endures but change, and as many growing economies in eastern Europe and Asia now seek to pursue the mixed blessings of widespread motor car ownership, some American and European cities are rediscovering public transport in an endeavour to create sustainable urban structures. In scores of cities planners are seeking to restore earlier more disciplined urban growth patterns by building new tramway and metro systems, and by revitalising suburban railways.

Traces of the patterns found in older industrial societies can be found in the relationship between Hong Kong's urban growth and its transport systems. However, it should be recognised that many other powerful influences have contributed to the shaping of contemporary Hong Kong, as discussed in Chapters 1 and 2. With this caveat, the following narrative sketches the relationship between Hong Kong's multifarious transport modes and urban development. Although each type of public transport is discussed chronologically, no major form of public transport has ever been abandoned in Hong Kong, so the various relationships described below are still evolving concurrently in a rather complex manner.

Water-borne transport

Hong Kong's large natural harbour favoured the territory's emergence as a major trading centre right from the beginning of the colonial era in 1841. Urban development was initially focused on the administration's headquarters in the City of Victoria (now the Central district) on Hong Kong Island.

The contemporary relationship between water-borne transport and land use is not harmonious. Apart from the progressive abstraction of cross-harbour patronage to new road and rail tunnels, ferry piers have suffered from numerous relocations arising from reclamation projects. This has removed the piers from activity centres and reduced the attraction of ferry travel. Moreover, most ferry passengers need to change to or from other modes to complete their trips, and the consequent cost and time penalties mean that it is often quicker, cheaper and more convenient to use land-based modes throughout the journey.

The Peak Tram

Opened in 1888, the funicular railway between Garden Road and Victoria Gap on Hong Kong Island was the first mechanised land-transport system in Hong Kong. Originally operated by the High Level Tramway Company, the 1.4 km line has been operated by the Peak Tramways Company Ltd. since 1905.

The land-use impact of the Peak Tram should not be exaggerated. When opened, fares were 30 cents uphill in first class, 20 cents in second, and 10 cents in third. Downhill trips were half the cost of going up, but in 1888 money terms these sums were substantial, and helped to preserve the Peak's exclusivity. Since then, the composition of the Peak Tram's ridership has been progressively changing; today less than 8 per cent of its 10,920 daily users are regular commuters, with trippers and tourists comprising the rest. A direct but localised land-use by-product of this clientele was the redevelopment in the mid-1990s of the upper terminal area to include the 'Peak Galleria' and the rebuilt 'Peak Tower' shopping and leisure complexes.

Hongkong Tramways

In contrast to the relatively exclusive Peak Tram, the electric tramway on the north shore of Hong Kong Island has manifestly contributed to the development of one of the most highly concentrated linear cities to be found anywhere on the planet.

Electric trams began running from Kennedy Town to Shau Kei Wan, and along a branch to Happy Valley, in 1904 (see Chapter 1). Since then the system's extent has changed little. However, residential densities have declined along the corridor, notably in Sheung Wan, whose density fell from an extraordinary 238,025 persons per km^2 in 1961 to 89,256 per km^2 by 1981. This reduction in residential density has been compensated for by the growth of commercial development in Central and Wan Chai. This generates many non-home-based trips on the tramway, which carried 312,720 rides a day in 1995, and remains the world's busiest tram route serving a single corridor (Figure 4.1)

Figure 4.1 Hong Kong tramways: North Point
Source: A.H.S. Cook and H.T. Dimitriou

Hong Kong's tramway has been amongst the world's most heavily used systems for many decades, but it has not followed systems elsewhere, either into oblivion or into the contemporary imperative to upgrade, although both suggestions have been made more than once. Despite its success, entreaties by the operator to extend the tramway on the Island or to build a new system in Kowloon (proposed in 1913) were eventually rejected by the government. In 1972 Hongkong Tramways proposed to build new systems in both Sha Tin and Castle Peak (Tuen Mun), and the latter proposal reached the state of detailed negotiations in 1979-1983. These negotiations collapsed, but the Tuen Mun system was subsequently built and operated by the Kowloon-Canton Railway Corporation, as discussed below.

Bus services

Relative to its size, Hong Kong could well be described as 'the bus capital of the world'. At the end of 1995, there were 4,834 registered franchised buses operating on 501 routes throughout the territory. These were supplemented by 333 private buses and 4,765 non-franchised public buses; some 500 of the latter served on 136 scheduled routes known as 'residents services'. In addition, there were 4,350 public light buses, 1,872 of which ran on a network of 267 scheduled routes, whilst the remainder provided non-scheduled services. Private light buses are used for various personal transport duties, and 2,585 of these were licensed at the end of 1995 (Government Information Services 1996 and Transport Department 1996)

The reliance on buses remains very strong despite the growth of rail services. In 1995, bus travel (including red minibuses, but excluding non-scheduled public and private buses and light buses) represented almost 52 per cent of all trips made on public transport. The density of the bus system reflects that of the urban form itself, and without bus operation on its present scale, Hong Kong's urban areas simply could not function.

Table 4.2 shows that the number of franchised buses relative to population has steadily increased in Hong Kong despite the rivalry of high capacity rail systems, minibuses and non-franchised buses. If bus capacity rather than vehicles alone were to be considered, the impact would be even more pronounced, as the buses of the 1930s and 1940s were relatively small single-deckers. By the end of 1995 double-deckers comprised 91 per cent of the total franchised fleet. Over half of the total fleet was made up of very large three-axle vehicles, each holding about three times as many passengers as the first-generation single-deckers.

Despite these achievements, the franchised bus companies have not always responded to increasing travel demand with requisite alacrity. In the 1970s the opening of the Cross Harbour Tunnel and the growth of the new towns in the New Territories both accelerated the demand for bus travel and increased average journey length. But the bus operators did not respond quickly enough, despite government's entreaties and (from 1977 onwards) the requirements of its annual 'route development programmes'. The plight of new town residents was particularly severe, with some commuters queuing well before dawn for buses to take them to work. The government had to intervene in 1981 by initiating memoranda of agreement with both major franchised bus companies to meet the public's demands (see Chapter 6 for more details of bus service planning).

The growth of rail capacity reduced the strain on the buses during the 1980s, although there was some irony when, in 1991, new Kowloon and Cross Harbour bus routes were specifically introduced to divert passengers from the Tsuen Wan Line MTR route under Nathan Road, which had reached its full capacity in the morning peak hour. Nonetheless, the bus industry in the mid-1990s was different to that of

Table 4.2 Registered franchised buses in Hong Kong

Year	Population (rounded to nearest 100)	CMB	KMB	Total*	Population per bus***
1936	989,200	**88	**111	**199	**4,966
1946	1,550,000	34	143	177	8,757
1951	2,015,300	151	**270	**421	**4,787
1956	2,614,600	200	450	650	4,022
1961	3,174,000	307	721	1,028	3,088
1966	3,732,400	495	1,055	1,558	2,396
1971	4,045,300	528	1,133	1,700	2,380
1976	4,438,000	702	1,700	2,446	1,814
1981	5,201,100	1,028	2,390	3,477	1,496
1986	5,488,000	1,020	2,731	3,811	1,466
1991	5,822,500	1,015	3,038	4,132	1,409
1995	6,189,800	881	3,506	4,834	1,280

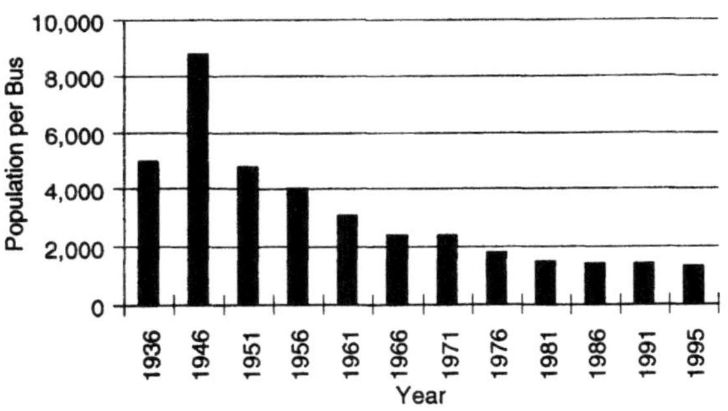

* *This includes New Lantao Bus Company and Citybus Ltd. buses where appropriate*

** *Estimated figures*

*** *These figures do not match exactly as population figures mostly relate to census dates in March whilst bus fleet totals usually relate to the year's end*

Source: Transport Department

even the previous decade. With the expansion of the franchised fleets and greater competition from other modes, productivity had fallen markedly. At the end of 1983 each licensed franchised bus carried, on average, 1,100 passengers daily. Twelve years later this figure had fallen to 781, despite a 39 per cent increase in the capacity of the average bus during the same period. Buses were therefore facing keen competition, which they endeavoured to meet by upgrading their networks with a wider variety of routes and a steady injection of air-conditioned rolling stock, albeit at the expense of rising fares.

There was competition within the franchised bus industry too, as a new franchised operator (Citybus Ltd.) had entered the scene in 1991 with a solitary route on Hong Kong Island. Subsequently this company gained a second franchise for 26 routes on the Island, which it took over from CMB in September 1993, followed by a further 14 in 1995 plus seven routes which it had initiated itself.

Whereas Hong Kong's bus industry still flourishes amazingly well by comparison with that of most other major cities of the world, the days of very cheap bus travel are clearly coming to an end (see Chapter 6). Nonetheless, unless present commitments to further railway expansion are met expeditiously, a proportionately increasing burden would again be placed on road passenger transport at a time when worsening road congestion also threatens to jeopardise satisfactory bus operation. Moreover, the bus operators in general, and the Kowloon Motor Bus Company (1933) Limited in particular, are saddled with labyrinthine bus networks which could benefit from some rationalisation. Although this necessity is recognised, it evades achievement, as the various district boards invariably oppose any endeavours to withdraw routes, even if they carry few passengers and are paralleled by other services.

Apart from franchised buses, the contribution of non-franchised bus and paratransit services in Hong Kong should be acknowledged. Non-franchised buses range from luxuriously-appointed double-deck coaches to the spartan single-deckers used for school and factory contracts. A particularly fast-growing element comprises the 'residents' services', which are essentially designed to provide peak hour travel between smaller housing developments and major commercial districts. These services supplement franchised routes, and offer point-to-point travel, usually in high-quality rolling stock. Although the franchised operators resent this form of competition, it might be preferable to regard it as an alternative to private car travel. As such, it offers a practical solution to customers' needs for comfortable travel in a very dense urban environment.

By contrast with high capacity buses, the 16-seat public light buses (PLBs) are not efficient users of road-space. This mode evolved from the illegal passenger vans which appeared in the New Territories in the 1950s (see Chapter 1). They were legalised as PLBs in 1969, but their growth was so rapid that a ceiling had to be imposed in 1976, when their numbers reached 4,350. Responding to market forces, the PLBs thronged the main corridors in direct competition with buses and trams. As major contributors to congestion, the government imposed restrictions on their activities on the one hand, and sought to provide a new role for them on the other, by developing a network of 'green minibus' services on licensed routes which were subject to fixed schedules and fares.

The green minibus network has slowly expanded over more than two decades to the point where the green vehicles comprise 43 per cent of the total PLB fleet. Green minibus routes have generally been devised to serve areas inappropriate to normal bus services for geometric or economic reasons, though to some degree they are now being developed as a premium quality alternative to ordinary buses as well. With their typically short routes (averaging 7.0 km) and semi-captive markets, the green minibuses are highly productive. At the end of 1995, each green minibus carried an average of 459 passengers a day. This was 59 per cent of the equivalent figure for franchised buses, despite the fact that on a vehicle-to-vehicle basis, the green minibuses had only 13 per cent of the capacity of the average bus.

The ordinary PLBs (or 'red minibuses') running on unregulated routes continue to thrive, although their network has altered somewhat in the face of Mass Transit Railway competition. By the end of 1995, the average red minibus was still carrying an estimated 354 passengers daily, considerably higher than the corresponding figure of 279 twelve years previously. Again, although inefficient in road-space terms, the PLB may be seen as a response to the need for relatively fast and convenient travel in a densely developed community in which private motoring is not feasible for most people. Thus, whilst minibuses may be decried as inefficient alternatives to conventional buses, let alone rail services, viewed as a collective travel alternative to the private car, they clearly have a role to play in sustaining the viability of Hong Kong as a high density city.

The Mass Transit Railway

The proliferation of private sector road passenger transport modes in post-war Hong Kong was not only expedient but also characteristic of the government's cherished laissez-faire policy. The government's only direct involvement in public transport was its own Kowloon-Canton Railway (KCR), but until the 1970s this had slumbered as a relatively minor mode, and proposals to extend the system to Central and Kwun Tong in Sir Patrick Abercrombie's 1948 Hong Kong Preliminary Planning Report were not acted upon (Leeds 1984).

In the 1960s, the territory's transport system had come under scrutiny in the U.K. Road Research Laboratory's Passenger Transport Survey (Road Research Laboratory 1967), which represented Hong Kong's first major traffic and transport study. This study recognised the potential inability of existing public transport modes to cope with demand, and recommended a 'new level' of transport. This recommendation was followed immediately by the Mass Transport Study and the Mass Transit Further Studies (Freeman Fox et al 1967 and 1970) which recommended a 52.6 km network of four heavy metro routes.

Despite some erstwhile local criticism about alleged capacity limitations, the MTR now has a directional line capacity of up to 85,000 passengers per hour, which ranks it as the most capacious metropolitan railway ever built. It is also the busiest metro in the world in proportion to its length. With such credentials, it would be very tempting to regard the MTR as a major force in reshaping Hong Kong's urban form (see Chapter 8).

However, the system does not appear to have had the same land-use impact as some other recently built or expanded metros, exemplified by the Hamburg U-bahn and, most notably, the Toronto Subway (Walmsley and Perrett 1992). The reason for this relates to the fact that the MTR was built to serve communities which were already very densely developed.

Summarising those few investigations which have been made of the MTR's land-use impact, Meakin (1989) concluded that whilst a trend towards clustering of new high-density developments has emerged around MTR stations, this effect has been constrained by the paucity of developable sites in the MTR corridors and other limitations, such as airport-related building height restrictions in Kowloon and eastern Hong Kong. The studies revealed that whilst commercial, residential and industrial 'density gradients' are detectable in the vicinity of some MTR stations, such gradients are less evident in primary commercial areas because corporate tenants with low rent-cost sensitivities concentrate closest to MTR stations. On the other hand, peripheral concentrations of density are found about 300 metres from stations in these primary areas.

Although (with some notable exceptions) the MTR may not have created readily apparent high-value clusters in the immediate vicinity of its stations, it is equally clear that the present pattern of activity in general (and commercial activity in particular) could not have been sustained without the passenger accessibility afforded by the MTR. It has allowed the business district of Central to flourish and build ever higher, and it has permitted the extension of business activity to the Queensway area and through Wan Chai to Causeway Bay, as well as westwards to Sheung Wan and across the harbour to Tsim Sha Tsui. Moreover, the MTR has enhanced accessibility to more peripheral localities, as witnessed by the growth of commercial activity in such places as Kwai Fong, Ngau Tau Kok and Kornhill (see Figure 4.2).

Figure 4.2
MTR and Metroplaza commercial development at Kwai Fong
Source: T.V Runnacles

In a theoretical study of Hong Kong's urban area, Cochrane (1982) showed that the acceptable density limits to sustainable residential and office land use (based on surface transport alone) had already been well exceeded even before the MTR was introduced. Clearly, the present scale of development could not have been achieved without the railway. Moreover, whilst the early research outlined in Meakin's paper (cited above) could find little evidence of high density clustering in the vicinity of MTR stations, this conclusion now warrants modification. The early 1990s saw the erection of some of the world's tallest buildings, including Central Plaza in Wan Chai and the Bank of China in Central (and many other marginally less striking commercial towers), all within ready walking distance of MTR stations. Such structures attest to the necessity of a high capacity rapid transit network.

No consideration of the MTR's impact should ignore the MTR Corporation's own property development role (see Chapter 6). The need for work sites and the construction of stations and entrances all required compulsory land acquisition (facilitated by the 1974 Mass Transit Railway (Land Resumption and Related Provisions) Ordinance) which subsequently provided the basis for joint-venture property development. The Corporation has retained management of nineteen developments, including three shopping centres, three office buildings and 28,729 flats. Amongst these developments are such major properties as the residential schemes at Kornhill and Heng Fa Chuen, and the office towers at Admiralty Centre. The new Lantau and Airport Railway is building upon this experience with its proposed 3.27 million m^2 of commercial and residential property development.

The Kowloon-Canton Railway

The Kowloon-Canton Railway (KCR) was opened between Kowloon and the Chinese border at Lo Wu in 1910, whilst the Chinese section from the border to Canton (Guangzhou) was opened in the following year. The KCR was a single-track route serving the rural New Territories (which had been incorporated into the Crown Colony of Hong Kong in 1898).

During the 1970s several piecemeal improvements were made to the KCR, including the relocation of its Kowloon terminus in 1975, partial double-tracking, signal modernisation, improvements to other stations, additions to freight handling facilities, a loop line to the new Sha Tin racecourse, and additional rolling stock. The 1976 'Comprehensive Transport Study' had suggested electrification, and the need to do this was confirmed two years later by the KCR's own consultancy study. This latter study anticipated that the new electric service would be able to handle 140,000 boardings daily, a great advance on the 39,700 carried in 1979. As it turned out, this was a considerable underestimate of the patronage actually carried in recent years.

Electrification was completed in three phases between May 1982 and July 1983, and involved the complete reconstruction of the railway, which became double-tracked along its full length and featured additional stations (some of which opened after electric services began). As reconstruction neared its end, the KCR itself was vested as a government-owned autonomous corporation on the same model as the MTRC.

Although the KCR handles both freight traffic and through trains to and from China (restored in 1979 after a 30 year interval), the East Rail Division (as the route is now defined) is essentially a rapid transit suburban line with a character not dissimilar to the MTR (see Figure 4.3). Patronage has increased steadily since electrification. In the last full year before electrification (1981) average daily ridership was

Figure 4.3
A KCR 'East Rail' electric train at Tai Wai (with the towers of Sha Tin visible in the background).
Source: A.H.S. Cook

46,912; by 1984 (the first full year of electric services) this figure had risen to 218,505, whilst in 1995 patronage had further increased to 640,559 boardings a day.

Like the MTRC, the KCRC has developed property over its railway premises, albeit confined until recently to developments over Sha Tin station and over the Ho Tung Lau depot immediately to its north. Elsewhere it is difficult to attribute any spontaneous land-use impacts to the KCR, simply because most of its route passes through planned new towns. However, the very sustainability of Sha Tin, Tai Po and Fanling/Sheung Shui, which had a combined population of over 950,000 in 1995, would be difficult to imagine without the KCR in its present form. Road travel, for long constricted by the Lion Rock Tunnel, has eased with the opening of the Tate's Cairn Tunnel and the Shing Mun Tunnels to east Kowloon and Tsuen Wan respectively, but the opening of these links (and the new bus routes that serve them) only briefly interrupted the growth of patronage on the KCR, which is the busiest single-route suburban railway in the world.

The further expansion and decentralisation of Hong Kong will be closely related to the KCR network's future growth in the New Territories, in which regard the commissioning of engineering studies for a 51 km 'West Rail' link from West Kowloon to Fanling, Man Kam To and Tuen Mun opens up major development prospects for the early 21st Century.

Light Rail Transit

Although the new towns are a major source of traffic to the KCR, MTR and buses alike, they were originally expected to be largely self-contained (see Chapters 2 and 3). In pursuit of this goal, a strong policy emphasis was placed in their design and planning on internal transport facilities. Roads were planned on hierarchical principles, and studies were conducted to recommend the most suitable internal public transport facilities. In the north-west New Territories, the first-generation new town of Tuen Mun was the subject of an innovative agreement in 1973 which provided for 'an exclusive and segregated right of way for public transport vehicles'. A year earlier Hongkong Tramways Limited had proposed 'modern tramways' for both Tuen

Mun and Sha Tin, and it was this same mode which was recommended in the 1978 Tuen Mun New Town Transport Study.

The proposed tramway, re-baptised as a 'light rail transit' (LRT) system, was to run on several routes in Tuen Mun, mainly on roadside reserves. An optional extension to Yuen Long (which was then about to receive official designation as a new town) was confirmed as part of the proposed network in 1979, whilst a branch to the joint-venture new town of Tin Shui Wai was added to the planned system in 1982 (Runnacles 1988).

In April 1979 negotiations to build and operate the LRT system began between government and the Hongkong & Kowloon Wharf & Godown Company Ltd., which, as the owner of Hongkong Tramways Limited, was regarded as the only company in the territory with appropriate expertise. After nearly four years of discussion, the Wharf Company pulled out of the negotiations in January 1983, largely because of differences of opinion about property development and forecasting uncertainties. After much discussion between the government and the various consortia which had previously submitted proposals to the Wharf Company, the KCRC was invited to take on the project. This the Corporation agreed to do in July 1984, and just over a year later a contract was awarded to construct the system.

The gestation period of the LRT system had been relatively long, and was subject to many uncertainties (Runnacles 1991). But construction was rapid, and the LRT system's first commercial services, comprising five routes, began in September 1988. By the end of that year daily boardings averaged 158,800, which, although a respectable achievement, was well below the KCRC's expectations. In subsequent years, the system has been extended from its original 23.35 km to 31.75 km, and the original 70 vehicle fleet has been increased by 30 additional cars, with 20 more for delivery in 1997 (see Figure 4.4).

Figure 4.4
Light Rail Transit (LRT) serving the Tuen Mun-Yuen Long corridor
Source: T.V. Runnacles

By the end of 1992 patronage had increased to 276,645 boardings daily. However, this brought criticisms of overcrowding, contributing to a decision to dismantle the network's protective 'Transit Service Area' (which had forbidden the local use of KMB's buses) from June 1993. But it appeared that this measure had little impact on LRT ridership, which continued to grow, and had reached 336,700 boardings daily by December 1995.

As with the KCRC's main line, it is hard to identify any spontaneous land-use impacts arising from the LRT, apart from its own property developments over the light rail vehicle depot in Tuen Mun and above three termini. Tuen Mun, Yuen Long and Tin Shui Wai are all planned developments, and the LRT system formed an element of that planning to the extent that the track reserves in Tuen Mun were in place many years before the first rails were laid. The LRT is a medium capacity system, and there is no discernible evidence that the western New Territories new towns are built to densities any different to those of the other new towns without light rail systems. Indeed, all new towns have lower densities than those of the old urban area.

Despite the lack of conclusive evidence that the LRT system has had any specific impact, it has enhanced the image of the new towns that it serves, which had always been one of the objectives of adopting the LRT scheme instead of conventional buses. Moreover, the adopted urban form of these planned communities recognises the provision of the LRT system, although Bristow (1989) argues that Tuen Mun could have functioned better if the LRT tracks had not so slavishly followed the road alignments. He also considered that it would have been better if the LRT network had been planned first, with the towns being designed around it. Of the three new towns the system serves, Tin Shui Wai comes closest to satisfying this desideratum.

The influence of Hong Kong's urban form on transport solutions

This necessarily abbreviated account has outlined the parallel development of urban expansion and transport services in Hong Kong, emphasising the dense and high-rise urban form that a dependence on public transport has permitted. But there is another aspect of this relationship which deserves some observation, and that concerns the equipment deployed by the public transport industry in response to Hong Kong's arduous demands.

Predictably, in such a dense and congested community, the most evident impact on public transport is the imperative to handle as many passengers as possible in the minimum possible space. For most surface transport systems this reflects the vertical planning that dominates Hong Kong's skyline. Double-deck vehicles allow the maximum concentration of passengers on the minimum achievable footprint and at the lowest cost per passenger place. Double-deck buses and trams (see Figure 4.5) were always preferred in the British Isles, but found only limited favour elsewhere for a variety of cultural and physical reasons.

The first double-deck vehicle in Hong Kong had been adopted by the electric tramway system in 1911, and double-deck cars shortly became standard on this network, as they still are. Today Hongkong Tramways has the only all-double-deck tram fleet in the world, and is one of only three of the world's 335 tram systems operating in mid-1996 to employ such cars. Suggestions to replace the double-deck

Figure 4.5
Double-decker buses and trams in Wan Chai
Source: T.V. Runnacles

trams by more conventional single-deck (or single-deck articulated) cars have been firmly resisted by the operator and government alike on the argument that longer cars of equivalent capacity would require longer stopping places, occupy more road-space, take more time to negotiate junctions and require larger depots.

In such an aggressively modern city as Hong Kong, it is more than a little surprising that such traditional trams have survived at all. Certainly there have been several suggestions over the last seventy years that the trams should be replaced by 'modern' and 'flexible' buses or trolley buses. However, such suggestions have always encountered capacity problems of some kind. A 1965 report examined the future of the tramway system (Public Works Department 1965) and concluded that replacement was not desirable, amongst other reasons, because tram-replacement buses could not be accommodated at kerbside stopping places in addition to the multitude of buses already using them.

A 1981 proposal to replace trams by trolley buses running in the centre of the road (like the trams) reached a negative conclusion because of the extra road-width needed to accommodate buses, which would not only have lacked the benefit of a fixed guideway, but which would also have been considerably wider than the trams (Transport Department, 1981). The 1.98 metre-wide HKT tramcars are now the world's narrowest, and one reason for the rejection of a 1980 proposal to replace them, not by buses but by 2.65 metre wide light rail vehicles, was because of the extra pressure on road space the new cars would have created (Martin and Voorhees Associates 1980).

Drawing from HKT's successful operation of double-deck tramcars, the north-west New Territories LRT system was also planned for double-deck operation, but neither the Wharf Company nor the KCRC was prepared to accept the risk and cost of specifying a new and apparently unconventional design, although headroom clearances still generally allow for the future use of double-deck cars. But even the single-deck cars that were purchased are sufficiently unusual to warrant comment. With passenger capacities ranging (according to type) from 205 to 226 per car, they are amongst the most capacious non-articulated cars ever operated on a light rail system. They are also distinctly long for a street-running rigid-framed car; with a body length of 19.4 metres, they are the second longest non-articulated light rail cars ever built, being only marginally exceeded by the 19.76 metre long cars used in Buffalo, USA.

The double-deck format took many years to spread to the franchised bus fleets. It was 1949 before KMB bought a double-decker, and CMB did not acquire its first until 1962. However, double-deck buses progressively became the norm, and by the mid-1980s it began to look as if both the major operators (KMB and CMB) would all but eradicate single-deckers. Since then there has been a minor reversal of policy, as both KMB and CMB have discerned a role for a proportion of single-deck air-conditioned vehicles on premium-fare routes of moderate patronage density. Nonetheless, the double-deckers continue to reign supreme, and since 1981 the majority of purchases have comprised three-axle vehicles, eleven metres and twelve metres in length, with capacities ranging from 135 to 171 passengers. At least one franchised operator has considered a four-axle 15 metre long double-decker, but such an immense vehicle is not currently permitted under the Road Traffic (Construction and Maintenance) Regulations.

Multiple-decking also characterises most of the territory's franchised ferries. Double-deck vessels comprise the majority of the conventional ferry fleet, but the outlying islands are served by a fleet of 13 triple-deckers with capacities ranging from 1,051 to 1,560 passengers. Even vehicular ferries are double-decked, with two-level loading/unloading ramps at their piers. Indeed, two of the vehicular ferries are triple-decked, with passenger accommodation on the uppermost level.

Brute capacity is a necessity on the territory's two 'heavy' railways. At the time it was opened the MTRC had the largest metropolitan railway vehicles ever manufactured, each car having a capacity of 313 passengers at retrospectively-determined acceptable comfort levels, the original specifications having assumed a capacity of 375 passengers per car. The emphasis in designing the MTR was truly on 'mass' transit: each car has only 48 seats, and the inter-car gangways within each train span almost the full vehicle width, thereby providing additional standing and circulation space. It is interesting to note that the standards used on the MTR have been closely followed on the metros subsequently opened in Singapore and Shanghai, and under construction in Guangzhou.

With the inexorable pressure for more capacity on the MTR (see Chapter 5), ways had to be found to raise the hourly directional passenger flow threshold, which had originally been set at 75,000 passengers per direction per hour on the Tsuen Wan Line. In 1993 it proved possible to inject one extra train, raising the capacity to 77,500 passengers per hour. The provision of new signalling on the moving block principle enabled a further increase to 85,000 passengers (or 34 trains) an hour in 1996, but beyond this volume station capacity becomes the limiting factor.

On the KCR the rapid growth of patronage in the 1980s demanded a doubling of train lengths from the original six-car formations to 12 cars. The original cars were also reconfigured to reduce seating and increase overall capacity so that a 12 car train comprising new or modified cars was able to accommodate 3,875 passengers at densities comparable to the accepted MTR standards. New cars ordered in 1996 feature five door openings per car on each side to increase passenger capacity by a further 15 per cent.

A commentary on Hong Kong's public transport/land-use interface

The various expressions of transport technology and land use are invariably locked together in a symbiotic relationship. In most mature cities there is clear evidence of this relationship in the juxtaposition of various urban forms. Thus, intensively developed central business districts and radial corridors of relatively dense activity reflect the rail-based pattern of urban growth, whilst lower density sprawl and major out-of-town commercial activities declare the impact of personal car-based mobility.

These patterns are less manifest in Hong Kong, where much of the present urban fabric reflects the vigorous growth of the last few decades. The constraints of natural topography, reflected in the high cost of land (and the cost of forming new land by reclamation and hillside terracing), have produced one of the densest and 'tallest' cities in the world. For most of its existence, Hong Kong's growth has been concentrated around the harbour, but from the 1970s onwards eight new towns have been created in the New Territories, providing nodes of high density development, some of which are relatively remote from the old urban core.

The government's influence on urban form cannot be understated, as it has a complete stranglehold on land disposal and redevelopment (see Chapter 2). Moreover, its planning has been essentially 'housing-led' since the 1950s, resulting in the creation of the original resettlement estates and industrial satellites, and subsequently the new towns. Current initiatives, derived from the cost-minimising Land Use Transportation Optimisation Study (LUTO) and now enshrined in Metroplan, are restoring the focus of new development to the harbour coastline, where three major reclamations (West Kowloon, Hung Hom Bay and Aldrich Bay) are already well advanced, with others now in progress or destined to follow (Central and Wanchai, Belcher's Bay and Green Island). Despite these initiatives, the government is paradoxically open to the criticism that its traditional laissez-faire policies have been characterised by a reactive (rather than a proactive) approach to problem-solving, coupled with extreme caution and an obsession with costs. Some regard may be given to these accusations in the transport context.

It has been noted earlier in this chapter that Hong Kong has traditionally relied, and still largely relies, on franchised buses for its passenger transport (see Chapter 6). Buses are admittedly a flexible public transport instrument, and the particular species of Hong Kong bus is also a formidable carrier. Nevertheless, the heavy dependence on buses led to serious problems in the 1970s when the franchised operators, themselves run by extremely cautious and conservative managements, failed to match the rapidly swelling demand for travel from the housing estates and new towns. To be effective, buses require unimpeded traffic flow on the roads, yet this has not always been the case (as witnessed by the chronic traffic paralysis on King's Road on Hong Kong Island in the late 1970s). A spate of new road-building during the 1980s improved traffic movement significantly, but there is an ever-present threat that capacity increments will fail to match the rising tide of road traffic. Inevitably, numbers alone dictate that it will be the bus users who will be the principal victims.

The government's present transport policy (Transport Branch 1990) re-affirms the need to expand rail systems, yet there has been a long history of dithering on railway investment which has arguably left Hong Kong with a smaller network than it deserves (Dimitriou 1994). Given Hong Kong's reputation as a 'can do' society, major public transport infrastructure projects have been typified by long gestation times. The tramway system on Hong Kong Island was first proposed in 1881, yet 23 years passed before it was opened, and the projected Kowloon tram system, proposed in 1913, was never built at all. In more recent times a short tramway extension to the Aldrich Bay Reclamation was proposed in 1985 and debated for seven years before it was also rejected, on the dubious grounds that 'buses and minibuses would be better'.

It is now generally acknowledged that Hong Kong would be unsustainable without the MTR, yet by the time its first line opened in 1979 surface travel was virtually at bursting point. The network bears some resemblance to Abercrombie's KCR extension proposals of 1948, although in its present form it owes its origins to the 1966 Passenger Transport Survey. Despite the success of the MTR, there are clearly missing links; in particular, the East Kowloon Line was regarded by professional planners as an imperative next step in the early 1980s, yet only now has a commitment been made to construct it (see Figure 4.6). In 1996 the MTRC completed a feasibility study and in 1997 obtained government approval for a line that would combine the existing Eastern Harbour Crossing rail tunnel with extensions from the northern end to Tseung Kwan O and from Quarry Bay at the southern end to North Point.

With a population of over 700,000, the north west New Territories is already one of the largest urban agglomerations in the developed world without an external rail link. As long ago as 1977-78 the Tuen Mun Transport Study (Scott Wilson Kirkpatrick & Partners 1978) recommended a fixed track link from the north-west New Territories to Tsuen Wan in addition to the proposed light rail system within Tuen Mun itself. In one form or another the necessity of such a link was re-affirmed in numerous subsequent studies, including the definitive Railway Development Study, yet no clear government support for such a link emerged until the KCRC undertook a feasibility study into what is now known as the 'West Rail' project in 1995. Even assuming this HK$75,000 million project goes ahead smoothly now, the railway will open a quarter of a century after it was first recommended.

Railway investment everywhere is beset by high costs, but in Hong Kong there is the added burden of the government's insistence that rail schemes should be wholly self-financing. Clearly, this places rail investment at a disadvantage compared with the expediency of running buses on common-user roads. For this reason even highly desirable schemes have to compete on a 'playing field' that slopes steeply towards the opponent's goal.

Although the Hong Kong Government prides itself on its lack of subsidy to public transport, there are in fact hidden subsidies of one sort or another. The rail

Figure 4.6 Tseung Kwan O new town - no MTR connection before the year 2001
Source: T.V. Runnacles

systems can claim government support in the form of initial equity injections (for the MTR and KCR) and the track-bed preparation of the LRT system. However, for the bus operators the free provision of high value terminus sites represents a major government commitment. Moreover, no true account is made of the track costs of bus operation in terms of highway damage or the value of dedicated bus lanes. If these costs were to be accurately assessed and charged, bus travel would become significantly more expensive. Although such charging would be politically unacceptable, accurate costing might nevertheless strengthen the case for railway investment, as advocated in Chapter 12.

A feature of recent years has been a number of private sector schemes for railway investments. No fewer than three separate submissions were made for light metro lines between Central and Aberdeen, whilst there have also been proposals for medium-capacity metros from Sai Sha to Diamond Hill and from Ma On Shan to Cheung Sha Wan. Yet another proposal envisaged an elevated light metro around the shores of the Kowloon peninsula, designated as 'Kowloon Sky Rail'. A common denominator of all these proposals was the linking of rail investment with property development. As such, the basis for financing these schemes would not seem to be very different to the property development activities associated with the MTRC, KCRC and LRT systems. Nonetheless, these schemes found little favour with the government; indeed, the three Aberdeen schemes, the Kowloon Sky Rail and the Sai Sha proposal were formally rejected, although a feasibility study of the Ma On Shan to Tai Wai line was undertaken in 1996.

Conclusions

Hong Kong is generally regarded as one of the world's most successful societies, and its public transport system is regarded as an equally outstanding achievement. There is probably no other community that can boast such a comprehensive public transport network, operating with minimal government interference, and either covering its costs or within reach of doing so. That such a dense urban development can function at all is entirely attributable to the remarkable scope and scale of the public transport system, and the implicit suppression of private motor car ownership and use.

Land use and transport would appear to be in harmony, but for all this, there is a lingering suspicion that the territory could be even better off, materially and in terms of the quality of life, if new investment were to be regarded more proactively, rather than being initiated as a desperate 'last resort' when existing systems reach saturation. Most of Hong Kong's transport history shows this to have been the case, and the future may not be so very different.

References

Barden, S.A. and T.V. Runnacles (1986) 'Transport in a high-density urban environment: the experience of Hong Kong'. *Transport Reviews*, 6(3).

Bristow, R. (1989) *Hong Kong's New Towns: a selective review*. Hong Kong: Oxford University Press.

Cochrane, R.A. (1982) 'The provision and allocation of highway space'. In *Transport into the 80s*, collected conference papers, 17-23. London: Thomas Telford Limited.

Dimitriou, H.T. (1994) 'Visions of a More Motor Vehicle Dependent-Free Strategy'. In *Moving Away from the Motor Vehicle*, CUPEM, University of Hong Kong. Hong Kong: University of Hong Kong Press.

Freeman Fox, Wilbur Smith and Associates (1967) *Hong Kong Mass Transport Study*. Hong Kong: Hong Kong Government.

Freeman Fox and Partners, Freeman, Fox, Wilbur Smith and Associates, et al (1970) *Hong Kong Mass Transit Further Studies*. Hong Kong: Hong Kong Government.

Giannopoulos, G.A. and G. Curdes. (1992) 'Innovations in urban transport and the influence on urban form. An historical review'. *Transport Reviews*, 12(1).

Government Information Services (1996) *Hong Kong 1996* Year Book Hong Kong: Government Printer.

Hall, P. (1984) *The World Cities*, 3rd edition. New York: St Martins Press.

Leeds, P.F. (1984) *The development of public transport in Hong Kong: an historical review 1841-1974*. Hong Kong: Hong Kong Government Printer.

Martin and Voorhees Associates (1980) *Island Corridor Strategy Implementation Studies*. Hong Kong: Highways Office, Public Works Department and the Mass Transit Railway Corporation.

Meakin, R.T. (1989) 'Hong Kong's mass transit railway: vital and viable'. In *Rail mass transit for developing countries*. Proceedings of a conference organised by the Institution of Civil Engineers . London: Thomas Telford Limited.

MVA Asia Ltd. (1993) *Travel Characteristics Survey*. Hong Kong: Territory Transport Planning Division, Transport Department.

Public Works Department (1965) *Report on the future of the Hong Kong Tramway System*. Hong Kong: Traffic Engineering Office, Public Works Department.

Road Research Laboratory (1967) *Hong Kong Passenger Transport Survey 1964-66*. Hong Kong: Hong Kong Government.

Runnacles, T.V. (1988) 'Tramway Developments in Hong Kong: Part 4, The Tuen Mun-Yuen Long LRT System', *Modern Tramway and Light Rail Transit*, 51(611).

——— (1991) 'The Planning of a Light Railway for Three New Towns in Hong Kong'. In *Urban Transport in Developing Countries: Lessons in Innovation*, ed. Heraty, M.. London: PTRC Research and Education Services Ltd.

Scott Wilson Kirkpatrick & Partners (1978) *Tuen Mun New Town Transport Study: Final Report*. Hong Kong: Tuen Mun New Town Development Office, Public Works Department.

Transport Branch (1990) *Moving into the 21st Century: the White Paper on Transport Policy in Hong Kong*. Hong Kong: Transport Branch, Government Secretariat.
Transport Department (1981) *Tramway Replacement by Trolleybuses*. Hong Kong: unpublished.
——— (1989) *Red Minibus and Green Minibus Operations 1988*. Hong Kong: Traffic and Transport Survey Division, Transport Department.
——— (1993) *Annual Transport Digest 1993*. Hong Kong: Transport Department.
——— (1996) *Traffic and Transport Digest: December 1995*. Hong Kong: Transport Department.
Vuchic, V.R. (1981) *Urban Public Transportation Systems and Technology*. Englewood Cliffs: Prentice Hall Inc.
Walmsley, D.A. and K.E. Perrett (1992) *The Effects of Rapid Transit on Public Transport and Urban Development*. London: Transport Research Laboratory, HMSO.
Xu, X.C. (1994) Written communication of 24 February 1994 from Tongji University, Shanghai.

5 Constraints and opportunities of planning the Mass Transit Railway

Timothy V. Runnacles, Nigel G. Wilson and Edward C.T. So

Introduction

A well-established range of modes is available for the development of a city's public transport network, including the various members of the motor bus family, electric trolley buses, tramway, light rail transit, monorail and guideway transit systems, suburban railways, and light and heavy metros. Of all these options, the most capacious and the most capital-intensive is usually the 'heavy metro'. By mid-1996 precisely eighty cities throughout the world were operating heavy metros: these varied in scope from the pioneering single-route metros of cities such as Calcutta, Miami and Tianjin to the mature networks of such metropolises as London, Paris, New York and Tokyo, each with a dozen lines or more and tracks stretching for hundreds of route-kilometres. In addition, rail transit in eighteen other cities was provided by lower-capacity, but fully-segregated, 'light metros', whilst four others were operating an urban metro service on modified suburban railway networks.

Although the name 'metro' is a generic label for urban heavy rail systems, other terms are widely used, such as 'underground', 'tube', 'subway' and 'U-bahn'. Hong Kong adopted the descriptive title of 'Mass Transit Railway' for its metro, and whilst this is usually shortened simply to its initials of 'MTR', most Hong Kong people know it simply as 'de tik', or 'underground iron'.

It is generally accepted that a heavy metro is capable of handling line flows of more than 30,000 passengers an hour in each direction. Stations are typically spaced a kilometre or more apart, and commercial speeds are quite high. By comparison, light metros are suited to lower capacity demands, and stations are usually spaced closer together. However, both light and heavy metros are fully-segregated from road traffic and pedestrians, unlike light rail transit and tramway systems. Whilst metros have strong technical affinities with some suburban railways, their tracks are dedicated to one type of traffic, meaning that metro trains have no need to share their paths with inter-city and freight trains. Virtually by definition heavy metros are limited only to the largest and densest of cities. Characteristically their construction is subject to three major groups of constraint, comprising those of policy, engineering and finance. Each of these constraints is discussed in this chapter, so as to explain both how the Hong Kong's MTR has evolved and where its future lies. This chapter concludes by examining the opportunities presented by mass transit railway planning and implementation.

For any particular metro the relative importance of policy, engineering and financial constraints can vary from nation to nation, and even from city to city, depending on local attitudes, legislative frameworks, the density of development and the financial strength of the individual authorities. The influence of these constraints will also vary over time in response to fluctuating economic circumstances and the evolving attitudes of the citizenry, administrators and politicians. However, before these circumstances are discussed in relation to Hong Kong, it would be useful to briefly introduce the MTR and its operations.

The Hong Kong Mass Transit Railway

Hong Kong's Mass Transit Railway Corporation (MTRC) was established in 1975 as a public statutory body 'to construct the Mass Transit Railway and to operate it having regard to the reasonable requirements of the public transport system of Hong Kong', and to 'conduct its business according to prudent commercial principles' (Hong Kong Government 1975). All of the issued capital of the MTRC is held by the Hong Kong Government. The MTRC constructed and now operates a 38.6 km railway system comprising the Tsuen Wan, Kwun Tong and Island lines. The Kwun Tong and Tsuen Wan lines were opened in phases between October 1979 and May 1982, including an immersed-tube link under Victoria Harbour between Kowloon and Hong Kong Island which was opened in February 1980; the Island Line was opened in 1985-86. In August 1989 the Kwun Tong Line was extended by 4.6 km from Kwun Tong to Quarry Bay via a second immersed-tube harbour crossing. This link was financed and constructed by the New Hong Kong Tunnel Company Ltd.; known as the Eastern Harbour Crossing, it features parallel road and rail tunnels.

The railway employs computerised operation; automatic control regulates the speed of trains on each section of track and governs their acceleration, braking and coasting. The ticketing system is also fully automatic, with entry and exit barriers at all stations. Passengers may use either single-ride or stored-value tickets, the latter being encoded with values from which each journey's fare is automatically deducted.

Each eight-car train can accommodate 2,500 persons, and the Tsuen Wan Line (the busiest of the three existing lines) can now handle 34 trains an hour towards Central Station during the peak period, offering a directional line capacity of 85,000 passengers an hour. On average 2.38 million passengers used the MTR every weekday in 1995, making it the most heavily-utilised metro in the world relative to its length. The MTRC is now expanding again: construction of the 34 km Airport Railway began in 1994, and it will open in 1998. Two services will be operated on the new railway, which will be marketed as two 'lines'. The Tung Chung Line will operate between its terminus in Central (to be known simply as 'Hong Kong') and Tung Chung station, providing a conventional commuter service, whilst the Airport

Express Line will terminate at the new airport at Chek Lap Kok. The Airport Express, stopping at only two intermediate stations, will be a premium-fare service aimed at serving air passengers and their escorts; thus it will be quite unlike the MTRC's other services (Black 1995).

In addition to the Airport Railway, a study of the feasibility of an extension to Tseung Kwan O was completed in 1996 (Mass Transit Railway Corporation 1996a) and government approval for this rail extension was received in 1997. The preferred alignment would extend from Po Lam terminus in Tseung Kwan O to North Point, and would include the existing Eastern Harbour Crossing railway tunnel. The extension from Quarry Bay to North Point would duplicate the Island Line between these two stations, thereby relieving the congestion-plagued Quarry Bay interchange. The existing Kwun Tong Line would no longer terminate at Quarry Bay, but would be diverted to partially parallel the new line, with another 'double interchange' at the new stations of Yau Tong and Tiu Keng Leng. Eventually the new line could be extended westwards from North Point along the Central and Wan Chai Reclamation to a terminus near the Convention and Exhibition Centre, to which point the Tung Chung Line could be extended eastward to permit interchange. By December 1988, a total of HK$26,000 million had been invested in the railway's original three lines. This was funded partly as equity from the Hong Kong government in its capacity as the sole shareholder, and partly from funds raised by a variety of financial instruments, including export credits, bonds, syndicated loans, floating rate notes and other capital market mechanisms. In 1994 the government made a commitment to inject a further HK$23,700 million as equity for the Airport Railway, leaving the Corporation to raise the remaining HK$10,300 million on commercial money markets.

In conjunction with its railway construction and operation, the MTRC has also led joint-venture developments of large residential and commercial properties above its stations and depots. Twenty-one residential and/or commercial properties were completed in relation to the three original lines. Amongst these, the Corporation retains some commercial properties for investment and also manages completed estates and offices, placing it amongst the largest of Hong Kong's estate management companies. In accordance with its statutory framework, the MTRC seeks to exploit its existing assets commercially; its station kiosks, shops, banks, advertising posters and communication networks (for pagers and mobile telephones) all produce useful revenues. The Corporation's entrepreneurial approach has already extended to the Airport Railway; by mid-1996 development consortia had already been awarded a major proportion of the 3.27 million square metres of developable floor space at and above its stations.

Policy considerations

The competitive climate of public transport provision

The overriding principle of Hong Kong's passenger transport policy is to satisfy a variety of travel demands by providing a choice of services. Accordingly, a remarkable diversity of transport alternatives has developed, including taxis, franchised and non-franchised public buses, private buses and minibuses, scheduled and non-scheduled public light buses, ferries, a funicular, trams, light rail transit, a metro and a suburban railway, as well as the inevitable private cars, vans and motorcycles. Most of these modes compete directly with one another, and each offers its own particular competitive advantages. Therefore the planning of a new mass transit railway route must take full account of the competition it will face.

Private cars and taxis provide a direct door-to-door personal service, whilst franchised buses and trams not only charge cheaper fares than MTR trains but also stop frequently to pick-up and set-down passengers. Residents' coach services may provide some special value-added features, such as the on-board breakfasts and newspapers that were offered on a Sha Tin residents' route in 1989-90. More consistently, school-contract coaches provide personal care when conveying small children to and from school. However, all of these transport services are vulnerable to delays caused by road congestion, often afflicting them with low commercial speeds and unpredictable journey times.

Ferries enjoy the freedom of the seas, and whilst seldom susceptible to delay, most of their passengers rely on land transport feeder services at one or both ends their marine voyages. Moreover, most piers are inconveniently located at a distance from the main population and commercial centres, and relentless reclamation has worsened this problem over the years.

MTRC policy

By operating its trains on totally-segregated tracks, the MTRC offers speedy and reliable services with an awesome capacity of up to 85,000 passengers an hour in each direction. However, the Corporation generally charges higher fares than other public transport services (see Table 5.1), whilst its trains are of limited appeal to people whose journeys start or end beyond reasonable walking distance from a station. Moreover, at some MTR stations the long walk between the street and platform

Table 5.1 Fare scales for MTR trains and Kowloon Motor Bus services, 1996

Journey distance not exceeding:	\$ MTR Fare (1)	\$ KMB Fare (2)
Urban Kowloon:		
3 km	3.9	2.2
5 km	4.7	2.5
9 km	5.8	2.8
12 km	7.0	3.3
15 km	8.4	3.9
18 km	8.4	4.4
Cross-Harbour:		
5 km	8.1	6.8
10 km	10.3	6.8
24 km	12.1	6.8

Notes: *(1) Represents adult stored value fares for the relevant approximate journey distances.*
(2) Represents fares for non-express and non air-conditioned services.
Source: MTRC and KMB Co. Ltd.

level discourages short journeys and inhibits the handicapped and the elderly from using the railway at all. (Since 1995 the Corporation has responded to the latter problem by aiming to make all stations handicapped-accessible by the end of 1996, whilst from its inception the Airport Railway has been designed to be accessible throughout).

For most of its relatively short existence, the MTR's air-conditioned environment and spacious trains have been able to offer more comfortable journeys than other modes of transport. However, the competing modes are now emulating the MTR, and air-conditioning is becoming commonplace on most forms of transport, public and private alike. Moreover, the steady increase in train loadings has conspired over the years to reduce the MTR's appeal to the point where some people may find it more amenable to travel by an alternative mode, such as an air-conditioned bus.

Clearly, the MTR now faces severe competition from other modes of transport. Its competitors are basically efficient and cost-effective, and charge relatively low fares for the benefit of the travelling public. Even so, whilst competitors may offer various advantages, the most important blessings, those of speed and reliability, remain the preserve of the MTR. Nevertheless, to some extent competition amongst different modes of transport is dictated by government policy. The major instruments of government control are shown in Table 5.2.

Although the Corporation is autonomous, government is extensively involved in the development of any new railway extension in terms of scheme approval, financial matters and its interface with other government projects. Moreover, once constructed, the multifarious aspects of railway technology and operation require inspection and safety-certification by the government's Chief Inspecting Officer of Railways before public service can commence. Whilst changes to existing MTR services and fares do not require government approval per se, proposed annual fare revisions are submitted to the Transport Advisory Committee and the Legislative Council Transport Panel for consultation prior to presentation before the Executive Council. Such fares proposals are expected to show due regard to market forces and political acceptability, but if the public interest obliges government to direct a departure from the Corporation's statutory obedience to prudent commercial principles, the MTRC would be entitled to full compensation of the sum involved.

All passenger transport operations in Hong Kong are expected to be self-financing. The revenue generated should be sufficient to cover all expenditure, including capital, running, replacement and interest costs, together with a return on the investment to the shareholders. The government does not offer any direct subsidy to transport operators for providing a particular service. However, from general (taxation) revenue the government does indirectly subsidise the running of the franchised bus and ferry services by constructing bus termini, roads and ferry piers. The government also waives the fuel tax and first registration tax for franchised buses, thus charging nothing for their use of the roads and bus-only lanes. By contrast, the MTRC is statutorily required to pay all of its costs, including those for the construction of stations, tracks, tunnels, and viaducts, as well as the costs of land acquisition and compensation: the government grants no concessions.

Table 5.2 Government controls on various transport modes

Mode (Operators)	Government control over supply		Government controls on price
	Number of vehicles/vessels	Routes/frequencies	
Franchised buses (KMB, CMB, Citybus, NLB)	Yes/No(1)	Yes	Yes
Green minibuses	Yes/No(1)	Yes	Yes
Ferries (SF, HYF, minor ferries)	Yes/No(1)	Yes	Yes
Tramways (HKT, Peak Tram)	No	Yes/No(2)	Yes/No(3)
Government-owned railways (MTRC, KCRC, LRT)	No	Yes/No(2)	No(4)
Public light buses	Yes(5)	No(6)	No
Taxis	Yes	No	Yes
Cars	No(7)	No	No(7)

Notes:
(1) *Because government controls the routes and minimum frequencies, there is only limited scope for the operator to determine the number of vehicles or vessels.*
(2) *New routes must be approved by government but frequencies are determined by operator.*
(3) *Ordinary Peak Tram fares are de-regulated, but concessions must be offered to monthly ticket holders.*
(4) *Except that fares on the LRT system have been kept artificially low as a result of compliance with District Board pressure.*
(5) *The combined total number of PLBs and GMBs is limited to 4,350 vehicles with 16 seats.*
(6) *Except that many roads and areas are prohibited to PLBs.*
(7) *Not controlled, but heavily influenced by government taxes.*

With the advent of the MTR, road congestion eased as large numbers of passengers diverted to the metro, thereby increasing the Corporation's share of the travel market (see Table 5.3). These trends not only confirm that the MTR is providing a superior transport service, but also that economic benefits are being gained by the community in terms of less road congestion and a better environment. Perversely, the resulting improvements in road speeds encourage more competitive and reliable road-based transport alternatives, so in this respect the MTR suffers from its own success.

The economic benefits attributable to the MTR can be estimated in monetary values. Examples include savings due to the deferment of new road construction, increases in effective working hours or leisure time, savings in road vehicle running costs, and reductions in environmental protection expenses. However, these benefits are excluded from calculations of the MTR's financial return because they do not accrue to the Corporation. It is noteworthy that a Hong Kong University student project (CUPEM 1993) has suggested that land auction tax, the transfer of development rights above MTR stations and income from land saved by not providing such alternative transport infrastructure as roads and bus termini should be used to finance railway construction.

Table 5.3 Percentage market shares for various scheduled public transport services since 1978 (the year before the introduction of the MTR)

Year	MTR	Bus	Ferry	Other rail systems	GMB	Total
1978	0.0	76.3	12.7	10.6	0.4	100
1983	19.0	61.5	6.4	8.3	4.8	100
1988	23.4	52.4	4.4	11.3	8.5	100
1993	27.5	44.6	2.7	15.7	9.4	100
1995	27.4	44.1	2.5	15.8	10.2	100

Source: Transport Department

Given the inherently high capital costs of the MTR and the difficulty of finding suitable alignments, it is virtually impossible to change the route of an operating railway in response to changing demand patterns. This is of concern in the light of the government's policy of developing new towns in the New Territories; this process is significantly lowering the population density in the older urban areas, and is thus weakening the potential demand for travel on the existing MTR system. Hence this policy may not only reduce custom on the existing MTR system but also seriously dilute any prudent commercial case for additional lines in the urban area.

Whilst the depopulation of the older urban areas threatens to deprive the MTR of patronage, the prospects for tapping the growth of the New Territories communities is hardly less fraught. The development of a new town normally takes over ten years to complete: it is because the population builds up so gradually that a railway may not be deemed essential during the initial years, when buses would suffice. But if the railway is not provided early in a new town's development, a comprehensive bus network will inevitably evolve, making it all the more difficult to attract bus passengers to the railway if and when it is eventually built. Furthermore, any proposal to withdraw bus services immediately following the opening of an MTR line would be politically unacceptable. Ideally, a balance should be struck between providing an MTR link to a new town at an early stage, when it would be less profitable, and later, when higher potential profits could be sapped by passengers' loyalty to competing bus services. The foregoing scenario summarises the dilemma which has confronted the provision of an MTR branch to Tseung Kwan O new town. The town's layout was defined in 1982 and included provision for a railway, but thirteen years of debate ensued before the decision to conduct a feasibility study was finally taken in 1995. Even assuming that government approval follows swiftly, it will be at least 2001 before the link is opened, by which time the town's population will have reached approximately three quarters of its projected capacity.

Engineering considerations

Despite the significant practical problems imposed by Hong Kong's challenging topography, heavily built-up urban environment and tangled infrastructure, the engineering constraints to mass transit railway planning are sometimes simpler to resolve than other less tangible features, which are more susceptible to change.

Construction options

The construction of a metropolitan railway can potentially encounter major conflicts. Even if a proposed metro has the unlikely benefit of a sponsor prepared to mobilise construction cost funds, and who may even be willing to promote it as a civic show-piece, pressure will inevitably arise to reach the most 'cost effective' solution. In this regard the principal determinant of cost for a railway route is its vertical alignment in the macro-sense, namely whether it can be constructed at-grade, or whether circumstances would oblige an alignment either above or below the ground.

Surface or 'at-grade' construction is by far the cheapest civil engineering solution, although the considerations of land costs, the interrelationship with streets and highways and the nature of the surrounding development may render this option impractical. Whilst at-grade construction, even 'on-street', may be appropriate for tram and light rail systems, the opportunities to build a heavy metro at-grade in a densely built-up urban environment are rare indeed. Nevertheless, every effort should be made both to discern opportunities for surface-level construction and to maximise its extent. Sometimes this can be achieved by constructing the railway in conjunction with a major highway. This principle has been successfully adopted for almost half of the 34 km length of the Airport Railway, whereby the tracks are either located under a dual-three-lane elevated expressway or (where the expressway is also at grade) alongside it, as shown in Figure 5.1.

Where a surface alignment is impossible, the next cheapest alternative is to elevate the tracks on a viaduct. However, elevated railways may present their own

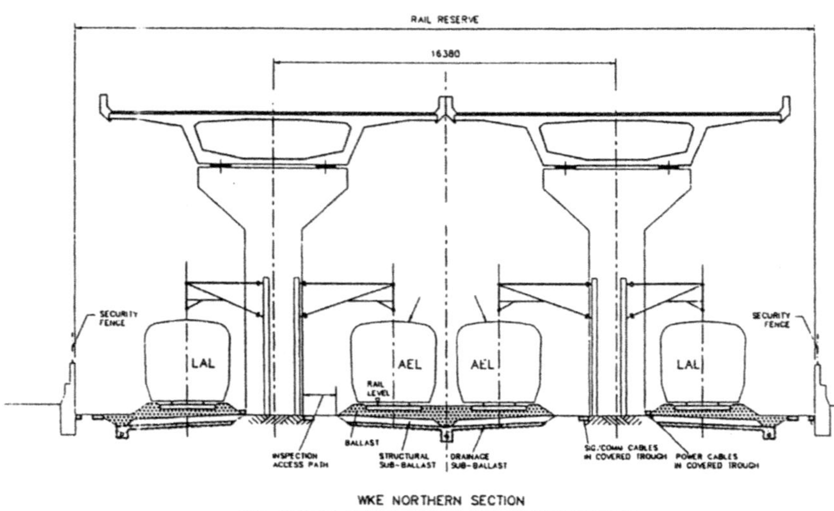

Figure 5.1 Typical cross-section of the northern part of the West Kowloon Expressway and the Airport Railway reserve

problems: land cost implications still apply, and there may be adverse effects in terms of visual obstruction and noise generation. These environmental impacts may dictate direct mitigation costs and could also require the provision of a development buffer zone, both of which would add to the railway's ultimate cost. However, there would appear to be no undue disbenefits if a viaduct can be located above the median strip of a dual-carriageway road. Compared with a surface alignment, the relative merits of this solution, with its lesser land-take, will doubtless depend upon a neighbourhood's ability to accommodate the combined transport facilities. Significant construction costs have also been saved on the Airport Railway by accommodating both the road and the railway on the same two bridges where they cross the sea channels between Tsing Yi and Lantau. This type of solution is only practical where road and rail projects can be planned in parallel, with the intention of securing the most cost-effective overall solution.

Underground alignments impose by far the highest costs, and geology is critical in determining how great these costs are. Tunnelling in soft ground, especially where it requires compressed air to repel water during construction, can be so expensive that it can only be justified over short sections of route.

Despite its great cost, underground construction in congested built-environments offers the attraction of creating an additional infrastructure asset without making any net use of land. Furthermore, although the construction period and the disruption to normal life may be longer than for surface or elevated lines, the completed railway is not only able to operate without creating any disturbance, but is also immune to adverse weather.

In the case of the MTR, distinctly different types of underground construction were used on the original three-line network. The Kwun Tong and Tsuen Wan lines made much use of 'cut and cover' construction, involving the excavation of streets for the railway right of way. After excavation, the railway trench was covered, first by temporary decking, and then permanently. By contrast, the Island Line was of bored-tunnel, or 'tube', construction. Of the two methods, cut-and-cover is much cheaper, although those who witnessed such construction along Nathan Road during the mid-1970s will recall the concomitant (albeit expertly choreographed) traffic detours and disruption to businesses.

Apart from being much more expensive, tube tunnelling minimises surface disruption but can produce a railway with very deep stations (Figure 5.2). The need to avoid disturbance was crucial when the Island Line was built. The railway had to be inserted beneath roads already so desperately congested that they were the target of overwhelming public criticism. At that time there were no alternative roads to accept diverted traffic, so the Corporation was well advised to do nothing to aggravate the situation.

The potential scope for underground alignments in an urban area is likely to be compromised by building foundations, especially where they are piled, as is almost invariably the case in Hong Kong's ground conditions.

In effect, it is almost impossible to establish any alignment that deviates from the routes of principal roads, and even then adjacent building foundations may well require to be underpinned and stabilised. The demolition of buildings is only pur-

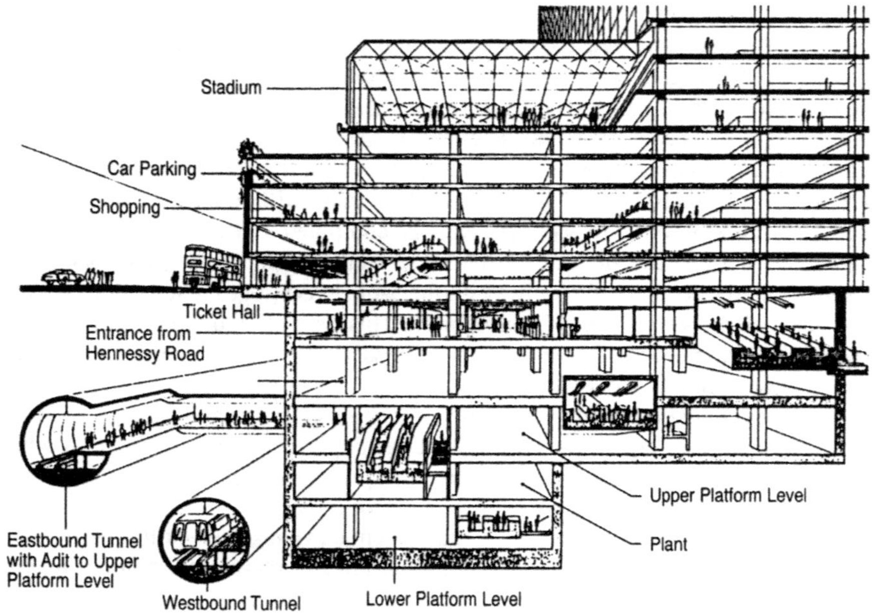

Figure 5.2 *Schematic cross-section of Wan Chai station and tunnels (The building above the station was not developed in the form shown on this concept drawing)*

sued in the most special and limited of circumstances, not only because of its associated cost and social implications, but also because lead times tend to be long and would almost invariably be further protracted by the processing of objections.

But not all sub-surface construction is beneath city streets. In some instances tunnels are necessary to combat geographical obstacles, such as hills and sea channels. The latter are of particular interest, as the MTR now has three routes beneath Victoria Harbour, each comprising immersed tubes laid in a trench on the sea-bed. The experience of the Eastern Harbour Crossing again demonstrated that the coordinated design of infrastructure can result in benefits to all parties. In this case the railway was aligned beside a dual-two lane highway and built as a single structure, as shown by Figure 5.3. This reduced disruption during construction and minimised the overall capital costs.

Moreover, the ability to link revenue from both motor traffic tolls and payment for the use of the railway enhanced this option's flexibility and appeal. Nevertheless, such an arrangement can seldom be commonplace, if only because of the different design criteria and land-use implications of road and rail alignments. Thus, joint construction of the Western Harbour Crossing road tunnel and the Airport Railway Western Immersed Tube was ruled-out because the road and the railway required separate landfalls on Hong Kong Island.

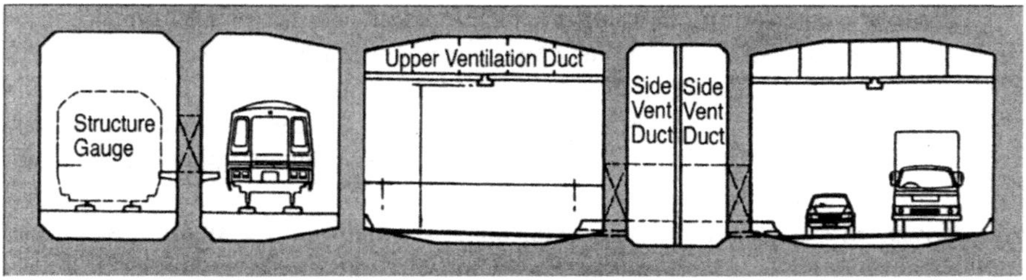

Figure 5.3 A cross-section of the Eastern Harbour Crossing rail and road tunnels (between Quarry Bay and Yau Tong)

The location of stations

Selecting the right station locations on a metro such as the MTR is critical to its success, for stations enable passengers to gain access to the railway and hence generate most of its revenue. Naturally, stations should be situated wherever predicted patronage volumes indicate the greatest potential, but this can create conflicts because these same localities are almost invariably characterised by the greatest land costs and the most acute traffic congestion.

Operational and alignment constraints inevitably feature strongly in the selection of MTR station locations. Unless a station can be built entirely under a public highway, or in a dedicated reserve, extensive land resumption will probably be required, with its concomitant costs and disruption. Although the powers incorporated in the Mass Transit Railway (Land Resumption and Related Provisions) Ordinance (Hong Kong Government 1974) provide a legal basis for such resumption, such powers would only be invoked in the present political climate if deemed essential in the interests of securing long-term community benefits. On the other hand, given that the most financially viable MTR routes within the Metropolitan Area have already been built (see Figure 5.4), additional lines may be less constrained in engineering terms.

As shown in Chapter 3, all of the new towns have been planned with some form of railway in mind, whether it be in the form of an MTR reserve (as in the case of Tseung Kwan O) or in some other form suitable for light metro or light rail routes. This means that the construction of such lines should be simpler and would encounter less disruption, as already witnessed by the construction of the Light Rail Transit system in Tuen Mun and Tin Shui Wai.

Railway planning in association with adjacent urban developments, particularly in the form of pedestrian walkways, can integrate and enhance transport and pedestrian movements. This requires both the planning constraints and engineering conditions to be clearly defined. In certain cases, it has even proven advantageous to build isolated sections of the MTR network prior to a confirmed decision to construct the line as a whole. This is quite usual when a railway is planned to pass under a highway, so as to minimise future disruption. However, in such circumstances the detailed alignment of the railway must be firmly accepted by all parties.

There are also examples of land-use developments being required to incorporate railway-related facilities prior to the commencement of railway construction. These are typically implemented by government at the development approval stage of planning, and are funded either by paying the extra costs directly, or by awarding plot-ratio building concessions to compensate for any loss of development space resulting from the railway works.

Railway passengers may either enter or leave a station on foot, or use another mode of transport to complete a longer journey beyond the railway network. Pedestrian access predominates on the MTR, and experience clearly shows that where both ends of a journey are accessible by foot, the MTR's market share of passenger transport may be as high as 60 per cent. However, the railway's share drops markedly when even one part of a journey requires a feeder bus. Because the bus network is so comprehensive and competitive, a single bus journey can surpass a combined bus and MTR trip in terms of both speed and economy. This, coupled with passengers' natural reluctance to change vehicles, emphasises the need to make interchange as effortless as possible. Indeed, the Corporation has striven hard to promote the development of feeder bus and minibus services to its stations, especially at termini where good intermodal connections can effectively extend the railway's influence. Such interchange stations must be planned from the outset with adequate space for terminating bus and minibus services, together with convenient pedestrian links between the road passenger transport services and the railway.

Hitherto, interchange between the railway and other modes has related almost entirely to buses and minibuses, albeit with the major exception of Kowloon Tong Station, which handles massive daily surges of commuters changing between the MTR and the Kowloon-Canton Railway suburban system. 'Park-and-Ride' facilities, which figure so prominently on many overseas metro systems are largely absent, possibly because land values in Hong Kong are so high that the commensurately exorbitant parking charges would defeat the purpose of parking a car for the

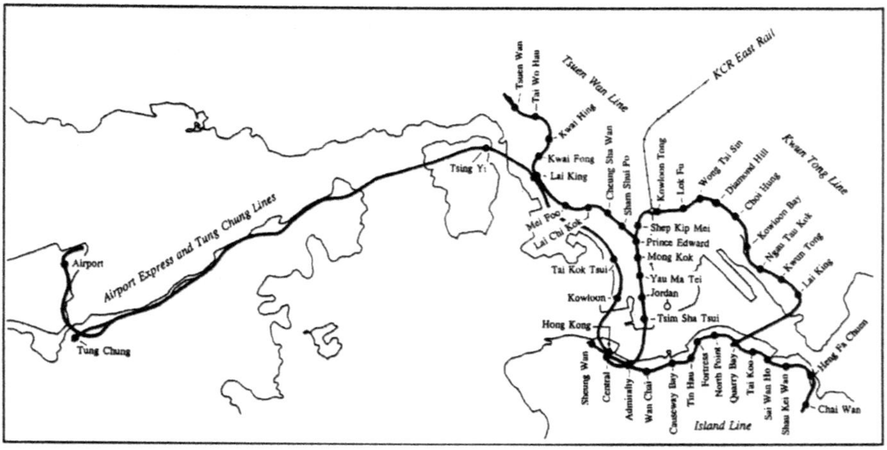

Figure 5.4 The existing MTR network, including the Airport Railway

day and completing the journey by train. Despite this generality, car parks were built during the construction of the Tsuen Wan Line adjacent to Tsuen Wan and Kwai Fong stations. Moreover, the Corporation has long aspired to develop a park-and-ride facility at Choi Hung Station to intercept affluent car commuters from the Sai Kung peninsula. Finally, after many years of negotiations with government, an agreement was reached to proceed with this project in 1995.

If road/rail station interchanges have made only a relatively modest contribution to the revenue of the MTRC's three urban metro lines, the Airport Express Line will present an entirely different picture. Because of its function as a dedicated link for air travellers, escorts and 'weepers and greeters', it has been forecast that 94 per cent of all its users will rely on motorised connections. Accordingly, the emphasis on providing good interchange facilities is correspondingly impressive. The three urban/suburban Airport Express stations at Hong Kong, Kowloon and Tsing Yi are each being provided with long kerb frontages where departing passengers can be 'dropped-off', plus particularly elaborate facilities within the station structures for arriving passengers to be picked-up. The pick-up facilities will include not only station car parks, but also multi-bay taxi ranks fed from taxi queuing areas capable of holding scores of taxis at any one time. To complete the picture, each station will also have termini to cater for hotel limousines and small buses (Budge-Reid 1996). In addition, the Tung Chung Line commuter trains will be supported by their own bus and minibus termini on sites adjacent to the stations.

Whilst physical measures can ease the disruption of changing from one mode to another, they do nothing to reduce the out-of-pocket cost of making two or more separate trips on different transport modes. In this regard, for quite some time the MTRC has been considering composite fares for combined bus and MTR journeys. This has been potentially feasible for several years on selected routes whose buses have been equipped with common-stored value ticket readers compatible with those used on both the KCRC and MTRC railway networks. Such equipment permits a composite fare which is less than the individual bus and train fares added together, thereby ameliorating the 'boarding penalty' that discriminates against all journeys by more than one vehicle (other than those made on the MTR itself and the Light Rail Transit system). In practice, no composite fares had been permanently introduced at the time of writing, even though the technical capability of the equipment has been proven. This was primarily because the proposed introduction of the 'Contactless Smart Card' technology by the end of 1996 was extending genuine prospects for innovative fare structures, especially as the MTRC and several other transport operators collectively hold all the shares in the company responsible for Smart Card distribution[1].

Environmental and urban development considerations

Although electric urban rail systems have generally been regarded as being amongst the most benign of transport modes, they can no longer escape the watchful eyes and acute ears of Hong Kong's environmental protectors. Indeed, environmental impact studies are now a prerequisite before agreement can be obtained to the alignment of any new railway, whether it be an extension to an existing network or a new private sector initiative. An environmental casualty in the latter category was represented by the ill-fated Kowloon 'Sky Rail' light metro proposal of 1990: the proximity of its proposed viaduct to neighbouring buildings meant that it failed to win government approval, following which the project was abandoned.

Noise is now recognised as a particular environmental concern of urban railways, and careful attention must be given both to the construction and operational phases of new lines to ensure that the requirements of the Noise Control Ordinance are met. Maximum permitted noise levels, as measured at the façades of 'noise sensitive receivers', must be adhered to, and this may well restrict the type of construction employed and the permissable working hours. In turn, these restrictions could both increase costs and lengthen the period of construction for new rail links.

But noise criteria not only limit construction techniques; they can affect the appearance and character of the finished railway as well. The demanding criteria regarding noise sensitive receivers may well dictate that preventative noise amelioration measures should be adopted. These could take the form of acoustic barriers to shield the buildings in question, and may range from relatively simple lateral barriers to the total enclosure of tracks (Figure 5.5); regardless of the measures adopted, they are likely to increase construction and operating costs. Moreover, such structures could well magnify the railway's intrusive visual impact, potentially leading to greater difficulty in obtaining approval for the proposal (although this has not been an issue up to now).

Where a railway passes through new areas scheduled for development or redevelopment, the concept of an essentially vacant 'buffer zone' can be applied. In this

Figure 5.5 The 'Dragon Box', or covered viaduct between Kwai Fong and Kwai Hing stations on the Tsuen Wan Line

Source: T.V. Runnacles

case the Hong Kong Planning Standards and Guidelines restrict the development of noise sensitive receivers within a zone up to 150m wide (Environmental Protection Department 1991). However, this creates the consequential difficulty that the railway becomes divorced from its potential passengers. Because of the probable reluctance to assign such a large buffer zone, pressure may arise to locate rail links underground, or to completely cover them, for strictly environmental rather than topographical reasons. Whilst this would significantly increase costs, it has been accepted that the MTRC could reasonably benefit from the enhanced land utilisation by granting part of the land so-released for the Corporation's development at full market value. The share of profits from such development would be used to offset some or all of the additional costs, including the extra cost of underground construction, as well as the higher operating costs imposed by the need for mechanical ventilation. In this way, all parties benefit: the railway operator has its extra costs covered, and its line is located nearer to the resultant catchment area; the joint venture party for the property development benefits from proximity to the railway; government benefits from the more effective use of its scarce land resources or a lesser demand for further reclamation; and the residents and passengers benefit from more convenient access to the railway.

Where a new town is planned with a railway in mind, development may be so concentrated as to provide the maximum potential 'walk-in catchment' area. However, whilst everybody cannot be located within the preferred 400 to 500 metres radius of stations, it has proven practical to locate non-intensive uses such as open space, breeze-ways and lower density residential development at greater distances from the proposed stations. In such cases, these remoter areas are served by feeder buses which terminate at the stations, thereby encouraging transfer between road and rail.

Financial considerations

The responsibilities of government

Railway construction is capital intensive and thus generally requires considerable investment. Therefore it must be viewed in the long-term, as it is not likely to represent an attractive short-term commercial venture. Consequently, any railway project and its associated funding must be framed from the very beginning in terms of its wider public benefit, and should therefore be assessed in comparison with other public sector projects.

Only government can make the broad-minded judgements necessary to decide whether the construction of a rail link meets the public interest and hence warrants public funds. This is because government not only controls, but also has responsibility for financing, the expenditure on the highway system, which constitutes the principal alternative to rail investment. A possible exception to this may be the private sector 'Build and Operate' projects for new highways, examples of which have been successfully implemented in Hong Kong in the form of road tunnels. Nevertheless, it has been invariably been demonstrated that a stable and well-defined economic and political environment, plus a viable project, are essential ingredients for this type of financing.

Government as a surrogate agent of development

The Hong Kong government derives considerable economic benefits from the improved mobility that a railway affords, not only in terms of passenger travel but even in terms of freight movement. This is because a railway reduces or defers the demand for new highway projects, enhances the attractiveness of otherwise relatively isolated new towns, reduces pollution from slow-moving vehicles and optimises use of the territory's scarce labour resources (Dimitriou 1994). Furthermore, other commercial companies, land owners and individuals benefit directly from railway construction, even if they do not use it. This is because a railway offers improved mobility, reduces the operating costs of road passenger transport vehicles, and enhances land values to the benefit not only of the real estate industry but also government, which can charge a premium on land-lease sales. In such instances, government is effectively acting both as a surrogate for these parties and as a beneficiary. These benefits are expressed through the increased growth of gross domestic product and increased profits tax, together with the general benefits accruing from enhanced land accessibility, improved labour mobility, freer-flowing traffic, the abatement of pollution and reduced fuel imports.

By the 1990s the Hong Kong government had a significant asset in its two railway corporations, namely the MTRC and the KCRC. The original investment of the 1970s and 1980s in the construction of the MTR and the upgrading of the KCR was an enlightened move and has enabled both corporations to attain a financial status that railway operators throughout the world can only envy. Bank loan arrangements by the MTRC have reinforced the attractiveness of Hong Kong's urban railway network to international financiers. The MTRC is the only business in Hong Kong that has obtained short-term and public long-term credit ratings from Standard & Poor's, Moody's Investors Service and the Japan Bond Research Institute. The MTRC continues to enjoy investment-grade ratings, with full access to the international capital markets.

Managing the MTRC's debt and equity situation

Out of necessity the MTRC is one of Hong Kong's largest borrowers: although its debt has progressively reduced over the years, it still stood at HK$14,736 million by the end of 1995. In response to this situation, the Corporation has pioneered many innovative debt instruments in the local and international capital markets. In its approach to the financial markets, the MTRC has been assisted by its international credit ratings and favourable market reputation. This has been demonstrated on several occasions over the years, and more recently, by the Corporation being regarded as one of the 'Best Asian Bond Issuers' in Euroweek's annual poll of 1995, whilst its 'Yankee' and HK$6,000 million loan facilities were ranked amongst the runners-up of the 'Best Asian Bond Deals' and 'Best Asian Syndicated Loans' of the year 1995 by Euroweek and Asiamoney (Mass Transit Railway Corporation 1996b). This advantageous position provides the Hong Kong government with financing options that are not usually available elsewhere for railway investment, namely the ability to accept a reduced initial return on the original investment by deferring dividends that would otherwise be payable on demand. The dividends so retained in the railway business can be used as a basis for borrowing, with an ensuing multiplier effect. The Corporation's proven ability to borrow significant sums on the international market means that it is only necessary for the government to fund a small proportion of a railway project's full construction cost. Nevertheless, it is still important that the government should fund at least a portion of the costs because this demonstrates the strength of its commitment to the project, quite apart from the direct financial benefits.

The amount of equity necessary to underpin any MTR expansion project depends upon several considerations. These include the level of competition for government funds, the government's own ability to borrow vis-à-vis that of the Corporation, the level of assurance that can be provided both in terms of loans and revenue projections, the permitted extent of competition amongst public transport operators, presumed interest rates and assumed cost-escalation, together with projections of the Corporation's financial situation without the extension.

Experience has shown that international financiers consider a debt-to-equity ratio of up to three-to-one to be acceptable for a well-defined and viable rail project. This is usually based on a secure existing revenue stream with demonstrable support from government. The support may variously be shown in terms of an equity commitment, an agreement to permit dividends to be retained for a defined period, and a firm transport policy and clear-sighted planning commitments, such as the granting of suitable land for railway construction and property development. That said, it was noteworthy that by the end of 1995 the debt-to-equity ratio had temporarily dipped to a mere 0.6:1; this was because the phased release of government equity for the Airport Railway was used to redeem some of the previously outstanding debts.

It is enshrined in the Corporation's enabling Ordinance that the appraisal of any new venture is based upon 'prudent commercial principles'. This particularly applies to the selection of fare levels, where the Corporation's ability to raise fares is generally expected to be consistent with inflation, thereby maintaining the real value of fares through time. Given this premise, it may be seen that an equity injection by government, accompanied by the deferral of dividends, can provide the Corporation with a significant funding ability for projects that government might have otherwise postponed. This was the principle adopted in respect of the Airport Railway project.

Despite the high capital cost of a railway, it does not necessarily follow that the government has to inject substantial equity or deplete its Capital Works Expenditure Fund. Instead, the Corporation can itself raise capital by taking advantage of both the injected equity and the dividends deferred from the shareholder's (government's) funds. Nevertheless, the government must be satisfied that it has set the correct priorities in drawing-down funds from the Capital Works Expenditure Fund, especially when the Fund may be subject to external influences or constraints, as was the case in the early 1990s when the maximum government borrowing permitted to remain by 1997 was limited by the Memorandum of Understanding between the British and Chinese governments. Additional factors that can influence government's financial involvement in MTRC projects concern not only the preservation of acceptable debt-equity ratios through a project's early life (when the interest burden is generally highest), but also the aspiration that net revenue should cover interest costs at all times. Any railway corporation (or other enterprise) undertaking a rail construction project in Hong Kong requires to be assured that its project has not only been rationally assessed but also has a high probability of being realised. Nevertheless, even after an agreement has been reached with government, assumptions about matters outside the railway developer's control may still go astray.

Such hazards could include unexpected variations in inflation, construction costs and exchange and interest rates, or perhaps a down-turn in the growth of base revenue arising from economic factors, or even from a shift in the government's transport policy. Should one or more of these circumstances adversely affect the financial outlook, bankers would have to be assured that the government (in its capacity as shareholder) can provide adequate cover to maintain the debt-equity ratio at acceptable levels. This is where the concept of 'callable equity' enters the picture; this is effectively a quantum of standby government equity that need only be summoned in response to defined criteria, such as an over-run of the agreed ratio of debt to equity. If this safety-net is not acceptable, it would be necessary to provide additional equity 'up-front' to cover the down-side risks.

The many financial instruments utilised by the Mass Transit Railway Corporation during the financing of its original three lines, and the world banking community's more recent acceptance of the terms of the Airport Railway's financing proposals, amply demonstrate the robustness of the funding concepts underpinning the construction of Hong Kong's metro.

MTRC opportunities

Passenger demand and supply

Hong Kong is one of the world's most densely populated territories. By the end of 1996 over 6.3 million people lived on 1,092 square kilometres, and whilst this was not by itself exceptional, it is the remarkable density of such urbanised areas as Hong Kong Island's north shore and the Kowloon peninsula that is almost unparalleled. These areas typically house over 26,000 people per square kilometre (see Chapter 4). With low car ownership levels of one car to every 20 people (in 1996), the territory relies strongly on public transport, and public transport movements are generally concentrated along a relatively small number of well-defined corridors. However, travel speeds by road are low, and reliability is compromised by widespread and sometimes unpredictable traffic congestion. However, with expansion of the road network limited by a severe lack of space, an off-street underground railway provides the ideal public transport solution.

Hong Kong's Metropolitan Area is not only very compact, but it is also lively and industrious. Unlike many cities, where public transport usage is concentrated within just a few peak hours, Hong Kong's citizens are seemingly in a state of restless ebb and flow from dawn until past midnight, seven days a week. They travel to and from social celebrations, visit friends and relatives, attend meetings, carry packages from one office to another, frequent cinemas, improve their minds at evening classes and devote much energy to eating out in the metropolis's innumerable restaurants. Other peoples do these things too, but in more spacious communities most of the associated journeys would probably be made by car. However, in a society where space rather than income limits the ownership and use of cars, public transport comes to the rescue, and the MTRC's fast and reliable segregated rail system clearly represents an ideal solution, notwithstanding its initial expense.

Hong Kong's economy continues to grow, with Gross Domestic Product having increased at an average annual rate of 7.5 per cent over the two decades up to 1995. One manifestation of this progress has been an apparently ceaseless growth in demand for passenger transport. Rising personal real incomes not only create more trips per person, but also a greater willingness to pay more for a faster, more reliable and more comfortable transport service. Moreover, average journeys are becoming longer as people relocate their homes to more congenial, but remoter localities.

Bus operating costs mainly comprise labour and fuel. Assuming that further gains in productivity have reached a plateau, labour costs must at least rise in line with wage increases. With real increases in wages, labour costs are rising faster than inflation. By contrast, many of the MTRC's costs, such as interest payments and depreciation, are not subject to inflation. Empirical data confirm that bus fares have increased much faster than MTR fares (see Figure 5.6), and that this trend is very likely to continue. It follows that MTR travel will become even more competitive over time as fare differentials converge.

The introduction of a metro in any city almost invariably redistributes people's movement patterns. Many of Hong Kong's estate agents believe that people will increasingly wish to live along rail corridors and would prefer to travel to places of work, education, recreation, shopping and entertainment that are located near to railway stations. These sentiments are manifested by numerous residential property advertisements and shopping centre promotions which extol the blessings of being 'near MTR'. Such publicity not only demonstrates the importance of the MTR in everyday life, but also insidiously influences the public's travel patterns.

A further advantage of rail travel compared with that by road is its lower accident rate per passenger kilometre. The availability of the MTR thus reduces overall traffic accidents. In addition, the MTR improves the environment in terms of reducing noise levels, air pollution and visual intrusion, especially where it operates underground. The development of a new town such as Tseung Kwan O also creates an opportunity for the Corporation to integrate a new rail link within the community. In such a new town, the stations can be placed at activity centres with direct and accessible pedestrian connections to all the major trip generators and attractors, whilst carefully designed interchange facilities will ease transfers between the MTR and bus feeder services, thus maximising railway patronage and enhancing its commer-

Figure 5.6 Changes in MTR and bus fares since 1980

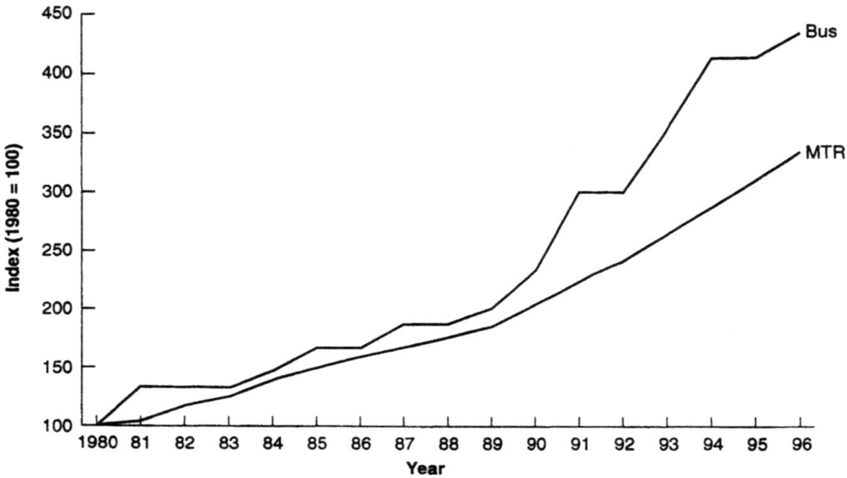

cial viability. New town development has relieved population pressure in the older parts of the Metropolitan Area and encouraged redevelopment at lower densities. However, this trend has not been altogether favourable for the MTR, a fact now appreciated by Metroplan, the metropolitan development strategy, which explicitly proposes higher density developments within about 400 to 500 metres of existing MTR stations, with lower density development being situated further away (see Chapter 8).

Development and land use

As a result of Hong Kong's experience, the benefits of railway development have possibly attracted greater recognition than at any time since the early pioneering period in the nineteenth century. This has manifested itself in several ways.

Firstly, the public has generally been extremely tolerant of railway construction, despite the undoubted inconvenience it has sometimes caused. Secondly, they have also shown themselves to be avid users of rail transport and are, by and large, prepared to pay an economic fare for the service provided. This represents a premium over the corresponding fares of competing buses and consequently it is no surprise that the average household income of rail passengers is higher than the median. Thirdly, the provision for railways to serve the territory's new towns and other major developments has almost become a planning prerequisite, subject always to demonstrations of sufficient passenger demand and financial viability. Nevertheless, virtually every significant urban development study incorporates existing or planned rail links as the primary mode of passenger travel. Figure 5.7 indicates the various rail

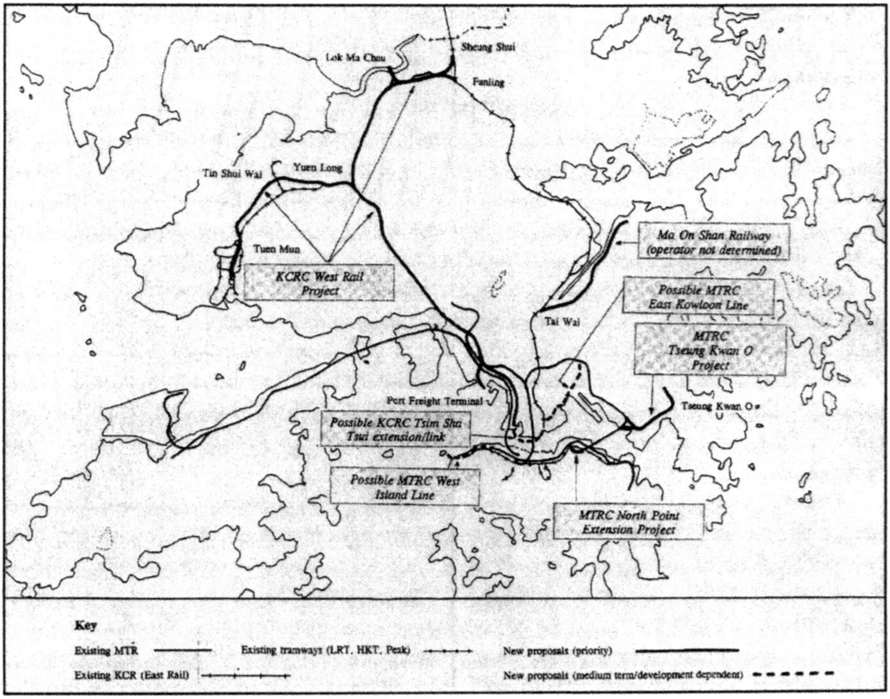

Figure 5.7 Development of territory's rail system proposed in the Railway Development Strategy

schemes identified in the Railway Development Strategy (Transport Branch 1994), not all of which would necessarily be MTR links per se.

Despite the pivotal role played by railways, experience has shown that a requirement remains for adequate alternative public transport modes to maintain widespread support for the railway. This principle of 'customer choice' has been adopted as a cornerstone of the government's transport policy (Transport Branch 1990) despite the adversity such a policy might inflict on the finances of a new railway.

Ever since 1976 various major companies have been keen to associate themselves with property developments integrated with rail projects. At the time of writing these developments comprised four projects on the Kwun Tong line, three on the Tsuen Wan line, twelve on the Island line, two on the Eastern Harbour Crossing extension of the Kwun Tong Line and five on the Airport Railway.

Additionally, the Kowloon-Canton Railway Corporation has completed or commenced four on its main line and another four on its light rail network (with two more planned). All these developments have been undertaken as joint-ventures, and have been successfully completed to the benefit of all parties. Significant contributions have also been made to the financing of MTR, Kowloon-Canton Railway, Light Rail Transit and Eastern Harbour Crossing projects by retaining a share of the profits on behalf of the railways. This reduces the extent of equity or debt financing that would otherwise be required, to the benefit of both the government and fare-paying passengers.

Perhaps inspired by the success of the two railway corporations, the Hong Kong Government has received several unsolicited private sector proposals expressing an interest in building and operating a railway together with related property developments. Options for the property elements could either allow the investor to develop his own land-holding at a site that would otherwise be poorly served by public transport, or provide sites over stations and depots by 'private treaty grants'.

However, the government has hitherto failed to accept the principle of granting land by private treaty to railway promoters other than the MTRC, KCRC and the New Hong Kong Tunnel Company Ltd. The government's policy is that a transport need must first be demonstrated before any related development can be considered. Indeed, should any private sector rail (or road) proposal look promising, the government would invariably gazette it for public tender to reap the benefits of maximum competition.

The main reason why Hong Kong's two railway corporations have been able to obtain approval to participate in real estate projects has been the recognition that prime sites with excellent transport facilities would be sterilised if the airspace above depots and stations was left undeveloped. The relatively large sites typically associated with railways have inspired government to consider them as 'Comprehensive Development Areas' and thereby secure preferential planning treatment for them, especially in environmental planning terms. Such developments may now be expected to incorporate noise amelioration measures and they may possibly feature covers over principal roads and the railway. By way of example, in 1993 the Kowloon-Canton Railway Corporation and the Environmental Protection Department agreed on a programme of noise abatement works at 18 locations, whilst in 1994 the MTRC started the construction of noise barriers at Heng Fa Chuen. In connection with the Lantau Port and Airport development, some 7,500 metres of noise barriers are being constructed for the roads, and parts of the railway track will also be covered (Environmental Protection Department 1995).

One area where the government has generally failed to support the benefits of railway usage concerns the provision of 'Park-and-Ride' facilities at appropriate stations. Even in the 1960s surveys were showing that the provision of adequate, convenient and suitably priced car parks at certain railway stations could encourage motorists to complete their journeys into busy commercial areas by train (Bailey and Dimitriou 1973). The virtue of the park-and-ride concept is that it still allows commuters to enjoy the convenience of motoring in rural or suburban areas whilst encouraging off-street travel through congested districts, thereby reducing traffic delays to the benefit of priority road users.

Commercial opportunities

Whilst most of the direct travel-related opportunities available for the MTR's development have been highlighted in the foregoing narrative, it should not be overlooked that additional commercial opportunities can benefit stations and nearby properties if the railway is meticulously planned and integrated with its surroundings. These include not only the proven attraction of developments which are close to stations, but also the willingness of many commercial developers to pay for additional or direct 'integrated' station entrances. Such entrances enhance the MTR's accessibility both to centres of activity and to surface public transport routes, and thus further enhance property values, thereby providing additional patronage for the railway.

Indeed, a major source of revenue enjoyed by the two rail corporations is the recurrent income from the management of their associated properties and the rental income from retained commercial premises, most of which have been built as part of railway developments. In the MTRC's case, this income represented about nine per cent the total revenue in 1995. Further evidence of the MTR's attraction may be found in the income derived from station advertising and kiosk rentals, which accounted for eight per cent of all revenue in 1995.

Conclusions

In the overall context of the opportunities generated by Hong Kong's transport system, it could be contended (as elaborated in Chapters 8 and 12) that the government is the principal beneficiary of urban rail development. This is because the government owns two railway corporations that not only have solid track records of successful operation and profitability, but which also account for a great deal of associated valuable property.

Furthermore, these assets have the potential to yield dividends through the respective corporations themselves. Taken together, these resources provide options for financing worthwhile new railways that are not so readily available to most of the world's governments. Given that many of Hong Kong's prospective railways should be capable of yielding an operating profit, even if full commercial viability may elude them, this means that the further development of Hong Kong's rail system is likely to be practical without recourse to government subsidies.

Note

1 The 'Octopus' Smart Card was introduced in 1997.

References

Bailey, S. and H.T. Dimitriou (1973) 'The Commuter and Park and Ride' in *Traffic Quarterly*, Vol. 27, No. 4.

Black, R.J. (1995) 'Airport Express puts quality first' in *Railway Gazette International*, Vol. 151, No.7.

Budge-Reid, A.J. (1996) *Hong Kong Airport Railway - the Transport Interchange Challenge*. Proceedings of the 'AsiaRail '96' conference held in Hong Kong, 25-27 March 1996.

CUPEM (1993) *Urban Planning Workshop II: Supporting Urban Development Strategies for Metroplan*. Centre of Urban Planning and Environmental Management, University of Hong Kong, Railway Development Group. Unpublished.

Dimitriou, H.T. (1994) 'Visions of a More Motor Vehicle Dependent-Free Strategy' in *Moving Away from the Motor Vehicle*, ed. H.T. Dimitriou. Hong Kong: Centre of Urban Planning and Environmental Management, University of Hong Kong.

Environmental Protection Department and Planning Department (1991) *Environmental Guidelines for Planning in Hong Kong*. An Extract from the Hong Kong Planning Standards and Guidelines. Hong Kong: Government Printer.

Environmental Protection Department (1995) *Environment Hong Kong 1995*. Hong Kong: Government Printer.

Hong Kong Government (1974) *Mass Transit Railway (Land Resumption and Related Provisions) Ordinance (Cap. 276)*. Hong Kong: Government Printer.

——— (1975) *Mass Transit Railway Corporation Ordinance (Cap. 270)*. Hong Kong: Government Printer.

Mass Transit Railway Corporation (1996a) *Tseung Kwan O Extension: Feasibility Study*. Hong Kong: Mass Transit Railway Corporation.

——— (1996b) *Annual Report 1995*. Hong Kong: Mass Transit Railway Corporation.

Transport Branch (1990) *Moving into the 21st Century*: the White Paper on Transport Policy in Hong Kong. Hong Kong: Transport Branch, Government Secretariat.

——— (1994) *Railway Development Strategy*. Hong Kong: Transport Branch, Government Secretariat.

6 Planning road-based public transport services

Ernest S. W. Lee and Richard T. Meakin

Introduction

This chapter introduces the planning framework and processes within which Hong Kong's road-based public transport services, accounting for 90 per cent of daily trips, are formulated.

Transport planning in the territory has, as explained in the previous chapters, been influenced by several factors and processes. Prior to the 1970s, when the development of Hong Kong was largely centred upon the northern shore of the Island and Kowloon, transport provision was largely bus-oriented, implemented mainly in response to complaints, rather than through careful planning. New demands were met by a mix of legal modes (mainly buses) and illegal modes which were mostly dual-purpose vans and pak-pai cars[1] .

In the 1970s, the need for public transport plans was prompted by the New Towns Development Programme and the Ten-Year Housing Programme (see Chapter 2) when the decision was made to build 'self-contained' and 'balanced' new towns. In transport planning, the emphases were on a system of railways as the principal mode, supplemented by a few bus-rail feeder-trunk services. The pursuit of the principle of self-contained development of new towns has resulted in a substantial (more than 90%) discrepancy between estimated and actual commuting trips between the new towns and the metropolitan area (Lee 1991).

In the 1980s, considerable changes to the public transport planning processes were necessitated by the advent of the MTR (see Chapters 4 and 5), the continued urbanisation of the New Territories, a growing affluence and the adoption of representative government. In 1984, new legislation was introduced which formalised and obliged bus operators to engage in systematic planning. A more dynamic approach in route planning subsequently evolved. The government also gave more attention to the minor modes, such as public light buses.

While public opinion has played an increasingly influential role in route development, some centrally-devised but less popular approaches, such as the public transport intermodal co-ordination policy (which favoured high capacity rail modes and restrained non-rail modes, and which claimed that interchanges in long trips were unavoidable) still influenced route development concepts. By the 1990s, road transport planning is mainly concerned with optimising the efficiency of networks and meeting customer satisfaction.

Types of transport plans

Urban transport plans in Hong Kong may be classified according to their functional characteristics into Strategic Plans, Regional Multi-mode Plans, Mode-specific Development Plans, and Route Development Plans. These are illustrated in Figure 6.1.

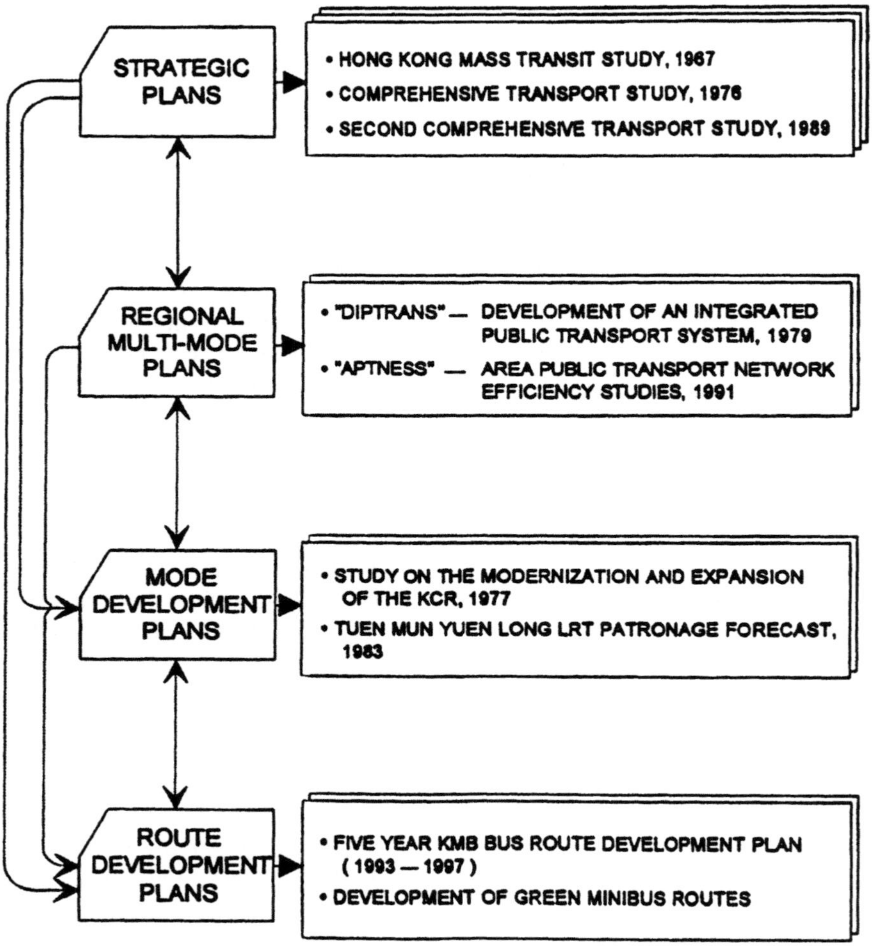

Figure 6.1 Examples of transport plans

Strategic Plans. These are structural plans which are prepared for broad, comprehensive development needs. They cover the entire territory. The planning horizon is normally 15-20 years, with more emphasis placed on planning for the immediate years. The number of traffic zones is limited in order to make modelling manageable and traffic estimates are made mainly for major traffic corridors. Recommendations are based on the justifications and viability of modes in terms of need, financial aspects and implementation phasing. The results are also used for the formulation of policy options.

Examples of such plans include studies for the MTR (Freeman Fox, Wilbur Smith and Associates 1967), the Comprehensive Transport Studies (Wilbur Smith and Associates 1976 and 1989), and the Railway Development Study (Transport Branch and MVA Asia 1993).

Regional Multi-Mode Plans. These plans are commissioned with specific briefs. They usually cover a planning period of 10-15 years. They may be regional or district-wide, covering several modes as deemed necessary as a result of structural changes in the transport network, such as new roads and tunnels (MVA Asia 1990) or the advent of new modes such as the MTR (Halcrow Fox and Associates 1979).

Development Plans. These are commissioned to review the role of existing transport modes, or to assess the viability of a new mode which is already committed. The planning horizon for these studies usually ranges from 5-10 years. Although the emphasis is on a particular mode, the impact on other competing and complementary modes may also be assessed.

The outcome usually includes patronage forecasts which are assessed in relation to ticket revenues, costs of operation, system characteristics, rolling stock procurement programmes, land requirements for depots and stations, right-of-way requirements and an implementation plan (Transmark 1977).

Route Development Plans. These are much more basic, detailed programmes of network changes. They may be schemes formed from experience, or plans prepared through a systematic modelling and planning process, taking into consideration anticipated land-use changes, travel demand, supply of services by other modes, costs, receipts and socio-political factors. They can be prepared by operators, or any members of users, or local politicians.

The planning horizon is less than five years, with detailed month-to-month proposals for the first year. The route development programmes of franchised buses specify routes in detail, including origins and destinations, frequency, hours of operation, fares and fleet allocation, and an implementation time frame (Kowloon Motor Bus 1994).

Functionally, there is not necessarily a direct or evolutionary relationship between the above types of plans. However, it is always good professional practice for planners engaged in Route or Mode Development Plans to cross-refer with strategic and regional plans wherever possible. After all, the end results should be a pragmatic compromise of different socio-economic, financial and political requirements.

The spirit and purpose of public transport planning

In the past, several principles, or approaches, have influenced public transport planning in Hong Kong. They are summarised below.

The Hierarchical Approach. This was used in formulating and developing transport plans in the late 1980s (Transport Branch 1989). Transport services are viewed as comprising a hierarchy of different modes, each with their unique positions in the public transport policy and system structure. Their relative positions and roles are determined on the basis of their operational efficiency, investment size, scale of operation, carrying capacity, social and environmental costs, and their effectiveness. Their relative positions are shown in Figure 6.2.

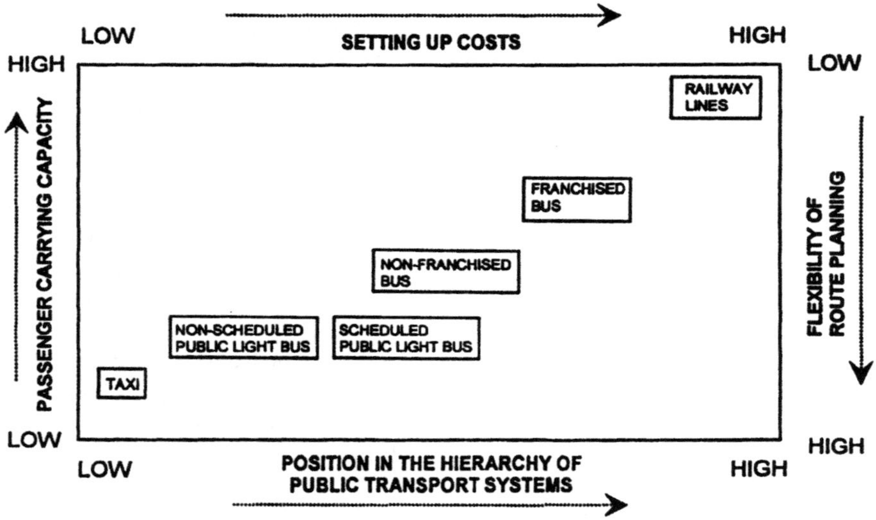

Figure 6.2 The relative position of public transport systems in Hong Kong

An important measure of the operational efficiency and hierarchy of a mode is its 'passenger-carrying capability' (PCC). This may be expressed as a function of the number of passengers carried per vehicle per car unit space, or in passenger-car-unit (PCU) terms[2] .

The concept of the PCU is extended to public transport planning so that the modes' passenger carrying efficiencies can be estimated (see Table 6.1). The term PCC is used here to show the relative efficiency of different passenger carriers, with reference to the private car as a base. The mode with the highest PCC value would be considered the most efficient. It follows from Table 6.1 that franchised buses are the most efficient, and the private car, the least efficient. The adoption of the above measure is justified because the scarcity of road space is a major constraint in transport planning in Hong Kong, and because franchised double-decker buses carry 12.5 times more passengers per unit of road space than the private car.

Table 6.1 Passenger-carrying capability of road transport vehicles

(a) Class of vehicle	(b) Passenger car unit (PCU)	(c) Average carrying capacity	(d) = (c)/(b) Passenger carrying capability (PCC) perPCU	(e) Relative PCC (Relative to the private car)
Private car	1.0	4.0	4.0	1.00
Taxi	1.0	5.0	5.0	1.25
Public light bus	1.5	16.0	10.6	2.65
Non-franchised bus (single-decker)	2.0	40.0	20.0	5.00
Non-franchised bus (double-decker)	3.0	102.0	34.0	8.50
Franchised bus (double-decker)	3.0	150.0	50.0	12.50

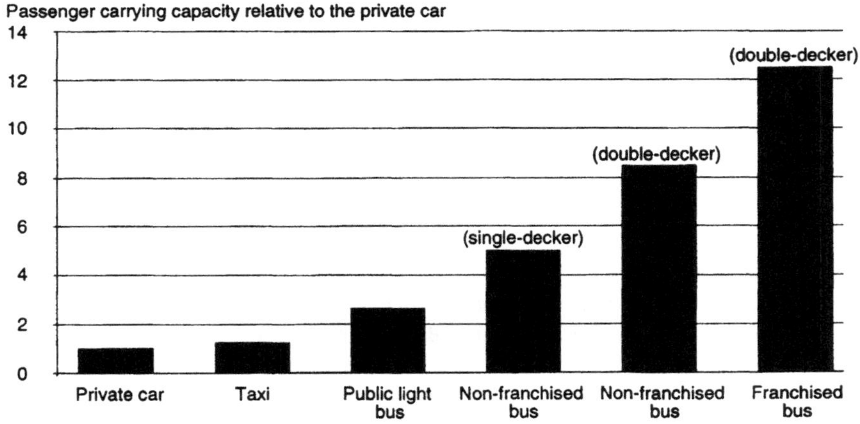

Source: Compiled from Transport Planning and Design Manual, Transport Department.

From this perspective, railways are at the top of the hierarchy as they have the largest carrying capacity per line. Since they are not affected by road traffic congestion, their regularity and punctuality can be better maintained (see Chapter 5). Below the railways in the hierarchy are buses. They include franchised and non-franchised buses, with the former operating on fixed routes and timetables, and enjoying a more significant position in the hierarchy than the latter. Further down the hierarchy are the 'minor' modes, such as taxis and minibuses.

The bus and minibus systems have their respective operating characteristics, costs and benefits. The benefit and attractiveness of the bus system lie in its flexibility. The network density and operating frequency can be adjusted easily to suit demand. Set-up costs are much lower than those of the railway. However, a major problem with buses is that they have to operate in a mixed traffic environment. Since they have to compete for limited road space with other modes they are more vulnerable to traffic congestion. The 'minor' modes can be even more flexible but are equally vulnerable to traffic congestion. Their smaller scale and lower set-up costs, however, enable operators to enter the market easily which means that the operating environment is more competitive than it is for larger buses.

The hierarchical approach to public transport provision has very much influenced the planning of transport facilities in Hong Kong. In the 1960s, when the urban road network was once again predicted to have insufficient capacity to accommodate the growing demand for travel, the construction of the MTR was recommended. The hierarchical approach was also extended beyond transport systems planning to service planning. Until the late 1980s, when the peak hour passenger level of rail systems exceeded capacity, planning for bus services has been conducted on the premise that they should supplement the railway services as feeders.

The 'Demand-based' Approach. This is an often-used approach in planning public transport services. The forecast of demand is crucial because one of the main objectives of transport planning is to match infrastructure provision with forecast traffic demand, at the right locations, at target times. Transport studies are typically undertaken to forecast travel demand systematically, to justify the provision of infrastructure facilities, and devise feasible schemes to allow existing transport systems to operate efficiently within operational limitations. Most public transport route development plans are in fact passenger demand-based plans.

The 'Service-based' Approach. This approach is service-oriented. It emphasises customers' choice of operators and modes, and their stated and revealed preferences. Plans conducted prior to the 1990s were concerned with the ability of transport systems to meet projected demand. The standard of service, comfort, choice of route and mode, interchange needs and interchange locations, were not all given due recognition and thus not reflected in the planning objectives. This changed with the study of the future of the Cross Harbour Ferry Services, which used stated preference methods in surveys and evaluation (Vaughan and Lee 1989).

Planning for franchised bus services

Licensing aspects

Planning for franchised buses in Hong Kong is governed by the terms of the franchise granted to the operators, and the Public Bus Services Ordinance (Chapter 230 of the Laws of Hong Kong). Both physical and service planning are involved. The former refers to the provision of depots and vehicles. The latter refers to the planning of routes, fares, and service-related facilities.

Up to 1975, bus route franchises in the territory were granted to operators on a geographical basis. In that year, in order to motivate bus operators to invest and expand their carrying capacity, a direct relationship between the level of investment by the operator (measured in terms of the average net fixed assets (ANFA)) and the profit entitlement of the operations (as a per cent return on the ANFA), was established by statute. This is also known as the 'Profit Control Scheme' (PCS). This practice lasted throughout the 1980s, despite public criticism. It was argued that, while the PCS provided an incentive for operators to invest in the fleet and depots, it gave them no guidance to identify and meet passenger demand, nor to utilise the assets acquired in a cost-effective way. The PCS has particular planning implications on fares, as well as routes which have to be cross-subsidised.

Notwithstanding the above, the PCS has served its function in encouraging operators to invest in the fleet, depot and ancillary facilities (Scurfield and Lee 1988). These were particularly relevant at a time (between 1982 and 1986) when the investment climate was at an ebb, and when the future of Hong Kong was a concern. By providing 'guarantees' of return on investment through profits related to the ANFA, the government could at least facilitate the private sector in securing the necessary financial facilities for investment on public transport at difficult times. In the 1990s, the positive aspects were no longer agreeable to the public. The franchise granted to Citybus Ltd in 1991 did not contain PCS provisions. In 1992, when the franchise of the China Motor Bus Co. Ltd was renewed, the PCS provisions were withdrawn.

Route planning

The bus route planning procedure has undergone several changes. In the early 1970s, this was done in an ad hoc, incremental manner. New routes, and improvements to existing ones, were introduced in response to demand or to suppress the demand for informal public transport modes, such as PLBs, illegal passenger-carrying goods vehicles and pak-pai cars.

The construction of the MTR in 1975 necessitated extensive road closures and traffic diversions. Route planning was short-term, mainly to facilitate road closures and traffic diversions. In view of the very tight programme of work, there was little participation from passengers. In 1977, the New Town and Public Housing Programmes began to accelerate. The need for expansion of bus routes to cope with internal movement of people had become more urgent and apparent. The first rudimentary bus route development programmes (RDP) were prepared by government planners. Their effectiveness was constrained by the reluctance of operators to commit investment on fleet expansion and renewal, probably due to the possibility of MTR competition. In any event, the law imposed no obligations for route planning by franchised operators.

In 1981, the requirement for the planning of franchised bus routes was formalised between the government and the operators. The Transport Department was required to produce for the franchisee a firm RDP for the following calendar year, and a tentative RDP for the subsequent year. In preparing the RDP, the Transport Department had to take account of a number of factors, including:

- outstanding RDP items not implemented in the previous year;
- public complaints on the level of bus services;
- the existing bus network;
- forecast population changes and housing developments;

- changes to other public transport services;
- recommendations of major transport planning studies undertaken;
- resources available for fleet and depot expansion;
- capacity of existing and proposed bus terminals;
- changes in bus operating speeds;
- changes in passenger demand and market share; and
- scheduling practice and fleet allocation.

As the RDPs were drawn up by government, they reflected prevailing policy guidelines and anticipated demands. Through cross-subsidy between bus routes, socially desirable but unremunerative services could be maintained, and the development of various public transport modes in Hong Kong could be coordinated according to policy, generally in a more effective manner. On the other hand, they lacked sensitivity to passengers' needs with regard to comfort and convenience, as well as to commercial considerations. Such an approach might be defended at a time when Hong Kong's urban structure was fast changing, when the demand for transport was increasing rapidly, and when the operators were not yet ready technically to prepare plans by themselves. However, RDPs produced by civil servants were no longer acceptable by the late 1980s. Since then, public transport operators have become much more technically competent in route planning.

Five-year plans

An amendment to the Public Bus Services Ordinance in 1984 transferred the responsibility for preparing the annual RDP from the government to the bus operators, recognising that it is the responsibility of the operators to identify and meet passengers' demands. Five Year Plans (FYP) are to be produced by the operators. The planning process is depicted in Figure 6.3.

Having initially scrutinised them, the Transport Department submits the Route Development Plans to the District Boards for discussion[3] . The whole programme and the comments received are presented to the Transport Advisory Committee (TAC), a board headed by a non-government chairman and comprising government and appointed members of the public to advise the government on policy matters relating to transport in Hong Kong. Outstanding matters of disagreement on the plan between the Transport Department and the operator are referred to the Secretary for Transport for arbitration. The Plan is 'settled' after deliberation.

These FYPs may be regarded as a form of contract between the government and operators (Meakin 1986). By including a new route proposal in the RDP, the operator effectively applies for a new route franchise. In agreeing to the FYP, the government is effectively making a commitment that the franchise to operate new routes will be granted to whoever has proposed the Plan. The transfer of responsibility for route planning from the government to the operators enabled the latter to take into consideration their commercial interests. Government's intervention was accordingly reduced. Local politics also comes into play when the RDP is tabled for discussion at District Boards.

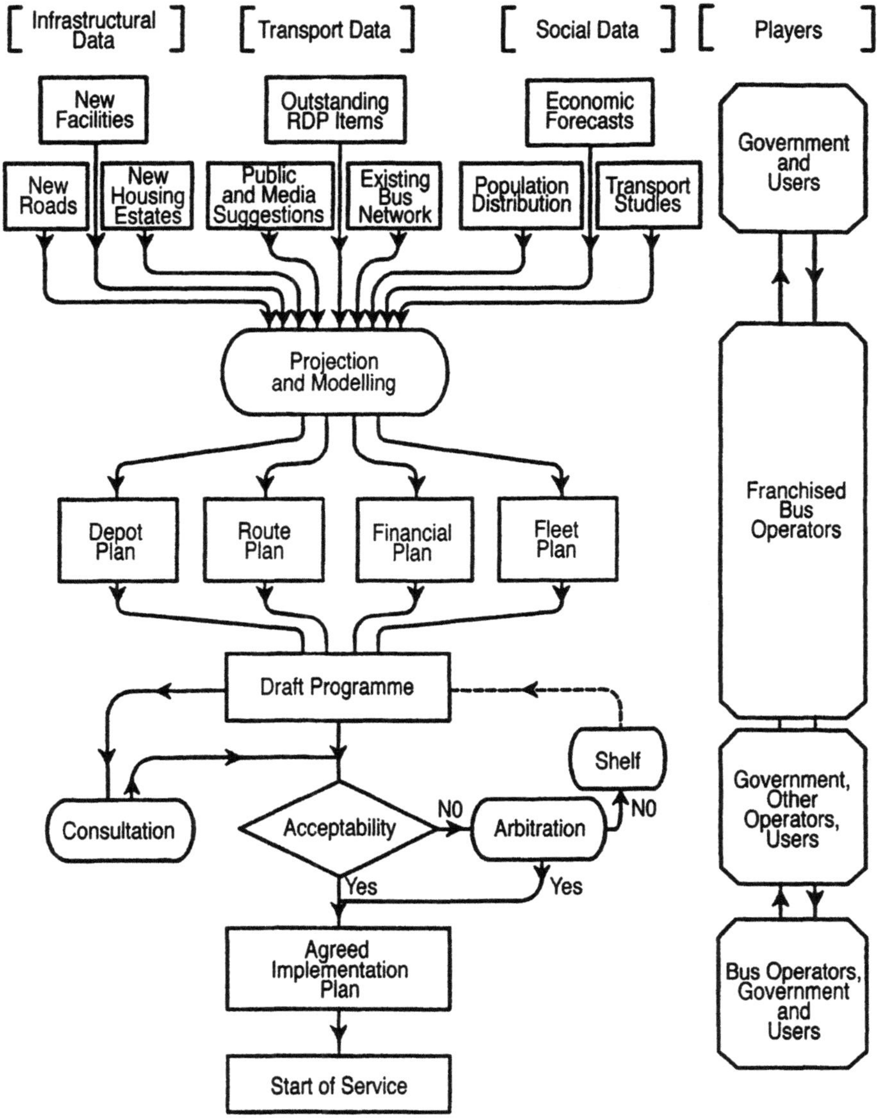

Figure 6.3 The planning process for franchised bus services, Hong Kong

Flexible route planning

Despite their codification in the law, some latitude is allowed during the implementation of Route Development Plans. The new cross-harbour air-conditioned route (No. 300) introduced in 1991 was a classic case. Similar routes were hitherto restrained because they were regarded as competing with the MTR, hence frustrating the hierarchical approach of public transport planning discussed earlier.

During the summer months of 1991, the over-saturation of the MTR system along the Nathan Road corridor in the morning peak hours was causing security concerns and this created an opportunity to use exceptional methods to develop new bus routes as a short-term relief. Air-conditioning and fare discounts were used so as to enhance the competitiveness of the new routes. To facilitate bus movements, new bus-only lanes were designated. New marketing tactics were employed. A new livery was introduced. These routes were inaugurated with a more commercial approach than would be the case for conventional routes.

The planning dilemma

The differences of opinion concerning feeder and trunk bus networking has constituted a dilemma. Although not expressly documented as a route planning standard in Hong Kong, a common planning guideline employed by consultants who designed public transport networks for new towns in Hong Kong in the 1970s, and later by the Transport Department planners throughout the 1980s, was the 'feeder-trunk' route concept. This was based on the premise that the provision of a feeder-trunk network is a more efficient way to serve diverse destinations where routes converge along a common corridor.

Feeder-trunk networks may take many forms (see Figure 6.4). They are planned to collect passengers from population sub-centres to a local town centre. In which case, passengers are fed to a few trunk routes via one or more interchange points. From these places onwards trunk routes will transport passengers to a down-town destination. This is the bus-bus feeder-trunk network form.

Another form is the bus-rail feeder which has been an essential part of Hong Kong's bus route planning policy for some time. Bus-bus feeder-trunk networking is justified by the elimination of route duplication along main transit corridors. Bus-rail feeders enhance the utilisation of the railways. When comprehensively designed, bus-rail feeders can improve the attractiveness of a railway system by making it more accessible to users.

The recommended bus networks in many public transport studies conducted in Hong Kong in the 1970s (Scott Wilson Kirkpatrick and Partners 1978; and Maunsell Consultants, Asia 1979 and 1980) were based on the bus-bus and feeder-trunk networks. It was argued that direct bus routes between new towns and the metropolitan area should be discouraged because they duplicated one another, and competed 'wastefully' with each other and with the railways.

The DIPTRANS (Development of an Integrated Public Transport System) Study (Halcrow Fox and Associates 1979) provided another example of plans which discouraged trunk routes that parallel the railway lines and encouraged rail feeders. Commissioned to integrate the MTR into the bus and ferry network, this study recommended that government should withdraw several existing bus routes which paralleled the MTR, curtail some trunk routes so that they would become MTR feeders instead of competitors, and introduce new MTR feeder bus routes.

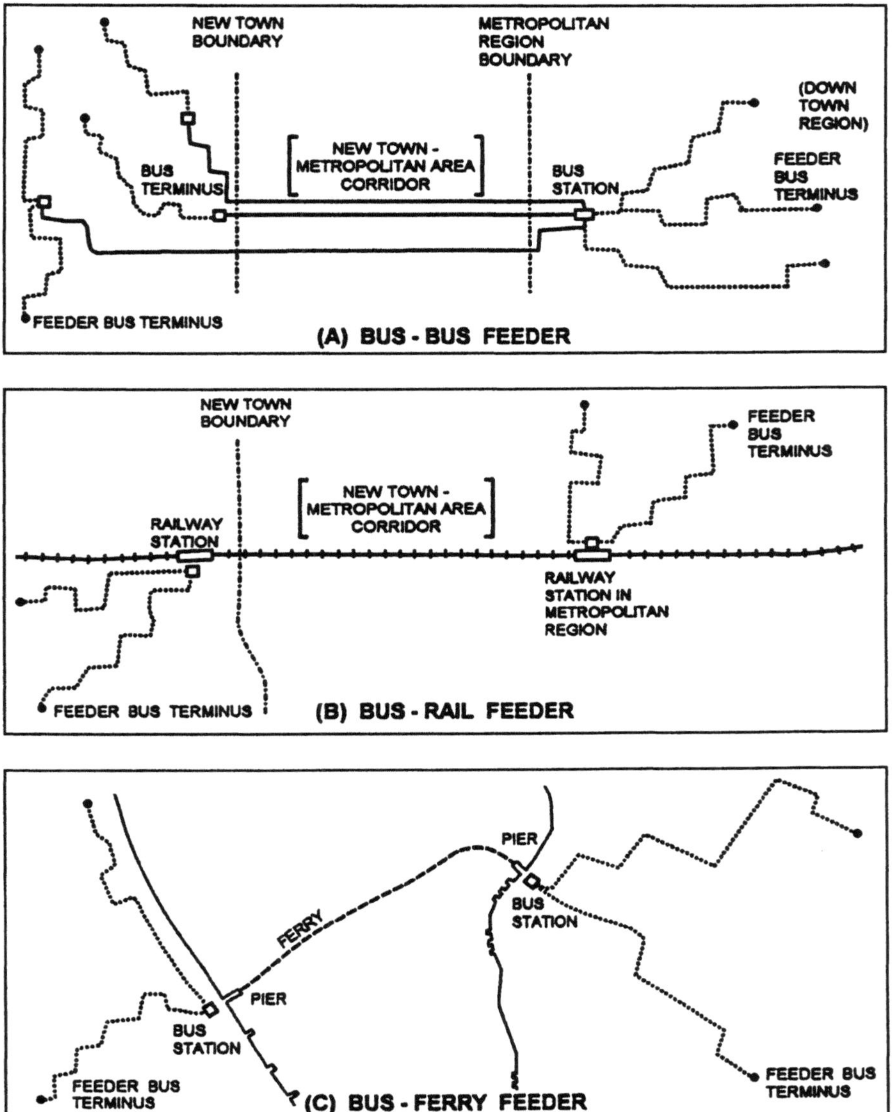

Figure 6.4 Types of feeder-trunk public transport network arrangements
Source: Lee, 1991

The DIPTRANS recommendations were, however, regarded by many as too protectionist, disruptive and socially unacceptable. Neither the government, nor the commuters, nor the operators endorsed them. Only a small proportion of the recommendations were implemented. Most of these plans were shelved. Nevertheless, the spirit of this study survived and was cited by the railway corporations in promoting railway feeder bus routes.

In the late 1980s, the potential efficiency of feeder-trunk networks was again emphasised. The principle was expanded in route planning guidelines which expressly restricted the number of direct (trunk) bus routes that might be operated from new towns to the metropolitan area. Feeder routes were encouraged. The bus companies continued to challenge the guidelines, on the grounds that:

- they were protectionist in favour of the railways;
- they were against the commercial interests of the bus companies; and
- there was a substantial demand for direct bus services.

Referred to as 'the public transport inter-modal coordination policy' (Transport Branch 1990), these guidelines had been used in the late 1980s by government planners to scrutinise the Route Development Plans of the franchised bus companies. Public consultation, however, confirmed the preference for direct bus routes rather than feeder-trunk routes. The latter require changing modes, and are hence more costly and less convenient. As a result of the saturation of parts of the MTR in morning peak periods, this policy was discarded quietly in 1991, only one year after its official proclamation in the 1990 White Paper.

A more recent attempt to restructure the bus network was the APTNESS (Area Public Transport Network Efficiency Study) exercise of 1990 (MVA Asia 1990). With a brief to reduce duplication of bus routes from new towns, this study considered options to reduce the number of trunk routes. After extensive consultation, only one pilot feeder route was implemented, and in a significantly modified form.

Characteristically, the planning and operation of feeder bus services in Hong Kong are different from those in many western cities where public transport is subsidised. In Hong Kong, separate fares are payable. Implementation of recommendations based on the feeder-trunk concept can therefore be frustrated by some unique factors. They include:

- the reluctance of passengers to interchange;
- the argument that bus operators should not deprive customers of a choice; and
- the non-availability of land for use as interchange stations for new feeder-trunk routes.

The last aspect of land availability is particularly pertinent, since the land use in new towns has been committed and it is extremely difficult to find land at suitable locations for conversion into public transport interchanges without disrupting other competing land use.

Planning for public light buses

Typology

Hong Kong's public light buses (PLBs) have a 16 per cent share of the public transport market. These 16-seat minibuses carry 1.7 million passengers a day (Transport Department 1997). Since 1974, the fleet has been confined to 4,350 vehicles most of which are individually owned. The moratorium of fleet size has created a premium price for the licence which, at the end of 1995, stood at about $2.0 million per vehicle. A new entrant into the PLB operation therefore has to be able to meet the hefty initial setting-up costs. Internationally, PLBs have been cited as a successful example of enterprising, market-driven, public transport (Rimmer 1986). This is particularly true in the recent decade, with their propensity to seek and exploit gaps in the transport network, including quality gaps.

Two types of PLB operations are in existence, both of which operate under a Passenger Service Licence under the Road Traffic Ordinance. They are distinguishable to the public by their colour scheme. Prior to April 1995 PLBs running non-scheduled services were identified by a red 'waist band' against a cream livery and scheduled PLBs by a green 'waist band'. To enable the display of advertising on the side panels of the PLBs the 'waist bands' have now become 'head bands'. PLBs with a red band are known red minibuses (RMB), and PLBs with a green band as green minibuses (GMB). RMBs are not regulated by the government in terms of routing and fare, while GMBs operate fixed routes prescribed by the government, according to a timetable and fare table.

The operation of PLBs has presented some policy dilemmas. They are popular because they run at frequent intervals and are air-conditioned. They provide guaranteed seating and stop on demand. On the other hand, their method of operation often creates traffic congestion and road safety hazards. In some cases, they frustrate the planning of systematic public transport modes.

Planning philosophy

Planning for PLBs is largely done by the government, guided by some principles promulgated in policy papers (Environment Branch 1979 and Transport Branch 1990). The fleet size and operating areas are restrained. This is sometimes referred to as the policy of 'containment'. It was explained that new demand corridors and routes should be offered and met by franchised buses which, as stated earlier, are more efficient road users. Accordingly, the red minibuses are denied access to new housing estates, tunnels and expressways, and are left to serve their established areas with a network of about 120 informal routes. The second principle is to manage their mode of operation, so as to limit the competition between PLBs and franchised buses and between red and green minibuses. Hence, PLBs are prohibited access to commercial districts and there are also stopping restrictions along the main urban corridors. Gaps along PLB clearways serve as de facto stops.

A third principle is to institutionalise and formalise their role. The objective is to remove PLBs from the main urban corridors, and disperse them to suburban, rural and feeder routes. This is sometimes referred to as the GMB Conversion Scheme.

Planning guidelines

The above are illustrated in the form of planning procedures (see Figure 6.5) and guidelines. The GMBs are to:

- have short route distances (say 2-6 km);
- serve as rail and ferry feeders;
- substitute bus routes which are unprofitable to operate due to insufficient patronage;
- serve areas not reached by conventional buses due to accessibility problems;
- replace suitable red minibus routes which are already in service, and which pose street management problems;
- provide a high quality (air-conditioned) service to reduce the use of private cars;
- provide an overnight service; and
- provide a less intrusive form of transport (cf. double-decker buses) in housing estates.

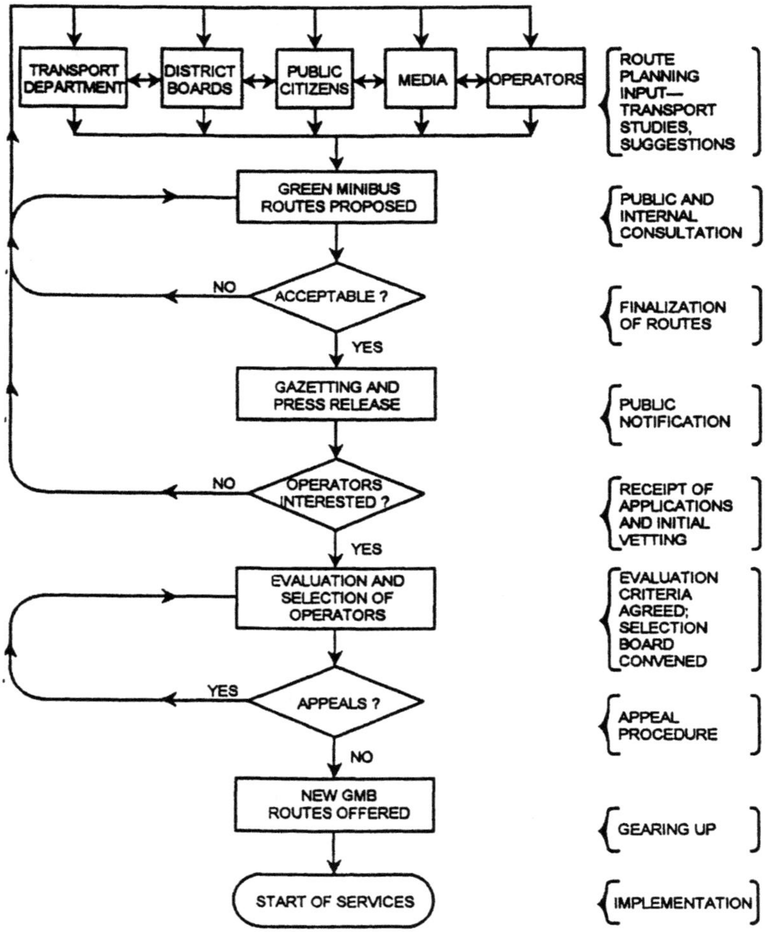

Figure 6.5 Planning process for new green minibus services

The planning dilemma

The number of GMBs grew steadily from 10 in 1973 to about 1,951 in December 1996 (see Table 6.2). It has become increasingly difficult to identify routes that will fit the above planning guidelines. The planners are, therefore, faced with several dilemmas. The first is whether the guidelines should be relaxed at the expense of erosion of the market share of franchised buses and residents' coach services, or whether the GMBs should be completely restricted to their complementary role.

Table 6.2 The development of scheduled public light bus (Green Minibus) services

	No. of routes	Fleet size
1973	1	10
1974	3	16
1975	6	34
1976	6	34
1977	8	45
1978	19	90
1979	45	178
1980	53	328
1981	57	374
1982	88	579
1983	109	822
1984	130	937
1985	137	1,050
1986	162	1,142
1987	173	1,222
1988	183	1,289
1989	184	1,295
1990	193	1,360
1991	207	1,470
1992	220	1,531
1993	231	1,620
1994	246	1,717
1995	258	1,845

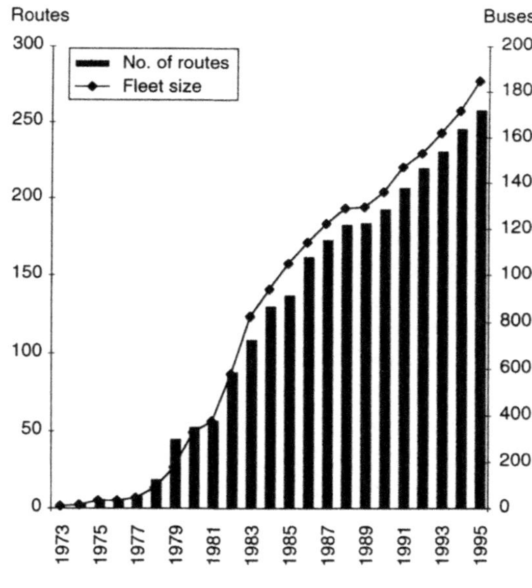

Source: Transport Department: Annual Departmental Reports and Statistical Digests, various issues.

The second dilemma concerns service. Experience shows that once RMBs are enfranchised, the operators easily become complacent and less responsive to market signals. Ironically, it has been found that the GMB vehicles are usually older, and operators are slower in reacting to market demand than their RMB counterparts (Lee 1989). A typical example is the conversion of the fleet to air-conditioning. The services on some routes had been disrupted due to labour disputes between the licence holders and drivers/sub-contractors. The third dilemma concerns the carrying capacity of the vehicle. Despite the legal stipulation that the PLB vehicles are permitted to carry a maximum of 16 passengers, the operators have requested an increase to 18 or even 19 which is physically achievable given the size of the vehicle. These applications have been turned down, one of the reasons being (apart from traffic management considerations) that an increase would affect the viability of competing modes. In spite of these dilemmas, in the past two decades, the policy towards the PLBs has changed, from prohibition and restriction, to recognition of their role as a complementary and supplementary mode of transport (Meakin 1993). It is likely that there will continue to be a role for PLBs to fill gaps in the territory's public transport networks.

Planning for taxis

Typology

In 1996, taxis were responsible for about 12 per cent of the public transport trips. While 70 per cent of taxi licences are held by persons who own one taxi only this was not the case in the early 1960s when large operators dominated the scene. Since the 1970s, the preponderance of company-owned and operated taxis has progressively been replaced by individuals, who own the taxi and drive one shift, leasing the other shift to another driver. This structure of the industry makes planning for and consultation with the taxi industry more difficult.

There are three types of taxis, their sub-division being related to their area of operation. Urban taxis (red/silver in livery) can operate throughout Hong Kong Island, Kowloon and the New Territories. New Territories taxis (green/silver) can operate in the New Territories except for Tsuen Wan, Sha Tin and Tseung Kwan O New Towns. The Lantau taxis (blue/silver) are confined to Lantau Island.

Each category of taxi is subject to quantity licensing (Lee 1993). The licences are made available by public tender. The number of new licences to be tendered is decided with regard to the supply and demand for taxi services, their financial viability, and comprehensive policy reviews by the TAC. The transfer of ownership of the taxi vehicle is volatile and speculative.

Planning issues

Planning for taxis encompasses several aspects:

The quantity. In view of the fragmented ownership of the fleet, should new licences, be issued to individuals, drivers, owners or companies, and should the transfer of licences be subject to conditions?

The use of licensing as a management tool. Should new licences continue be issued, and if so, how frequently and how many? Should a premium be charged on the tender and should government claw-back the increasing premium rather than letting it benefit speculators?

The role. Should taxis be regarded as private transport, a personalised form of public transport, or simply a mode of public transport? In 1979, the White Paper on Transport Policy (Environment Branch 1979) considered taxis as an inefficient user of road space. Arguably, they are even less efficient than private cars, because they operate many empty trips while plying for hire. On the other hand, they are the only mode of public transport that comes close to the level of service of a private car. By meeting the demand for trips that are urgent, or difficult on scheduled public transport, taxis help reduce private car ownership and usage.

The fares. Should fares be pitched to reflect the level of taxi service? Should there be a relationship between taxi fares and those of other public transport modes?

Zoning. Should taxi operations continue be restrained by their boundaries? Should the Lantau taxis be allowed to expand their operating boundary to cover the New Airport? Should zoning be abolished altogether?

Planning for residents' coach services

Typology

Residents' coach services are a form of unfranchised bus service exempted from the provisions of the Public Bus Services Ordinance. They are operated under a Passenger Service Licence. In 1995, they had a small market share of about one per cent. The development of residents' coach services was not rapid until 1990 (see Table 6.3). In that year, following the industrial actions in CMB, which affected the provision of franchised bus services on Hong Kong Island, the government started to use residents' services to foster competition in the bus industry, to serve as a 'safety net', and to reduce the reliance on franchised bus companies. The government also seeks to reduce the growth in the assets of franchised bus companies by reducing bus fleet growth, to decrease the pressure on fares.

Although some forms of non-franchised bus service were already in existence in the 1970s, they could not, until 1982, collect fares separately from passengers (Lee 1990). Residential coaches are expected to offer high quality services. Capacity is reserved for residents; routes and timetable reflect the residents' preference. Vehicles used are often air-conditioned, single-deck coaches, without provision for standing passengers. Fares are higher to reflect the premium service offered. Most residents' services are operated during peak hours when demand for travel is at its highest, and when the available franchised bus services are insufficient to satisfy demand. They fill market niches in terms of meeting network peak capacity or quality that cannot be fully satisfied by franchised buses or are uneconomical for PLBs.

Planning issues

Residents' bus services are planned, administered by and operated for a dedicated group of passengers, who are residents of a housing estate or estates. The initiative to provide a service comes neither from the government nor from bus operators.

Table 6.3 Development of residents' coach services

	No. of routes	Vehicle allocation
1982	15	35
1983	23	54
1984	20	55
1985	19	68
1986	22	85
1987	23	85
1988	22	91
1989	26	105
1990	53	190
1991	66	303
1992	80	361
1993	91	386
1994	101	418
1995	136	516
1996	184	567

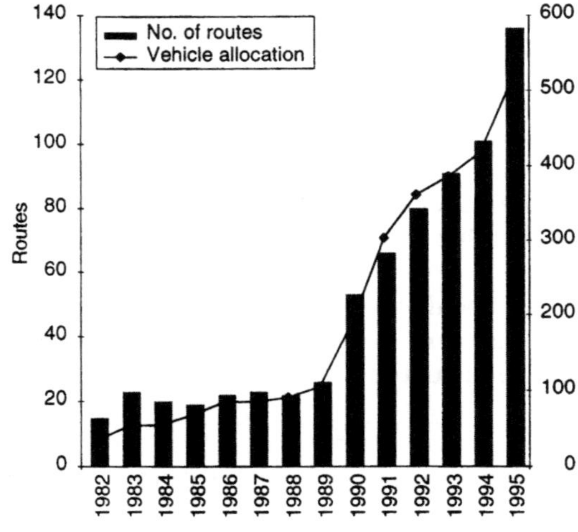

Source: Transport Department and Information Services Department

Instead, the management of a residential development, or the residents' organisation of an estate (or estates) is the applicant. They have the authority to hire and fire the operators.

While the fares of such services are determined jointly between the estate management and service providers, the routings, terminals, time of operation and frequency are regulated by the Transport Department. From the planning angle, residents' services are expedient ways of providing scheduled bus services without going through the arduous processes of plan preparation, costing, and public consultation.

The planning (or scrutiny of plans by applicants) of residents' services is similar to that for green minibuses. The planning guidelines are also broadly similar. In fact, many residents' services originated as proposals to link up large residential estates with market places, railway stations, and downtown commercial districts. Some could equally be operated using minibuses but in many cases coaches are used in lieu of PLBs because of the lower overhead costs of operation, and their larger carrying capacity. Since there is no market premium for a bus licence, it is more economical to operate a new route using buses than using PLBs.

In recent years, new routes or niches in Hong Kong's public transport market which could viably support new bus services are increasingly difficult to identify. Hence, route planning for new residents' bus services must inevitably be aimed at a different, more affluent market sector so as to realise latent demand, and divert passengers from existing franchised bus routes. Keen competition therefore exists between residents' coach operators and franchised bus operators.

Residents' bus services may therefore be viewed as an expedient way to introduce scheduled bus services without the need for a franchise. The close involvement of operators and users in the planning, route selection, and fares negotiation, makes the operation very much more responsive to users' needs. The operators must provide cost-effective services so as to retain the endorsement of the users, without which they would lose their right of operation.

The land-use and transport planning interface

The interface between transport network and land-use planning takes place at several areas (see Chapter 3): at the stage of land-use zoning, site search, development control and street management. Ideally, a proper land-use/transport study and land-use planning exercise should have due regard to the transport requirements before land-use plans are finalised. Land reservations, including the location, size and availability dates, should be made in conjunction with transport development programmes which are tied to target population intake time frames. The access routes to these places should be straightforward, avoiding unnecessary 'dead mileages' for public transport vehicles. Bus-only roads and/or lanes should be planned where possible. Land requirements for depots and terminals should be projected, justified and made in the outline zoning and outline development plans. These types of planning efforts are especially important for new towns where land use could be planned from the top beforehand.

The opportunity should be taken to include and integrate provisions for transport facilities, such as interchanges, bus bays, taxi and PLB stands, into new developments upon application from developers. This will ensure good and healthy urban design, help enhance the value of the property, and facilitate use of the interchange by pedestrians and vehicles. In built-up areas where transport interchange facilities are insufficient, the opportunity should be taken in any urban renewal and/or redevelopment projects, by inclusion of the required transport interchanges in the redevelopment plans. The requirements should be written into the lease conditions, where possible, so as to ensure compliance and facilitate enforcement. This approach will optimise land-use and transport requirements, and is especially suitable for city centres where land is a scarcity and price is at a premium. The technique is to place the transport interchange at the ground level, with pedestrian access either at-grade, through subways and/or footbridges. Commercial and residential developments, or government/institutional and community developments can then be built above. Examples are the Central Bus Station underneath Exchange Square in the Central District, and transport facilities at major MTR and KCR stations (see Figure 6.6).

Site searches in Hong Kong for additional land to accommodate transport requirements will become increasingly necessary. This process starts when the requirement for new transport facilities is projected either as a result of planning shortfall or of unanticipated changes in land-use intensity. Site search will then be activated at two levels. For short-term requirements (which may be satisfied elsewhere in the longer term) temporary modifications to land use can be made, such as the conversion of amenity plots into bus bays, playgrounds into termini, so long as local objections can be contained or removed. If the demand is perceived to be permanent, site search should be conducted at the land-use zoning level, that is, in the form of modi-

Figure 6.6 Commercial development above rail stations
Source: E.S.W. Lee, 1996

fying the ultimate land use in the town plans. In this case, applications to the relevant sections of the Town Planning Ordinance for permanent changes to the land use will be required.

At the operational level, it is inevitable that some on-street transport interchange facilities will be required. This is particularly so in old, inner urban areas where land has already been developed, where amenity areas suitable for conversion into off-road transport interchanges are limited, and where new transport services have become essential. Site search into the availability of kerb-side spaces for temporary use will be necessary. Bus-only lanes and bus-only gates should be provided to facilitate access and by-pass congestion areas. The interests of transport provision, traffic circulation, land-use control and environmental improvement may have to be compromised. It has to be appreciated, however, this course of action is unsatisfactory, and should be avoided as far as possible. At all levels, it is imperative that environmental protection measures are taken into account when drawing up transport plans, and/or modifying land-use plans. Good access control and traffic management should be introduced so that unnecessarily tortuous routings are avoided. Covered, integrated public transport interchanges should be provided with ventilators, preferably with monitoring devices to enable the quality of air to be checked.

The way forward

Current practices and policies of road transport planning in Hong Kong reflect the interaction of institutional, political, social and financial activities underlying transport developments over the last 30 years. In future, the number of modes and their complexity will continue to grow, corresponding with the changing social, political and economic developments in the territory.

The fine-tuning of institutional arrangements to accommodate these changes also needs to continue so as to facilitate public transport plan preparation and evaluation. The stipulated procedure for bus route developments could be made less stringent. Despite the flexibility discussed earlier, some are still too rigid, and influenced by legal, bureaucratic, political and financial considerations. They take insufficient account of public choice. It may also be added that, in past years, there has been a general lack of innovation, imagination and commercial sensitivity in some bus planning exercises. This could have been a consequence of the protection from competition conferred by the franchise system, as well as the lack of foresight, authority and stamina in route planning and service monitoring. All these need to be revamped.

The withdrawal of the franchise of 26 CMB routes in 1993 was widely supported by the public. It will add new impetus to the planners' efforts if new route franchises will continue be granted to new operators, particularly those with good track records of providing residents' services. When more franchised operators are allowed on the road, there could be 'competition on the road' - a less desirable practice - as well as difficulties for the government in finding enough depots and exclusive termini for the new operators.

While competition from PLB and non-franchised bus operators has, to some extent, moderated the rigidity of franchised bus route planning practice, and exerted pressure on franchised bus operators in making their services more attractive, further action is needed to perpetuate the monitoring of planning efforts of franchised operators. There is still room for monitoring by customer groups. These initiatives are partly reactions to challenges from the PLBs and residents' coach services.

As regards taxis, any policy decisions on the issue of new licences should have regard to the ability of the trade and of society to absorb the impact; the prospect of relaxation of zoning restrictions also deserves closer examination.

Urban public transport planning will continue to call for an integrated approach, encompassing the contributions of the land-use, transport and environment planners as advocated elsewhere in this book. Increasing attention should be devoted to the need for environmental improvements while executing transport plans and programmes. Short-term, personal interests may have to be compromised in favour of long-term community interests, particularly in areas of environmental concern. These compromises will not be easy, but government planners should have the expertise as well as the stamina and political will to charge forward.

In the years to come, with increasing public participation from a more fully representative government, public transport planning in Hong Kong will become increasingly sophisticated, and politically aware. The transport planners will, as a result, however, be in a more invidious position in fine-tuning and balancing the interests of various sectors of society.

Notes

1 The term pak pai refers to those private cars used illegally in the 1970s for hire and reward. Literally, pak pai means white licence plates. In the early 1970s, licensed hire cars bore red licence plates while the plates on private cars were white.
2 The term PCU is normally employed in traffic flow calculations, wherein the size of different classes of vehicles is expressed with reference to that of a private car. In Hong Kong, the standards adopted are: the PCU of a private car is 1, that of a public light bus is 1.5; a lorry has a PCU of 2; a bus 3, and so on (Transport Department, Hong Hong Government, undated).
3 District Boards are statutory bodies formed under the District Boards Ordinance, Chapter 366 of the Laws of Hong Kong. There are altogether 18 District Boards. From 1995 to the 'Handover' members were all elected.

References

Environment Branch (1979) *The White Paper on Transport Policy*. Hong Kong: Government Printer.
Freeman Fox, Wilbur Smith and Associates (1967) *Hong Kong Mass Transport Study*, Hong Kong.
Halcrow Fox and Associates (1979) *DIPTRANS Study Report: Development of an Integral Public Transport System, Hong Kong*. Transport Department, Hong Kong.
Kowloon Motor Bus (1933) Co. Ltd. (1994) *Five Year Plan (1995-1999)*, Hong Kong.
Lee E.S.W. (1989) 'Planning and control of para-transit services in Hong Kong'. *Transport Reviews*, Vol. 9, No. 4.
——— (1990) 'Formalising Informal Transport: Paratransit in Hong Kong' in Heraty M. (ed) *Developing World Transport*. Grosvenor.
——— (1991) 'Idealism and Pragmatism - Lessons from New Town Public Transport Planning in Hong Kong' Unpublished PhD Dissertation, The University of Hong Kong.
——— (1993) 'Vehicle and Driver Licensing' in L. H. Wang and A. G. O. Yeh (eds): *Keep a City Moving: Urban Transport Management in Hong Kong*. A.P.O., Hong Kong.
Maunsell Consultants Asia (1979) *Sha Tin New Town Stage II Engineering Feasibility Study Supplementary Public Transport Study*. Public Works Department, Hong Kong.
——— (1980) *Tai Po Fanling Shek Wu Hui Supplementary Public Transport Study*. Public Works Department, Hong Kong.
Meakin R.T. (1986) 'Assessment of Government's Supervision of Franchised Bus Operations'. Unpublished M.So.Sc. Dissertation, University of Hong Kong.
——— (1993) 'Management of Taxi and Minibus Services' in L.H. Wang and A.G.O. Yeh (eds): *Keep a City Moving: Urban Transport Management in Hong Kong*, A.P.O., Hong Kong.
MVA Asia (1990) *The Area Public Transport Network Efficiency Study (APTNESS)*. Transport Department, Hong Kong.
Rimmer, P.J. (1986) *Rikisha To Rapid Transit. Urban Public Transport Systems and Policy in Southeast Asia*. Australia: Pergamon.
Scott Wilson Kirkpatrick and Partners (1978) *Tuen Mun New Town Transport Study. Final Report*. Hong Kong.
Scurfield R.G. and E.S.W. Lee (1988) 'Alternative means of encouraging public transport provision'. *Proceedings of Seminar H/J, PTRC 16th Annual Meeting*. England.
Transmark (1977) *Modernisation and Expansion of the Kowloon-Canton Railway*. Environment Branch, Hong Kong.
Transport Branch (1989) *Moving into the 21st Century. The Green Paper on Transport Policy in Hong Kong*, Hong Kong: Government Printer.

Transport Branch (1990) *Moving into the 21st Century. The White Paper on Transport Policy in Hong Kong*, Hong Kong: Government Printer.

——— (1992) *Consultative Paper on Taxi Policy Review*, October 1992. Transport Advisory Committee, Hong Kong Government.

Transport Branch and MVA Asia (1993) *The Railway Development Study*, Highways Department, Hong Kong.

Transport Department (1997) *Annual Transport Digest*, (and previous years). Hong Kong: Government Printer.

——— (undated) *Transport Planning and Design Manual.*

Vaughan, B. and E.S.W. Lee (1989) 'Study of Cross Harbour Public Transport Services, II' Paper presented to the 17th Annual General Meeting of PTRC, England.

Wang, L.H. and A.G.O. Yeh (eds) (1993) *Keep a city moving: Urban Transport Management in Hong Kong*. Asian Productivity Organisation, Hong Kong.

Wilbur Smith and Associates (1976) *Hong Kong Comprehensive Transport Study*, Hong Kong Government Printer.

Wilbur Smith and Associates and Transport Department (1989) *The Second Comprehensive Transport Study*. Transport Department, Hong Kong.

Part Three
POLICY AND PLANNING RESPONSES

Policy and planning responses: Editors' Overview

Harry T. Dimitriou and Alison H.S. Cook

While it cannot be denied that Hong Kong has encountered serious land-use/transport problems as a result of its fast growing economy, its high population density levels and limited territorial space, of all the cities in the world facing difficulties, it is uniquely placed to provide a visionary response. There are several reasons for this. First, unlike most other cities, Hong Kong does not have a hinterland of any significance to finance and support. This leaves the public purse with higher levels of public revenue to spend on itself than comparable cities (with the possible exception of Singapore). Second, Hong Kong has benefited from a competent administration cadre during its colonial years of governance supported by a talented entrepreneurial class - a combination not often found elsewhere[1]. Third, no other city in the capitalist world possesses the kind of control over its land use as the Government of Hong Kong. The city enjoys this advantage by virtue of almost all land in the territory being Crown Land (i.e. land in public ownership). This places the Hong Kong Administration in a position to confer leasing rights to the private sector from which it collects substantial rent and development premiums. Fourth, few cities in the world subsidise the housing of their inhabitants to the same degree as Hong Kong. About half of the territory's inhabitants reside in public housing, and because such housing, is often provided in new towns or estates that are equivalent to city populations in other parts of the world, theoretically this enables government planners to coordinate land-use and transport planning efforts more easily (Figure 1). Fifth, and last of all, the motorcar culture in Hong Kong has not yet established firm roots among the mass of its population despite the recent concern shown by government and academics about the rising levels of traffic congestion attributed to this mode of transport (Transport Department 1994). This situation is in part a result of the excellent public transport services but also attributable to the high cost of owning and operating a private car, and the limited road and parking space available[2].

Establishing the degree to which the Hong Kong Administration, its planners and business community have managed to rise successfully to the challenge of developing a new vision for the territory to tackle its land-use and transport problems effectively is one of the underlying themes of the third part of the book. How far the

Figure 1 Integrated transport, retail development and housing: Sheung Shui
Source: A.H.S. Cook

above cited advantages could and should have been better exploited is a matter of conjecture, but there has certainly been no shortage of efforts (both by the public and private sectors) to effect change. A proliferation of plans and projects since the 1980s have had a dramatic impact on Hong Kong. Whether it is reasonable to expect these to address both the problems of today and the future effectively is quite another matter, as the discussions in the following three chapters suggest.

On a broader front, the limits to which the Hong Kong Administration can predetermine its future, when in reality its very existence is hypersensitive to changes in the international economy by virtue of the territory's function as an entrepot, need to be recognised. However, it is significant that the development experience of the territory, and the analysis of both this experience and its vision(s) for the future offers an important insight into a phenomenon that extends well beyond the territorial boundaries of Hong Kong. This is pointed out in Chapter 8 but is also argued by Cuthbert (1993) who in Chapter 2 investigates the complex question of whether the kind of urbanisation experienced today is part of a new social order altogether different from that of the past, or whether society remains essentially unaltered by the new urbanisation dynamics and new forms of industrialisation, technologies and social relations. Emphasising the relevance of Hong Kong's experience to this penetratingly important question, Cuthbert claims that:

> *The paradox (in Hong Kong) is that many of the accepted characteristics of post-modernism and post-modernity appear exaggerated rather than diminished: space-time compression, ideological discourse, deconstruction, internationalisation and other existential truths are all central to the lives of Hong Kong people. Symbolically, the city lives under the volcano in several dimensions, in its relation: to China, to international trade, to emigration (and immigration) and other factors. Hong Kong's industrial paradigm has demonstrated to the world a seminal model of capitalism in action. It offers laboratory conditions to examine the central features of post-industrialism and post-modern space*[3].

Although one cannot deny that both the administrators and business leaders of Hong Kong are understandably hypersensitive to changes in the international economy, the policy and planning responses outlined in the following three chapters suggest that the planners have treated the sea-change in the dynamics of urbanisation - where the organisation of production at a global scale has come into being with an accompanying change in human relations much influenced by the information revolution (see Poster 1990) - as merely a set of trends that should be linearly projected and addressed. Thus, rather than formulate a different and more innovative vision by, for example, exploiting some of the inherent advantages cited above and looking more towards integration with mainland China, the territory proposed (and is now in the process of implementing) a technocratic pragmatic and undemocratic development process based largely upon projection of past trends. This strategy has been supported by accomplished and internationally recognised project management expertise, and a partnership that is as much driven by the private sector as by the public purse. The implementation results to date are impressive but nevertheless have implications that have not been fully appreciated and which are only beginning to emerge.

While the technical and (short term) economic logic of the planning responses adopted are well explained by the numerous plans and programmes launched by government, the broader reasons for the choice of the vision(s) they promote are much less clear. This is particularly brought home by their areas of omission which to the critical observer suggests an underlying political agenda not hitherto publicly explained, and/or too limited an appreciation of the kind of international dynamics outlined by Cuthbert, even though Hong Kong has been thrust to the forefront of the financial world. For example, despite the global significance of the information revolution and the strategically important technological developments they help to generate, Hong Kong's plans (unlike those of Singapore) have offered very little to incorporate such aspects into their future vision(s). A similar observation can be made with regard to addressing the prospects of 'High Tech' industrial development in the territory, notwithstanding the belated decision to go ahead with a newly proposed Science Park for Hong Kong (Hong Kong Government 1995) and investigations now underway into how best to redevelop old industrial areas (Planning Department 1993).

The Hong Kong government has instead pursued a strategy relying heavily on mega transport infrastructure projects (Figure 2) as a means of reducing the economic and political uncertainties it confronts in a highly volatile and transitional environment. The aims of this strategy have clearly been an attempt to pre-empt and prevent, as far as possible, a different future unfolding in the territory to that projected[4]. In some sense (although not entirely), this approach reinforces Ulrich Beck's premise (Beck 1986) that in the future there will be a growing emphasis in policy-making on the minimisation of risk rather than the maximisation of opportunities as the true cost of failures increasingly comes to light. In retrospect, it can be argued that the Hong Kong strategy has proven very successful in helping to retain prosperity in the territory at a time of great political and economic uncertainty while much of the rest of the world was in depression[5]. How sustainable this strategy will prove

Figure 2
Hong Kong Island West: Route 7 under construction
Source: A.H.S. Cook and H.T. Dimitriou

to be into the next millennium is, however, quite another matter (this is an issue addressed by Chapter 12). This strategy has pre-determined the future of more than the territory's transportation system which may in turn cement the territory into a soon to be outdated past vision as an old fashioned heavily infrastructured and (inevitably) more polluted port city. The outcome is likely to be the generation of traffic and environmental impacts that critically impinge on Metroplan aspirations of creating a better quality environment 'in which to live and work'. It is a dilemma that the future leaders of Hong Kong's Special Administrative Region and the PRC, in conjunction with the 'makers and shakers' of the territory's private sector will have to grapple with and respond to, soon if not now[6].

Against a backcloth of the above issues, Part Three focuses on the three following areas of concern:

- The main concepts and objectives underlying the modelling of the land-use/transport interaction at the territorial level (Chapter 7) and their implications.
- The relationship between transport planning, the environment and land-use planning for metropolitan Hong Kong (Chapter 8) as treated by Metroplan.
- Developments in urban traffic management, how these are being adjusted to the environmental protection objectives and constraints of the territory (Chapter 9).

Barry Robb and Alan Kam (both senior professionals of the Transport Studies Group of Hong Kong's Planning Department at the time of writing this book) in Chapter 7 provide a valuable insight into how transport planning and modelling has contributed to the broader aims of the territorial development strategy and plans for the metropolitan area. Conceptually, the authors acknowledge that although the conventional kind of land-use/transport modelling employed in Hong Kong possesses well known limitations, the political value of such exercises for the proponents of certain courses of action in the territory has proven quite high. Notwithstanding this, the authors claim that the problems associated with the various land-use/transport model forecasts made over the years do not lie with the models themselves but in how they have been used and the validity of the future that they have been presumed capable of forecasting.

Robb and Kam commence their contribution from a theoretical perspective arguing that the historical roots of the land-use/transport modelling process used for the territory are derived from a context that is so geographically and politically different from Hong Kong that the results produced for the territory should have been treated with more caution. Following an account of the various land-based transport studies conducted, the authors offer an extensive review of the concepts that underlie the land-use/transport modelling exercises employed. This refreshingly candid review addresses a number of important issues, including the problems of sectoralism versus integration, the suspected possibility of the existence of links of this planning process with the motor vehicle lobby, and the relevance of the Von Thunen's model (as the basis of land-use/transport modelling) to the advent of motorisation. Robb and Kam conclude by offering a way ahead for future land-use/transport planning for the territory. This clearly differs from the government's supply-based approach which, among other things, is seen by the authors to completely divorce transport from land-use planning. Little real insight is offered into the suitable intensity of urban development that should be pursued.

As authors of Chapter 8, the editors of this book investigate whether Metroplan offers any major advances in the way it has tackled the land-use/transport/environmental interface as a development strategy for metropolitan Hong Kong. The chapter points out that with its harbour focused expansion strategy, Metroplan reflects previous TDS recommendations and marks a major change in direction of policy from earlier government attempts to concentrate urban expansion in the New Territories under the New Towns Programme. The interwoven dependency of Metroplan's transport infrastructure proposals with preceding transport studies and TDS is emphasised and examined by the authors and the inevitability of this questioned. Attention is drawn to the underlying economic structural change of the territory resulting from the rapid decline of the manufacturing sector and the concurrent rise in the service sector, and how these changes call for a new planning vision rather than one based on more of the same.

A chronology of plans, studies and schemes significantly affecting Metroplan is provided in Chapter 8 together with a fairly detailed account of their relevance to the metropolitan strategy. In particular, the impact of the PADS Study and the two comprehensive transport studies are appraised. Considerable attention is paid to the planning process and framework employed by Metroplan, the principal issues raised and the development opportunities offered. The mechanistic treatment of the land-use/transport interface is attributed to TDS, CTS I and II, and the option evaluation process employed is criticised for being too restrictive. The reasons for the limited choice of options are explained, as are the planning principles adopted in the development of the selected option. It is at this level that the authors encounter the most innovative proposals of Metroplan, particularly in efforts to reduce urban densities and restructure urban areas with the aid of reclamation, the proposed use of the MTR as a focal point for future redevelopment, the introduction of development statements and use of urban design guidelines. The opposition to some of these ideas and the criticism of the cosmetic treatment of public participation employed, are also discussed.

While acknowledging the critical need for a metropolitan development strategy and applauding many of its new features, the editors conclude that Metroplan ultimately has no substantial vision of its own by virtue of its subservient relationship to preceding plans and its limited real concern for the environment. The authors also consider that Metroplan places the territory's transport, real estate and financial interests above quality of life interests of the inhabitants of Hong Kong. On the economic space-life space spectrum (Friedmann 1988) the provision of economic space has been an over-riding and explicit aim of the Hong Kong government (Ng and Cook 1997). As a result of its non-statutory status, as well as the refusal of government to set up a Metropolitan Development Agency to oversee the execution of its recommendations, the strategy lacks political clout.

In Chapter 9 the growing awareness of the need to link improved urban traffic management efforts with environmental protection measures in Hong Kong is outlined by Kelvin Leung and John Spiers, consultants with MVA Asia, one of the leading transport consultancies in the territory. Commencing with an elaboration of the interface between traffic and the environment, Leung and Spiers explain how urban traffic management measures are increasingly being adjusted to incorporate environmental concerns in urban transport planning in Hong Kong and why there is a growing (albeit belated) appreciation of the need to plan for pedestrian access and circulation. The problem of dealing with qualitative and quantitative standards in environmental appraisal is highlighted as one of the difficulties of reconciling the conflicts of traffic management aims with environmental protection - given the former's inherent greater concern with vehicles and the latter's greater concern for people (also see Chapter 10). While the authors recognise the world-wide dilemma of meeting the mobility needs of the city (and its economy) and the preservation and enhancement of the environment, they claim this is made particularly problematic in Hong Kong by its high population densities and the scarcity of land available for development.

Following a review of the three major components of traffic environmental impact studies: noise pollution, air pollution and road safety, the authors discuss changes in perception that have taken place in the territory since the 1980s, when Hong Kong became increasingly affluent and the concern for the effects of global warming more widespread. Leung and Spiers call for further collaborative efforts among developers, traffic engineers and environmental specialists to produce 'acceptable compromises' to enable Hong Kong to continue to prosper. In this context, they acknowledge the particularly keen debate over the applicability and suitability of international environmental standards locally. Following on from their account of typical traffic management measures currently employed in Hong Kong and the emerging environmental concerns as traffic growth increases, the authors examine the various responses that have been introduced. This is done both in general terms and in the context of a case study of the Green Island Reclamation Scheme. Leung and Spiers conclude that 'if motor vehicle growth and usage are allowed to continue at the current rate of about five per cent per annum, then the likelihood of any significant improvement in the environment throughout the territory is very small'. They also warn that Hong Kong is unlikely to achieve any real progress without significant cost and disruption, hinting that the territory is about to enter into a new phase of development warranting a very careful balance between the extremes of total environmental protection and unrestrained transportation development.

Notes

1 Although the entrepreneurial skills are expected to be unaffected by the political transition in 1997 it is widely believed that the administrative capabilities of government will dramatically alter after this date. This fear is partially being borne out by the high levels of resignation of senior and middle management civil servants - both expatriate and local Hong Kong Chinese (see Fitzpatrick 1995 and Chapter 12).

2 This case was first presented by one of the editors in a Toyota Quarterly Review entitled The Wheel Extended (Dimitriou 1994).

3 The title to Cuthbert's article (Under the Volcano) is taken from Malcolm Lowry's book of the same name. While the concepts of post-modernism and 'the fully fledged capitalist state' as presented by Cuthbert clearly attracts (and warrants) a wider discussion than offered here, his main points regarding the relevance of Hong Kong's 'exaggerated' and 'compressed' experience to other world cities remain very valid.

4 This was perhaps more true at the time Metroplan was prepared and TDS initially developed than today where, for example, there appears to be a rapid reappraisal of the needs to strengthen rail transport links, and associated development, with mainland China.

5 It is widely suspected that the reason the port and airport projects were given the green light had as much to do with the need to meet projected traffic growth as to offset the uncertainties confronting the Territory at the time by generating a public works programme that attracted major employment and investment opportunities much in the same way as Roosevelt introduced the 'New Deal' in the USA to stimulate the economy and offer greater political and social stability.

6 There are signs of this already taking place within the private sector in the form of a number of disparate studies being undertaken (see Enright et al 1997), and increasing talk of the need for a regional port and airport strategy for Southern China. The public sector has similarly begun to look at areas of concern previously ignored (see Planning Department 1995).

References

Beck, U. (1986) *Risk Society: Towards a New Modernity* London: Sage Publications.

Cuthbert, A.R. (1993) 'Under the Volcano: Post-modernism Space in Hong Kong', Unpublished Paper, University of New South Wales, Sydney April 1993.

Dimitriou, H.T. (1994) 'Bending, Under the Strain: Operating and Implementing an Integrated Transport System in Hong Kong', *The Wheel Extended*, No.88, A Toyota Quarterly Review, Tokyo.

Enright, M., E.E. Scott, and D. Dodwell (1996) *The Hong Kong Advantage: A Study of the Competitiveness of the Hong Kong Economy*. Hong Kong: Oxford University Press.

Fitzpatrick, E. (1995) 'Mandarins in Disarray: Constant Shuffling Threatens Morale and Efficiency' *Windows Magazine* Hong Kong, 22 September.

Friedmann, J. (1988) 'Life Space and Economic Space: Contradictions in Regional Development' in Friedmann, J. *Life Space and Economic Space: Essays in Third World Planning* New Brunswick (USA) and Oxford (UK) Transaction Books.

Hong Kong Government (1995) *Hong Kong: Science Park Study* Segal Quince Wickstead (Asia) Limited, September.

Ng, M.K. and A.H.S. Cook (1996) 'Are there Feasible Alternatives to the Reclamation-led Urban Development Strategy in Hong Kong?' Occasional Paper No. 132, Department of Geography, The Chinese University of Hong Kong.

——— (1997) 'Reclamation: an urban development strategy under fire'. *Land Use Policy* Vol.14 No. 1.

Planning Department (1993) *Study of the Restructuring of Obsolete Industrial Areas.* Final Report, Townland Consultants Limited and Roger Tym and Partners Limited.

——— (1995) *The Final Technical Report on the Territorial Development Strategy Review.* Part One: General Overview. Government Printer: Hong Kong.

Poster C.D., and A. Kruge (eds.) (1990) *Community Education in the Western World* London: Routledge.

Transport Branch (1994) *Report of the Working Party on Measures to Address Traffic Congestion*, Hong Kong Government, November.

7 Modelling land-use/ transport interaction at the territorial level

Barry D. Robb and Alan W. Y. Kam

Introduction

This chapter outlines the main concepts and objectives of modelling land-use/transport interaction at the territorial level of Hong Kong first introduced in Chapter 3 and explains some of the processes which have been used.

Every model, by definition, has a limited conceptual validity. With a simple model the limitations may be obvious to all, but as the complexity increases the limits of validity can become obscure. To be of practical value a land-use/transport model should be designed to address issues and confront problems which lie within its conceptual limitations. One needs to recognise, however, that such a model can also have what might be called a political value for the proponents of certain courses of action. In this sense, the value is endowed through the presumption of a conceptual validity in the face of possible political opposition. Given such circumstances it is sometimes difficult to recognise the full scope of the problem that needs to be confronted.

Origins and development of land-use/transport models

It can be argued that land-use/transport models have their roots in central place theory which was developed in the nineteenth-century to describe the gravitation of commodities to urban area markets across a featureless German plain in a closed economic system. The notion of a closed system stems from the period when the Germanic Principalities were unpopular with their neighbours. Von Thünen was one of the scholars who illustrated how these principalities could become economically self-sufficient and self-contained (Von Thünen 1875).

The territory of Hong Kong does not closely replicate these conditions. However, given the 1951 international embargo against trade with the People's Republic of China (PRC), the territory developed over subsequent decades without what might be described as 'normal' trading relations with its rural hinterland and neighbouring cities of mainland China.

Techniques derived from the Von Thünen model were used in the 1950s and thereafter to estimate private motorcar travel generated by suburban growth around American cities[1]. The concepts underlying these techniques were fairly simple inasmuch as the suburban residents were presumed to be 'economically captured' by the central city, and travel by the private motorcar was assumed to be a virtual certainty. Advances in digital computing through the 1960s facilitated the development of these and other models to describe inter-modal transport activities.

The advances in techniques made it practical to simulate all forms of mechanised travel in a city. The resulting new generation of traffic models was perceived as a universal forecasting tool which could be calibrated to reflect local travel characteristics in any city, or wider territorial area, where the boundaries were simple to define. They appeared at face value to offer comprehensive and rational guidance on the levels of city-wide traffic activity in future years, illustrating how new urban expansion areas were likely to interact with the existing distribution of land uses as urban population grew, economic development occurred, or some combination of these developments was realised (Creighton 1970, Dickey 1975, Black 1981, and for a critique of these models, Dimitriou 1992).

With these models the general approach to urban transport planning changed subtly from a simple probabilistic forecast of future expectations to an approach that determined capital investment in transport infrastructure needed to accommodate the forecasts. The change was inspired largely by the corporate interests of industry influencing the publicly-managed transport sector in a pluralistic political environment. These econometric models of land-use/transport interaction (which they ostensibly were) described the costs of urban travel and the benefits which could accrue from capital investment in improved transport infrastructure. They have become powerful tools to support investment proposals in active competition with other public sector projects for the allocation of limited capital resources. In this regard, the models not only have the advantage of offering supposedly rational economic guidance but are overtly based on socio-economic growth aspirations which are seen as politically persuasive in a liberal, democratic, materialistic society.

Introduction of land-use/transport models to Hong Kong

The British colony of Hong Kong has never been a democracy. In a classic sense, the economic development of the territory has been governed by a liberal, paternalistic administration (see Chapter 2). The paternalistic aspects are manifest as strong deterministic responses to perceived problems. The liberal aspects, however, are displayed by a certain willingness to listen to pluralistic views before determining a response. By the late 1960s, incremental development of the traditional urban areas of Kowloon and Hong Kong Island had reached the stage where further urban expansion was spatially constrained. Under the pressure of population growth, fuelled by immigration from the PRC, the government response was the preparation of plans to build new towns in various locations in the predominantly rural New Territories. A ten-year programme to build these new towns was formally initiated in 1973 (see Chapters 2 and 3).

In formulating plans for these new towns, pluralistic views on the spatial requirements for all the normal urban socio-economic activities were sought from the various (separate) sectors of the administration, including housing, commerce and industry, buildings and lands, education, roads and transport. Although the new towns were consciously planned with lower land-use densities than those prevailing in the traditional urban area, the new towns were conceived in essentially the same image and land-use mix as the metropolitan area.

In 1973, the government embarked on a very fast-track programme to urbanise the territory within 10 years. At the time, Hong Kong was still seen as a closed economic system with clearly defined points of import and export. The economic environment in Hong Kong and its territorial development plans appeared to some to provide an ideal opportunity for the application of the latest generation of comprehensive land-use/transport models being used elsewhere in the industralised world (Wilbur Smith and Associates 1976, and Transport Department and Wilbur Smith and Associates 1989).

The territorial model was seen to have different values to different sectors of government. It was seen to offer a practical value in its claimed capability to forecast travel demands by different modes in all the existing and planned transport corridors. It also offered a political function in illustrating how the deterministic socio-spatial development plans for the new towns could be expected to operate both economically and efficiently.

While the practical value was promoted as the primary professional objective of the land-use/transport model, the political value of the model, appearing to provide rational support for a definitive course of action, was an important, if less prominent, objective.

Territorial transport planning studies

Background

The post-Second World War population in Hong Kong grew at such a rate that the government gave high priority to housing, education, water supply, health and industrial development. With the rapidly increasing rates of urbanisation and the relative concentration of employment in the main urban areas, the need for systematic planning of transport infrastructure became obvious (see Chapter 3). The government subsequently commissioned a series of territory-wide transport planning studies (outlined below), commencing in the early 1960s with plans for the development of comprehensive road and public transport systems in Hong Kong (Freeman, Fox, Wilbur Smith and Associates 1967 and 1968).

Hong Kong Passenger Transport Surveys

In August 1964, the Hong Kong Passenger Transport Survey Unit (PTSU) was formed jointly by the British Road Research Laboratory and the then Public Works Department of the Hong Kong government. The major responsibilities of the PTSU were to conduct surveys in the territory for identifying current travel patterns, and to build up within government a permanent organisation to maintain up-to-date survey data. Two series of surveys were carried out between 1964 and 1966 (Dalby 1967):

- Network surveys (between 1964 and 1966); and
- Origin-destination (O&D) surveys (in 1965).

Network surveys were conducted by field observation and included the following:

- a street inventory of the physical characteristics of the road system to establish its capacity to accommodate traffic flows;
- motorcar and public transport travel time surveys;
- a survey on the expenditure on travel;
- surveys of traffic flows, traffic composition and vehicle occupancy in the road network;
- surveys on waiting times at public transport stops and termini; and
- a parking survey of the number of kerb and off-street parking spaces available, the number of motorcars parked legally/illegally, and the duration of parking.

Such surveys provided information to establish the characteristics of traffic and the ability of different parts of the transport network to accommodate it. The O&D surveys were designed to measure existing movements of a representative sample of people and goods vehicles using the transport network, particularly with regard to the origins and destinations of the journeys, the purpose of the trips, the land use at each end of the trip, and the mode of travel. This information was obtained by interviewing a sample of motorcar and motorcycle users at their homes; bus and tram users at transport stops; and Peak tram, railway and ferry users during the course of their journeys. Data for the taxi users were collected by means of a postcard survey. Commercial vehicle interviews were also conducted at points on the road network in the urban area and on vehicular ferries.

The information collected in these surveys subsequently formed the basic data for two territorial transport-planning studies described below. In 1967, the Passenger Transport Survey Unit was made a permanent organisation and renamed the Traffic and Transport Survey Unit. It was responsible for reviewing traffic and transport requirements, for keeping survey data up-to-date, and for forecasting traffic and transport trends in Hong Kong.

Hong Kong Mass Transport Study

The O&D surveys in 1965 indicated that approximately 81 per cent of all trips were made by public transport, of which about 61 per cent were undertaken by bus, 5 per cent by ferry, and 15 per cent by tram (see Table 7.1). The Kowloon Canton Railway (KCR) accounted for less than one per cent of trips.

One feature of public transport in Hong Kong which contributed to its high profitability and low fares was the comparatively uniform spread of patronage (over hours of the day, days of the week and months of the year) compared with most other

Table 7.1 Daily person trips by mode of transport, 1965

Mode	No. of trips	%
Car and motorcycle	410,663	14.6%
Taxi	127,995	4.5%
Bus	1,705,577	60.6%
Tram	421,568	15.0%
Ferry	132,689	4.7%
Train	17,938	0.6%
Total	**2,816,430**	**100%**

Note: *Many of the trips recorded in the surveys used several modes of transport, but in order to avoid double-counting during analysis a trip recorded on a particular mode of transport was rejected if the person making it reported using another mode which ranked higher. For this reason, figures for junior modes were less than the numbers of persons who used those modes for some part of their trips.*
Source: *Dalby, E. (1967)*

cities. Mass transport in Hong Kong was, furthermore, considered the most efficient means of moving people, although it should be emphasised that the limited available road mileage in Hong Kong required such efficiency. Since such a large proportion of travel was made by public transport, the government decided in early 1965 to conduct a detailed study to improve and expand the public transport system. The Mass Transport Study (MTS) was to start at about the same time as the basic travel data were made available from the passenger transport surveys.

Consultants (Freeman, Fox, Wilbur Smith and Associates 1967) were appointed to investigate and evaluate different forms of public transport systems and recommend a functional plan of public transport development for implementation by stages to the year 1986 (see Chapter 5). Three public transport systems were proposed to provide the basis for the study of alternative options. The first system consisted primarily of improvements and extensions to the existing surface public transport system with no rapid transit elements. The second system included extensive rapid transit facilities supplemented by bus, train and ferry services. The third system lay between these two, having a smaller proportion of rapid transit and relying more on surface vehicles. As these systems were designed to serve the same residential, industrial and commercial areas, the principal variations among the alternatives considered related to the type and level of public transport services.

Consistent with similar studies conducted elsewhere, the amount and nature of future travel was presumed to depend largely on population, employment, income and car ownership levels. One of the major requirements of the MTS was, therefore, to derive relationships between the movements of people and goods in Hong Kong, and the social and economic activities of the community, and then to use these relationships (collectively called a transport model), to predict future travel patterns.

The first step of the transport model used for this study was to predict the amount and nature of travel. This was measured by the number of 'generation' and 'attraction' trip ends in an area, defined respectively as the total number of trips 'originating from' or 'destined for' that area. The number of trip ends depended largely on socio-economic factors such as population, employment, income, car ownership, and the nature and intensity of land uses, including housing type and commercial, industrial and institutional uses. Data obtained from the passenger transport surveys were aggregated by traffic zones. The size of the zones reflected the level of detail required in the subsequent analysis. Category and regression analyses were carried out on the aggregated data to derive the trip generation and attraction rates. Using planning parameters projected for the future years in each traffic zone, the total number of trips generated from and attracted to each zone was estimated.

The conventional 'four stage' transport model consisting of trip-end estimation, modal split, trip distribution and assignment estimations was applied in the MTS to predict travel characteristics and project patronage for alternative public transport systems for the design year. The study led to the recommendation of four mass transit railway lines (see Chapter 5), three of which are now in operation, namely:

- Kwun Tong Line,
- Tsuen Wan Line, and
- Island Line.

Hong Kong Long-Term Road Study

During the period 1954 to 1966, it was found that, as a consequence of rising personal incomes, both vehicle ownership and public transport patronage in Hong Kong had increased at a faster rate than population (see Figure 7.1). For car and non-car owners alike, the total number of trips generated increased with personal income. Dispersal of residential development and employment also contributed to the territory's growth of road traffic. The resultant road congestion was seen as a serious and growing problem in many parts of Hong Kong. It was therefore necessary to estimate road needs and the distribution of future population, motor vehicle ownership and public transport services, and prepare a realistic plan for road development consistent with these.

The Hong Kong Long Term Road Study (LTRS), initiated in December 1966 to investigate the need for road network development to the design year 1986 and pre-

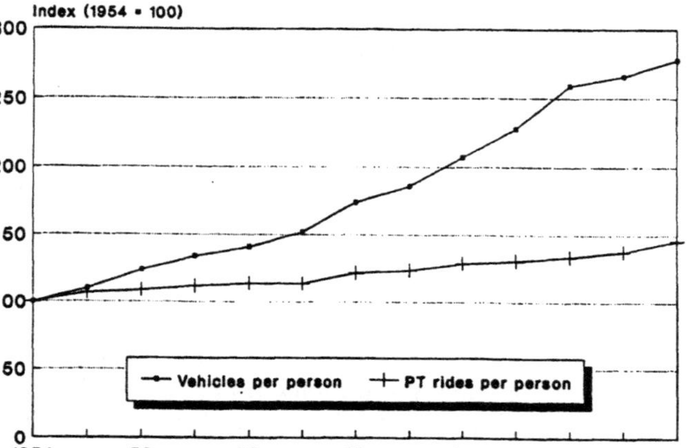

Figure 7.1 Trends in vehicle registration and public transport ridership, 1954-1966
Source: Transport Department

pare a programme for staged implementation of the proposed road works (Freeman, Fox, Wilbur Smith and Associates, 1968), drew basic data on travel characteristics from earlier passenger transport surveys, and forecast area development and mass transport utilisation based on the MTS. The same 'four stage' planning process was employed as in the MTS. Modal split and trip distribution were, however, both affected by the presence (or absence) of a rapid transit system. Hence, two basic forecasts of vehicle trips were prepared in the study corresponding to the two alternative conditions. In addition, a third set of trip projections was made to test the effect of a sharply reduced level of future car ownership such as might result from a very high tax on vehicles, in the absence of a rapid transit system.

A tentative road network was prepared for Hong Kong based both on experience and a judicious examination of the forecast development pattern. Assignments of the three sets of projected vehicle trips to the tentative network highlighted the corridors where vehicular volumes would be large and additional capacity required. These corridors were investigated and proposals developed to provide an improved, integrated road system.

As earlier noted, both the MTS and the LTRS were conducted with their own specific objectives, namely to investigate the requirements of public mass transport and road infrastructure respectively, and to establish a development programme for each. A fixed land-use development pattern was assumed for both studies. The only variable was the public transport system in the MTS, and the road network in the LTRS. Possible changes in land-use development arising from changes in the transport system were not examined, nor was the interaction between the public transport and road networks.

The first comprehensive transport study

Estimates of travel demand in Hong Kong in the above two studies were made on the basis of a number of planning assumptions, including assumptions about population and employment distributions and economic growth, which could not be predicted accurately and were subject to unexpected changes. There were, for example, indications in the early 1970s, that the territory's economic development, housing programme and population growth in the New Territories were significantly different from what had been earlier assumed. In addition, as the less difficult road projects were progressively completed, further expansion of the road system became more expensive. In particular, the continuation of unrestricted car ownership and usage negated the efforts to increase road capacity through new road construction. Accordingly, it was considered that a comprehensive territory-wide study of transport requirements was essential for Hong Kong and that the planning of new facilities would have to be complemented by appropriate transport policies.

Consultants were subsequently commissioned in 1973 to conduct the first Hong Kong Comprehensive Transport Study (CTSI) (Wilbur Smith and Associates 1976). The primary purpose of this study was to determine the measures required to achieve and maintain an acceptable level of mobility for passengers and freight up to 1991 (also see Chapters 3 and 8). It was the first full review of long-term planning and development of transport systems and policies in Hong Kong covering both road and public transport networks. The transport model used for this study attempted to predict the effects of transport policies on travel patterns and transport characteristics, as well as land use, thereby extending the basic model earlier developed under the MTS.

A series of special surveys was conducted for CTSI to provide the data for developing a comprehensive transport model capable of simulating the very complex travel demand relationships and resulting transport system usage which existed in Hong Kong. Of these surveys, the Home Interview Survey conducted in 1973 was the most extensive ever undertaken in the territory for transport planning purposes. It included a sample of about 25,000 households, stratified by income and area to ensure that sufficient usable interviews were obtained from high income groups. With any large-scale home interview survey, under-reporting of trips is unavoidable. To correct for this, comparisons were made with traffic counts at screenlines, and selected trip and modal factors were derived for applying to the trip matrices in the Home Interview Survey.

For predicting the future travel pattern in Hong Kong under different land-use and transport assumptions, a series of sub-models was developed. Typical trip-makers in Hong Kong, unlike their counterparts in many large cities of the industrialised world (where such studies had previously been conducted), could choose from a number of travel modes for most of their daily travel needs. Because several of these modes operated in direct competition with each other along the same transport corridor, a multiple choice of mode was available to the trip-maker along these routes.

The CTS modelling technique sought to reflect the manner and magnitude of trips made under varying sets of conditions by the use of the following:

- a vehicle ownership model;
- a trip production model;
- a trip attraction model;
- a multi-mode trip distribution model;
- a modal choice model;
- a sub-modal split model;
- a vehicle occupancy model;
- a peak period model;
- goods vehicle trip end and distribution models; and
- a traffic assignment model.

The above have since become the basic territory-wide transport modelling procedures in Hong Kong (commonly referred to as the CTS model). The outline procedures for application of the CTS model is illustrated in Figure 7.2.

In any city urban travel patterns are generated by the needs of people to move to and from places of their daily activities. Many such daily movements are repetitive and regular. The travel decisions that underlie these movements are the result of a number of factors which transport models deal with in a somewhat simplistic manner. In the case of CTSI, the principal objective of the study was to treat in aggregate form what in reality were individual trip-making decisions. The CTS model, there-

Figure 7.2 Outline procedures for application of CTSI Model

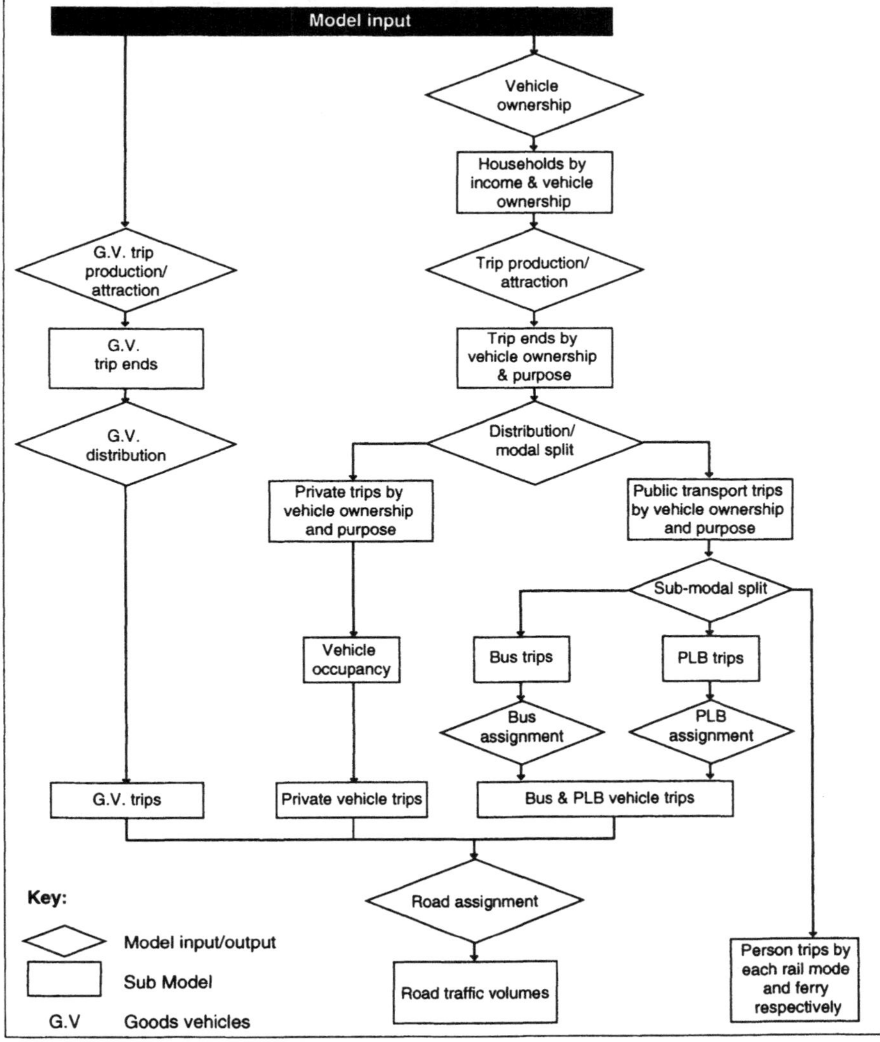

fore, represents the aggregate response patterns of sub-groups of trip makers faced with what are considered similar trip-making conditions and travel choices. The study's model was based on the premise that behavioural patterns for specific trip-making sub-groups would remain constant over time, and that changes would occur only in the population within each sub-group, as well as within the environment in which the facilities and services are provided.

In order to provide a sound basis for transport planning in Hong Kong, three alternative future land-use patterns were considered by CTSI, each assuming different areas of major urban growth. The three corresponding geographical distributions of population and employment (with major growth in North Eastern New Territories, Western New Territories, and existing urban areas respectively) were postulated using the same territorial total population forecast for the design year 1991. Three alternative transport systems comprising rail public transport network and road network developments at different scales were proposed for satisfying transport needs. The performance of these and the three alternative land-use assumptions was tested by applying the CTS model. Transport conditions in the intermediate years were also investigated. The analyses included detailed evaluation of a considerable number of transport issues and various alternative policies, such as restraints on the ownership and usage of private cars and taxis. These results laid the ground work for the government's transport policy promulgated in 1979 (Environment Branch 1979).

The Second Comprehensive Transport Study

Since the completion of CTSI in 1976, Hong Kong had continued to expand rapidly, both geographically (with the further development of the New Territories) and economically, with GDP real growth rates averaging 8.4 per cent per annum between 1976 and 1986. This expansion resulted in a demand for much greater travel. In order to cope with this, an average of some HK$3 billion was spent every year between 1976 and 1986 on the construction of highways and railways (see Table 7.2). While there was a large number of potential projects to choose from for the further expansion of the transport system, there were also numerous practical limits to both financing these projects and the capacity of the construction industry to execute them. New transport projects, therefore, had to be planned carefully in order to obtain maximum benefits from any new investment.

Table 7.2 Investment in highways and railways 1976-1986

	Highway	Railway	Total
1976	0.21	0.89	1.10
1977	0.30	1.13	1.43
1978	0.42	2.28	2.70
1979	0.60	2.91	3.51
1980	0.72	1.82	2.54
1981	0.95	2.24	3.19
1982	1.51	3.67	5.18
1983	1.50	3.41	4.91
1984	1.27	3.23	4.50
1985	0.95	1.48	2.43
1986	1.33	1.10	2.43
1976-1986	9.76	24.16	33.92
Average	**0.89**	**2.19**	**3.08**

(HK$ billion, current values)

Note: *Government rail equity is excluded.*
Source: *Transport Department and Wilbur Smith Associates (1989)*

In 1986, the Hong Kong government decided to conduct the Second Comprehensive Transport Study (CTSII). This was done, firstly, to establish detailed road infrastructure and public transport development programmes up to the year 2001, and secondly, to develop an analytical tool capable of assisting in the formulation of transport policies able to reflect quickly the effects of changes in policies on the transport infrastructure of the territory. The study was conducted by government, in-house, with specialist assistance from the same consultants (Wilbur Smith Associates) who helped prepare CTSI.

CTSII saw a change of emphasis from CTSI. Whereas the emphasis of CTSI was on 'demand-based' planning that of CTSII was on 'resource-based' planning. Transport projects were prioritised in CTSII by evaluating their costs and benefits. Transport policies were investigated for reducing travel demand to fit the capacity of planned transport networks. Because of the uncertainty associated with future developments in the territory, CTSII also marked the beginning of a continuous monitoring process. While the study had prepared an investment plan to cover the entire planning period of the study, the emphasis was very much on identifying projects for implementation within the more immediate future.

The study was intended to be updated regularly to incorporate transport network changes and socio-economic developments in the territory. Projects not committed for short-term implementation were to be reviewed, re-evaluated, and their priority in the programme reassessed. New projects would similarly be appraised as they arose. Another important element of the study was the analysis of transport policy issues, the findings of which were seen as a useful tool to manage future transport demands to acceptable levels.

A computerised transport forecasting model lay at the heart of the CTSII analytical process (see Figure 7.3). Based on an up-dated inventory of the transport system, and projections of the territory's development, the model estimated daily travel demand by mode, and forecast corresponding traffic congestion levels and traffic speeds. From this, it was possible to analyse and subsequently incorporate new transport projects or changes in transport policy into the transport system, and compare projected traffic flows with and without these projects or policies.

The model used by CTSII was, as already indicated, originally developed in 1973 as part of the CTSI Study for Hong Kong. An initial important task of CTSII, however, was to overhaul the earlier model and adapt it better to its new role. This process included:

- adjustments to reflect the most recent data on travel patterns;
- the incorporation of improvements in modelling techniques as developed in Hong Kong and elsewhere;
- the reorganisation and simplification of the model structure to reduce computer running times; and
- an expansion of the model to include economic evaluation procedures.

It was intended that the study, as far as possible, would rely on existing sources of data. The main database employed for person travel characteristics, for example, was derived from the 1981 Transport Characteristics Survey (Transport Department

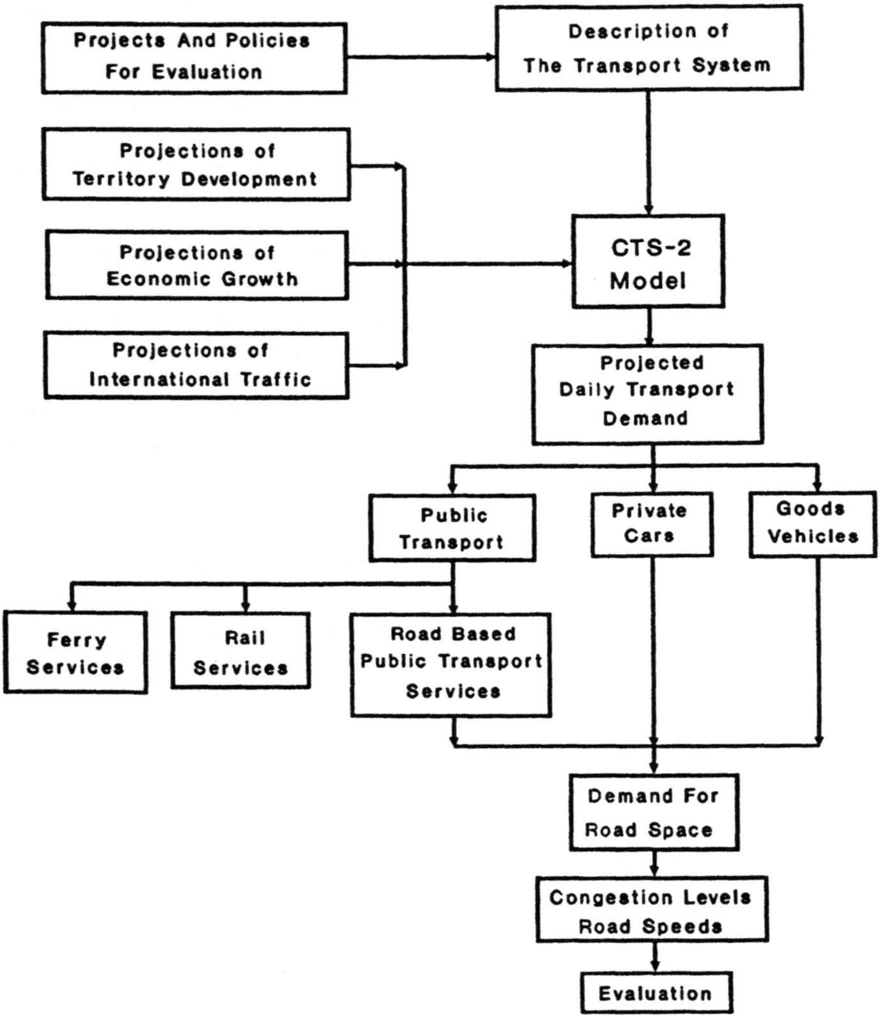

Figure 7.3 Analytical process of the CTSII Model

1983). The prime data source was a household survey of about 28,000 households similar to that conducted for CTSI. In addition the detailed surveys carried out for the 1984 Trucking Industry Study (Wilbur Smith and Associates 1984) provided comparable information for goods vehicle transport. Both surveys were updated using current 1986 statistics on traffic and public transport operations. Four additional special surveys were also carried out for CTSII to supplement these data. They included:

- a taxi survey;
- a survey of light vans, passenger vans and goods vehicle registered vans;
- an area speed-flow survey; and
- a parking accumulation survey.

The transport projects and policies examined in the CTSII Study were evaluated from several different viewpoints, including:

- operational evaluation;
- environmental evaluation;
- distributional evaluation;
- economic evaluation; and
- financial evaluation.

Projects were selected for inclusion in the recommended investment programme based on the estimates of funds available and the priorities indicated by the study's evaluation. Evaluation priorities were modified where necessary, as in the case of the need to make changes according to logical sequences of construction.

Land-use data projections are basic input requirements in any comprehensive transport studies. This task is never an easy one (as confirmed in Chapters 2 and 3) as there are so many factors that can influence the projections. A central land-use development strategy was adopted by the CTSII Study for the main series of transport tests, and evaluation of transport projects and management policies. This permitted the various analyses to be conducted at a sufficient depth by limiting the changes in land-use assumptions. Sensitivity tests of minor variations in the land-use assumptions were also carried out. It was found, however, that they would not affect the major recommendations of the study.

Territorial transport planning

The basic approach

The basic approach to transport planning for the territory can be summarised by reference to four distinct phases:

- data collection;
- model building;
- travel forecasting; and
- transport plan evaluation.

A traffic survey is first made of existing travel demand, and relations between the present demand and the existing land-use and transport system are analysed to enable a mathematical transport model to be developed. The forecasting phase then uses the transport model to estimate the future territorial transport demand based on projections of population and other planning and transport facilities data, and proposals for the area. The evaluation phase assesses the results of alternative plans and policies to establish how they would satisfy pre-defined social, economic and transport operational objectives.

In the early period of territory-wide transport planning for Hong Kong, considerable effort was invested in detailed surveys of the transport network characteristics and trip origin and destination patterns. These provided the basis for developing a conventional four-stage transport forecasting model. It was assumed that the time dependency of travel characteristics mainly arose from the variation of causal factors with time (Choi 1982), so that:

travel characteristics at time t = function of (causal factors at time t)

This assumption in effect removes the time dependency of the relationship between travel characteristics and their causal variables, enabling the determination of this relation from suitable data in one particular 'cross-section' in time. Earlier transport studies were mainly concerned with assessing major infrastructure schemes on the basis of a specific land-use projection.

Since the early 1970s, transport planning in Hong Kong, and elsewhere, was increasingly influenced by the systems approach to planning (Creighton 1970, Dickey 1975, and Stopher and Meyburg 1975). Within the territory, CTSI was used to examine the effects on travel characteristics of different transport policies as well as land use. Notwithstanding this, the projection of travel demand was still made on the premise that currently observed travel patterns and characteristics would persist in future. In some areas, changes to model parameters were used to simulate envisaged changes in the socio-economic characteristics, and investigate consequent changes in the projected travel patterns. Between CTSI and CTSII, a large number of land-use/transport studies were conducted in Hong Kong for new town developments and major transport corridors (see Table 7.3), all of which affected the territory's future in some way. By and large, these studies follow the same general CTS process outlined earlier. In other words, alternative land-use or network patterns were set up for each of the study areas, and the resulting travel behaviour predicted. The preferred option was then selected according to a set of predetermined evaluation criteria.

Table 7.3 Main categories of planning studies between CTSI and CTSII (1976-1987)

1	A series of sub-regional studies associated with the development of the Territorial Development Strategy, e.g. the Harbour Reclamation and Urban Growth Study (1983).
2	Studies of the transport requirements of New Towns, e.g. the Junk Bay Transport Review (1984).
3	District studies to develop local transport plans within the context of the overall transport strategy, e.g. the Western District Traffic Study (1987).
4	Planning studies of specific routes and corridors, e.g. the Feasibility Study of Alternative Transport Development in the Sha Tin - Tsuen Wan Corridor (1981).
5	Public transport studies, e.g. the series of Development of an Integrated Public Transport System (DIPTRAN) studies to forecast public transport requirements in conjunction with the development of the Mass Transit Railway (1979-83).

By the mid 1980s, resources placed important constraints on the feasible increase to the capacity of the territory's transport system. As a result, the economic use of the transport system emerged as increasingly crucial. Reflecting this, an economic evaluation sub-model was specifically developed for CTSII in order to determine the economic and financial performance of various transport options. The results were used to prioritise proposed transport projects and policies for implementation. This in turn, led to a transport infrastructure and policy development programme being established, which would be regularly updated to take account of actual changes in development.

The transport model first developed in the MTS, and expanded and further developed in CTSI and CTSII, had originally been a forecasting tool for road-based infrastructure planning. The use of this kind of model in Hong Kong, in retrospect, reflects the fact that it was developed at a time when roads were not yet fully developed in the territory, and when they were both easier to provide and less expensive to construct. Under these circumstances, roads were more readily acceptable to government as 'solutions' to transport problems.

Although the transport data employed for both CTS studies mirrored reasonably accurately the travel usage and performance of the road system of the time, in the case of CTSI it was at a time when the urban railway systems in Hong Kong were not yet in place. The Transport Characteristics Survey conducted in 1981, on which the CTSII transport model was developed, did not reflect the full use of the urban railway system which has subsequently developed. In particular, the continuous shift to urban railways from other road-based public transport modes in the mid 1980s in the face of growing urban road congestion was not analysed nor simulated in sufficient detail in CTSII to incorporate all the effects.

The Land-use/Transport Optimisation Study (LUTO)

As Chapter 3 confirms, transport planning until the late 1970s was traditionally carried out in Hong Kong in parallel with, but separate from, land-use planning. Transport plans were thus conceived once major land-use development plans were formulated in outline, and details progressively developed in accordance with these land-use plans. Transport infrastructure development in the territory, as a result, has typically lagged behind land-use development, thereby contributing to traffic problems.

As the intensity of development increased, the cost of constructing additional roads to new urban centres to cope with the rising travel demand became progressively more expensive. This was primarily due to limitations of space, and increasing social and environmental constraints. While traffic management measures of varying degrees of severity were introduced to help alleviate traffic congestion problems, the 'solution' to these problems in reality would lie in tackling problems at their source.

In the early 1980s, it was recognised that resources imposed an increasingly severe limit to the feasible increase in the capacity of the territory's transport system. It was also acknowledged that because travel demand was a direct result of the interaction of land-use/transport systems, any effective planning approach to this field of concern had to integrate the planning of these two components. A hierarchical three-level approach to territorial development planning was thus developed (see Figure 7.4, and for detailed discussion of these levels see Chapter 3).

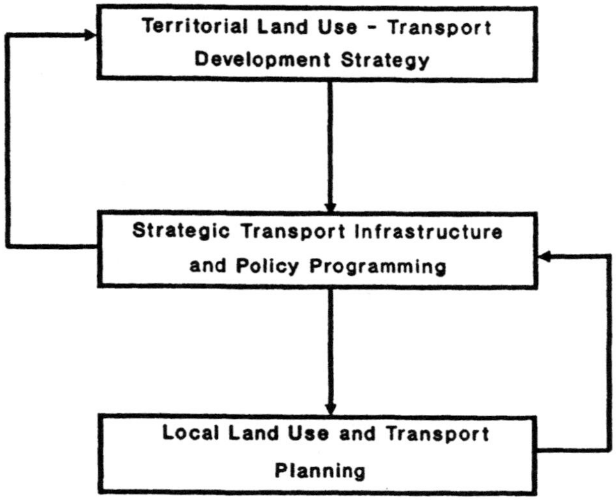

Figure 7.4 Three-level hierarchy of development planning

At the top level of this approach was the formulation of a territorial development strategy of land use and transport, employing a land-use/transport optimisation (LUTO) model (Choi 1984). Within this planning framework, a CTS type model was used at the second level. This second level model was principally intended as a tactical transport planning tool for the development of strategic territory-wide transport infrastructure and policies. The third level model looked at a much more detailed planning of land use and transport at the local level. It was appreciated that at all levels feedback would be required from one to another.

The first step of the LUTO Study, which was conducted in-house within the Lands and Works Branch of the Hong Kong government between 1981 and 1984, was to ascertain the development potential of various parts of the territory. With this in mind, planning studies were carried out for each sub-region of the territory to identify the maximum capacity for development from various planning, environmental and transport perspectives. Such studies were also carried out in order to provide plans for achieving different development levels in the sub-region together with the corresponding costs.

The LUTO Study ultimately aimed to arrive at an optimal pattern of future territorial development which would minimise total land development and transport costs[2]. A generalised mathematical model was developed for this purpose which consisted of three main components:

- a strategic land-use/transport model;
- a cost model; and
- an optimisation model.

The first of these models was, yet again, a simplification of the CTS model but with a coarser zoning system and set of transport networks. The mathematical relationships between the traffic flow characteristics and the socio-economic variables were also greatly simplified by virtue of the computational effort required by the optimisation algorithm. The principal use of this model was to provide the optimisation model with 'influence coefficients'[3] which characterised the transport influence of the land-use variables.

Various cost items of land resumption, engineering development, and provision and operation of transport facilities were analysed. These were expressed as mathematical functions of population and serviced land areas in each zone of the territory. The mathematical relationships were derived from the sub-regional studies to form the cost model. The optimisation model incorporated a solution algorithm based on mathematical programming and operated on a set of analytically formulated objective functions and constraints. Conceptually it simply assigned development to the sub-regions according to the least marginal cost, subject to various constraints.

In this context, the marginal cost was the average cost of providing the necessary serviced land and transport infrastructure in the territory to accommodate a unit of additional population or employment in a given sub-region. In this way, the optimal global land-use development pattern, together with the corresponding transport network, was determined with respect to a number of objective functions, subject to given constraints. The results were then used as an aid to the formulation of Hong Kong's long term development strategy for both land use and transport.

While roads in the urban areas of Hong Kong had become increasingly difficult to provide, as urban areas continued to develop and as physical and environmental constraints increased, the construction and operation of the MTR in the early 1980s succeeded in maintaining good levels of overall mobility. So much so that questions had arisen as to whether the economic growth of Hong Kong could ever be sustained without this mass transit system. The Metroplan Study (Planning, Environment and Lands Branch 1990) carried out in 1987, contains a further development and application of the integrated land-use/transport planning process (see Chapter 8). This study acknowledged the important role of the MTR by placing a major emphasis on urban passenger railways to help achieve the goal of restructuring both the metropolitan area and the territory.

A review of the concepts employed in land-use/transport modelling

Sectoralism versus integration

Fundamentally, humans tend to have a natural desire to break a problem down to a level where the identified issues are readily amenable to a subjective consensus - i.e., a decision-making level within the experience of the majority. This concept is reflected within the Hong Kong government by the creation of separate sectoral authorities (see Chapter 11). Although these administrative sectors often function at a disjointed level of decision-making, each with its discrete goals, policies and programmes, they are the conceptual basis of the style of management and planning of urban development in Hong Kong which, as a result, is inherently accommodated by the urban development process.

Arguably, none of the governmental sectoral areas of concern, or the issues they generate, could physically exist in isolation. This is so because urban development is a dynamic, socio-spatial phenomenon comprising many intangible qualities, the components of which must ultimately interact in mutual support to survive and develop. At some stage, therefore, the important sectoral issues of Hong Kong need to be conceptually drawn together and consensus reached on their importance. The method(s) employed to do this, to reduce conflict between the various sectoral subjective opinions, and to 'deconstruct' the problem at hand, involves the development of an integrative model.

The integrative model

The selection of a suitable model concept depends on the nature of the 'problem' confronted and on the perceived issues it is designed to tackle. The function of any model is to provide a convenient way of sampling, examining and rationally explaining interactive issues. This is done with a view to promoting objectivity and guiding potentially divergent opinion towards consensus on how to resolve the 'problem'. In the context of land-use/transport planning, the 'problem' is the allocation of urban space between competing sectoral interests.

With advances in the power of computers and computer software, it has become possible to build increasingly complex numerical representations of mechanised travel and land-use changes in large urban areas. As with most other major cities, however, the metropolitan area of Hong Kong was largely built before the era of computerised analytical techniques of urban development and transport planning. The basic spatial structure of the city - i.e., its division of usable land between discrete packages for private land use and network of streets for public mobility - had already been established well before the employment of these computerised models. Outline development plans for new towns had also been drawn up and agreed between the various administrative sectors of government without the 'advantage' of computational techniques. Instead, consensus had been reached using a different kind of model. This was the model of the 'orthogonal plan'[4] with spatial allocations that reflected agreement among the various sectors.

During the 1960s, when the early transport planning studies were undertaken in the territory, the spatial structure of Hong Kong's urban area already existed, as did the distribution of land uses. The space allocated for movement was configured as streets with raised side-walks. This was not because there was any significant amount of mechanised travel but simply because this geometric configuration was drawn from British transport planning and engineering practices of the time. Inherent in these were concepts of equity of access from private land to public mobility space, and concepts of safeguarding public health and welfare which had evolved over a century of learned experience of building industrial cities. There is no evidence that the application of such basic concepts was questioned in Hong Kong. Rather, passenger transport, road development plans and land-use structure were all perceived by the sectoral authorities as unrelated elements of urban development which could be reasonably and satisfactorily addressed separately.

Given the entrenchment of this 'disjointed' perception of land use, and its relationship to transport and other sectors, it is in retrospect, a little surprising that a new comprehensive approach to urban transport planning was attempted in the early 1970s. For conceptually, this new approach was the philosophical antithesis of previous transport planning efforts, with all aspects of mechanised transport demand seen to stem from the nature and intensity of land use. The basis of the new approach was an empirical observation of traffic generated by all economic categories of land use. The new model became a numerically computed abstract construction of mechanised travel demand, expected to occur in the event that the assumed growth in economic land use was realised. Initially, the purpose of this type of model was to reveal the spatial consequence of mechanised travel in the territory in order that adequate spatial provision could be allocated in the orthogonal plan.

Essentially the integrative model was and still is the orthogonal plan. Although the land-use/transport model may give the appearance of being an integrative model, in reality its objective is focused on justifying the allocation of space to the provision of roads. In this respect, it is far from being an integrative model. However, the model has significant political value for the roads sector if other land-use sectors are convinced that it is in their own interests to give up land-use space for the provision of roads.

Links with the motor vehicle lobby

It has been claimed (Ziv 1977) that the motives behind the construction of such models, in reality, stem from an alliance between the motor vehicle production industry (which required road space to be built to encourage the sale and usage of motor vehicles) and land owners on the periphery of urban areas (who could foresee substantial increases in land values if the peripheral areas were made accessible by motor vehicles to the central urban areas). Unlike most other Western cities, however, Hong Kong's urban area is not seen to be surrounded by opportunity space for suburban development. Nor does the territory have a domestic vehicle production industry.

The acceptance of the CTS type land-use/transport modelling in Hong Kong, therefore, cannot be attributed to these suspected links as neither of the alleged fundamental motives existed in the territory. One can only surmise that there were no market forces promoting this kind of model's adoption but simply a technocratic attempt by the combined roads and transport sector to justify the allocation of an extremely scare resource in Hong Kong, (i.e., land) by borrowing what appeared to be the latest concepts and techniques from British and North American transport planning practice (Dimitriou 1992).

The 'four-stage' transport model adopted by the various transport studies in the territory may be described in concept as a rationalised attempt to break down human behavioural travel habits into components which could be separately observed, analysed and empirically correlated with specifically definable input assumptions and land uses. Theoretically, each individual component (or stage of the model) required consensus on issues which had been simplified to a level which facilitated agreement among concerned parties within the roads and transport sectors. Thereafter, the re-aggregation of the components is computed using statistical econometric techniques.

Von Thünen's model and the advent of motorisation

As already indicated, major aspects of the CTS model as applied in Hong Kong are in fact derived from Von Thünen's early nineteenth-century concept of central place theory, whereby it was assumed that the probability of two land uses interacting would decrease as a function of the distance between them. This concept was enhanced by Pareto in the late nineteenth century (Pareto 1896 and 1909) who described this 'natural' distribution of land use as an 'optimal balance' between the resource value of the land (used for the various urban activities), and the resource cost of essential interaction.

In the early twentieth century, empirical observations (Ricardo 1912, Lösch 1954, and Christaller 1966) further substantiated the utility of these theories. Pareto's work, conducted before the era of motorised transport, focused on the location of land use when the opportunity to interact (principally on foot) had little spatial consequence. The fundamental concept was how land uses should be distributed in an orthogonal model.

With the introduction of motorised transport, however, increased speeds raised average trip distances. Despite this, the principles of central place theory remained valid, with the difference that vehicles required more space and a particular configuration of space for mobility purposes. It was thus believed that if an economic distribution of land use could be created in an orthogonal model, all that was required of the new breed of transport planners was to focus on the 'operational adequacy' of the roads in order better to accommodate the mechanised interaction seen to be essential to the survival of these land uses.

Used in this way, the land-use/transport model employed to determine spatial consequences has no control function and would, therefore, theoretically invoke the infinite expansion of cities and the widening of all roads, especially so if encouraged by market forces associated with selling more motor vehicles and urbanising more rural land. For Hong Kong this is clearly absurd.

A century ago, without any knowledge of the growth of motorised travel, Pareto postulated a concept of optimality which is again today proving valid for reasons which he could not have foreseen. As cities over-expand, the relative low cost of peripheral land no longer offsets the consequent cost of travel. The consequent cost is not merely the financial cost of vehicle ownership or the perceived cost of mechanised travel, but the community and other external costs stemming from the spatial and environmental impacts of motor vehicles in cities. In Hong Kong, this realisation has not yet had a major influence on the technocratic transport planning processes (Dimitriou 1994).

Although the CTSII study process was similar to that employed by its predecessor, the objectives of the Hong Kong administration in the meantime have changed from determining spatial consequence to determining justifications for financial resource allocation. To the technocrat with no external pecuniary motives, the 'four-stage' transport model, based as it is on econometrics, appeared to be an ideal tool to explain and justify the allocation of resources to new roads and road improvements. Here, statistical interaction between land use and transport was described in terms of economic cost, and any operational advantage to be gained by capital investment in transport infrastructure could be readily expressed in terms of economic benefit to the community. Using this technique, it was believed practical to substantiate capital resource allocation and examine the relative merits of different investment options, prioritising and programming projects seen to be in the public interest.

Limitations

Modelling techniques that aggregate what are in essence disjointed elements perceived by the administration to be important, are essential planning tools which raise consensus to a higher decision-making level. It must, however, be recognised that these techniques have conceptual limitations, and the higher decision-making level consensus is only likely to be valid if the objectives lie within these limitations. The transport planning objectives of the present generation of land-use/transport models in use in Hong Kong have been variously described as:

- substantiating the allocation of spatial resources for transport infrastructure; and
- prioritising the allocation of financial resources among competing transport infrastructure projects.

These are the objectives of the transport sector and are, arguably, entirely valid from an internal sectoral perspective. Unfortunately, however, what is supposedly both patently and rationally true to the transport planning technocrat may not be seen in the same light by others affected by the proposals recommended. Concepts, such as limiting the geographical expansion of urbanisation or judging the relative merits of allocating urban land to transport (rather than to some other activity), remain unaddressed. Consequently, the present generation of land-use/transport modelling techniques employed in the territory are in fact inadequate and cannot be legitimately used to guide a higher level of consensus-making on urban land-use/transport interaction issues.

Notwithstanding this view, others perceive the territorial CTS model to use generalised land uses as the principal causal factors generating travel. Because, on a numerical basis, the land-use distribution inputs driving the model can be changed, this leads to the presumption that the model is valid and reactive (especially through its economic functions) to changes in socio-economic development. On this basis, the land-use/transport model can, arguably, provide useful comparative guidance on the preferred territorial distribution of land use.

As stated at the beginning of this chapter, the new generation land-use/transport models were perceived as universal forecasting tools. These models, nevertheless, had to be calibrated to simulate the prevailing travel characteristics and associated causal factors in individual cities. Even in such circumstances, however, the assumption that travel characteristics at a particular point in time ('t') are a function of the causal factors of that same point in time, is only a reasonably valid hypothesis for the period over which the relationship can be expected to remain stable.

There are many cities in America and Europe where development pressures are low and spatial constraints on travel characteristics are of no consequence. In such cities, the structural distribution of land use and the typical centrality patterns of transport interactions between these land uses have remained stable for decades. In such situations, a properly calibrated model can be expected to remain valid until such time as something intervenes to bring about structural change. If any structural change occurs, e.g. in the size of the population and/or in the distribution of economic activities and/or in the opportunities to travel, which results in a change to the centrality characteristics, then one is effectively dealing with a different city. Any previously employed land-use/transport model would thus require recalibration of the relationships linking the travel characteristics with the causal factors at the new time ('T') to restore the validity of any forecasts.

In this sense, the application of the territorial CTS model (calibrated at time 't') to significantly different territorial development scenarios (at some future time 'T') as part of the TDS formulation process for Hong Kong as outlined in Chapter 3, is not strictly valid. This is somewhat akin to applying the model of one city to a series of different cities.

Although the territorial travel characteristics would change under each of the prospective development scenarios of the TDS process, thereby generating new and possibly conflicting centrality patterns, what the model will in fact attempt to simulate is the amount of personal travel likely to be generated by the scenario examined. It will assign this travel to the original centrality patterns of the territory on the basis of modal choice characteristics prevailing when the model was calibrated. Yet, many of these forecasts may no longer be realistic, even though it might be reasonable to assume that the amount of personal travel generated will be in the same order. This is because some of the future population may simply travel to different destinations by different modes.

The value of the kind of model discussed for a TDS scenario evaluation lies in recognising where people will be able to travel between, and more importantly, the locations for interactive land uses where for some reason socio-economic interaction poses insurmountable difficulties. Tackling such complexities requires good professional judgement entailing a complex time-consuming study or, if need be, a cursory analysis of the major drawbacks of a particular development scenario.

Returning to the period of time over which a calibrated territorial model can be expected to remain valid, it is necessary to recognise that over the last two decades Hong Kong has enjoyed a highly dynamic period of economic development as pointed out in the Introduction to this book. Within the metropolitan area in particular, land-use intensities have become especially high and travel opportunities spatially very constrained. These characteristics are quite unlike the relatively stable, spatially unconstrained cities of North America and (to a lesser extent) Europe, where the validity of this generation of models was hypothesised and, over time, statistically accepted. The vibrancy and adaptability of the Hong Kong economy therefore suggests that the validity period of a calibrated model such as those outlined earlier is limited to a few years rather than the decades enjoyed elsewhere.

Even though this period of validity may be limited, in comparative terms the model nonetheless has a practical and a political value to the transport sector which must promulgate cost-effective operational policies and allocate limited financial resources among competing infrastructure projects. As such, however, used deterministically for these purposes, the function of the model is constrained to being a management tool rather than the forecasting tool which it purports to be under other developmental circumstances. This raises the question of whether the modelling technique has any further practical, or political value. The answer to this is in the affirmative, but again the limitations must be recognised.

As described previously, the centrality characteristics of a TDS development scenario can be analysed on a judgemental basis in considerable depth. The model relationships between travel characteristics and the causal factors can then be adjusted to simulate how the future territory (and its urban areas) might operate. This is a 'probabilistic approach' which relies both on the professional judgement of the analyst and on some prior political consensus on the nature of the territory's future as it should develop. Although the resulting forecasts thus contain an element of determinism and are, to a degree, self-fulfilling prophecies, they are nevertheless valid. The limiting issue under these circumstances, therefore, is not the validity of the forecasts, but whether or not the political objectives were adequate.

This concept of undertaking a quasi-calibration of the model is time-consuming and would not normally be undertaken during the TDS formulation process, but would await selection of a preferred development strategy. Arguably, this is too late to influence strategy formulation. However, it provides the transport sector with a planning tool to 'prioritise' and 'programme' the transport infrastructure components of the territorial development strategy, over a much longer period of time than the limits of the absolute validity of the original model would permit.

There is, though, a danger in selecting the preferred development strategy in advance of the quasi-calibration. This lies in the possibility of political determinism overriding prudent professional judgement. When first introduced into Hong Kong, the land-use/transport modelling process was applied in a very deterministic manner to a set of complex developmental circumstances, involving a territory-wide new town development programme. 'Blue-print' plans for the new towns provided land-use distributions as inputs to the model. This land-use mix was conceived in the same image as the existing metropolitan area.

Consequently, the new towns were expected to develop their own internal centrality patterns and be socially and economically self-contained. This notion went considerably beyond the conceptual validity of even a probabilistic approach. No matter how ideal the land-use mix might have appeared at the outset, given the separate government sector policies and market forces, the 'planned' conjunction of events was highly improbable. In terms of forecasting 'subsequent' events, this model did not, therefore, prove to have the validity of a self-fulfilling prophecy. Arguably, however, the model did have significant political value for the proponents of the new town development programme.

The 'subsequent events' referred to above proved to be an acceleration of the housing programme through the 1970s, and a decline in the demand for light industrial land in the 1980s. These left the new population 'adequately housed' but without a local employment base. This produced travel demand patterns very different from those characterising land-use/transport interactions in the metropolitan area. Rather, the travel patterns were more akin to those of the American residential suburbs with the new population remaining economically tied to the central-city job market. This is an alternative territorial centrality pattern which could equally well have been modelled. As events have unfolded, this might have proved to have had greater practical value but it is questionable whether this scenario would have been 'politically acceptable' at the time.

In a situation of this kind, where political idealism supplants probability, nobody wins. The public blames the transport planners for failing to provide the transport-needed infrastructure and services. The transport sector blames the land-use planners for failing to correctly estimate the land-use needs and pursuing the appropriate planning concepts. The land-use planners are then left looking to the housing, industry, commerce, and lands authorities for a lead. These authorities, on the other hand, typically, simply respond/react to supply and demand pressures affecting their sector.

The way ahead

The hypothesis underpinning land-use/transport demand modelling is that the generalised categories of socio-economic land use will generate some immutable quantum of mechanised travel demand. Hong Kong is faced with overwhelming circumstantial evidence in the form of successful metropolitan redevelopment intensities that this simplistic hypothesis ought to be rejected.

This is prompted by attempts to assess the traffic impact of metropolitan redevelopment. Given that the metropolitan road networks operate at their practical capacities throughout much of the day, adding some new development and some additional quantum of travel demand should result in the networks being overloaded. There is a school of thought that redevelopment permission should be withheld on these grounds. Overloading a network, however, is an abstract concept. In reality a road network cannot operate above its practical capacity. There is therefore no observable evidence after the event of a redevelopment to support the view that development permission should have been withheld. In effect, the impact of redevelopment is not appreciable in traffic operational terms and is manifest by the imposition of a marginal diminution of travel opportunity on all land uses within the area of influence of the redevelopment.

This has given rise to the introduction of supply- rather than demand-modelling in Hong Kong wherein the physical operational capacity of an area road network is accepted as a control over the nature and magnitude of demand that can be generated in, or attracted to, the area. While this approach offers some compelling logic and is very useful for examining traffic patterns in a local area, unfortunately it completely divorces transport from land-use planning and offers no insight into the most suitable intensity of urban development. Nonetheless, network supply modelling, particularly within the metropolitan area where opportunities to increase network capacity are strictly limited, seems to be a rational component of the way ahead.

Fundamentally, interaction between land uses does not take the form of passenger car units which is the basis for expressing mechanised travel demand in the CTS modelling system. Interaction involves the movement of individual people, some conveying goods, who are all, to some degree, capable of perceiving their travel opportunities, and from their experience deciding rationally, or intuitively, why, where, how, and when, to travel.

This suggests that the way ahead for modelling land-use/transport interaction should be comprised of three integral components:

- a personal demand model;
- a road network supply model; and
- a rail network supply model.

To arrive at any defensible position over the most suitable intensities of urban development, and indeed the structural distribution of interactive land uses throughout the territory, requires the introduction of some quantifiable concept of sustainability.

This, therefore, is both the technical and the political problem that needs to be confronted. Modelling must produce meaningful, conceptually valid and politically acceptable indicators for decision-makers.

In transport terms the sustainability of urban development seems to imply the achievement of, or at least moving in the direction of, a satisfactory balance between:

- community/public investment in the opportunity for individuals to meet their compelling travel needs;
- the conditions experienced by individuals in fulfilling their travel needs; and
- the resulting impact on the community of how individuals meet their travel needs.

To address this balance, which might be described as the modern day equivalent of Pareto's Optimality, will require new judgemental criteria beyond conventional transport network performance indicators. It will be necessary to derive mutually compatible, explanatory indicators to describe:

- compelling travel needs;
- tolerable travel conditions;
- acceptable community impact; and
- public investment returns.

The objectives of a new territorial land-use/transport model will be to:

- gauge the practical limits of geographical expansion of the urban development structure throughout the territory; and

- compare the relative merits and demerits of differential levels of intensification of urban development within the constrained geographical structure and the practical supply of transport opportunity.

To explain sustainability in an environment of saturated supply needs a new language of planning indicators including:
- usability;
- tolerability;
- adaptability;
- behavioural discontinuities; etc.

While empirical determinism will undoubtedly remain the methodological basis for management models, probabilism is the only satisfactory basis on which to model conflicting social and economic issues. Such new indicators cannot be established as absolute standards. These can only be statements of each political quality objective, quantified in mutually compatible terms to describe limits and trade-offs.

Since Hong Kong now faces a future in which growth of interaction with its hinterland, the Pearl River Delta (PRD) area, is inevitable, the sustainability issues of the growth of port cargo throughout, of the capture of PRD-generated cargo by new road links to Hong Kong, and of local air quality, pose problems that have to be confronted in conjunction with domestic journey-to-work difficulties.

Conclusions

In retrospect, it can be convincingly argued that problems associated with the various land-use/transport model forecasts which have been made over the years in Hong Kong did not lie essentially with the territorial transport model itself, but in how it has been used and the validity of the future that it had presumed capable of forecasting. The fundamental purpose of any model is to provide some valid insight into the future which will not prove, in retrospect, substantially incorrect. Those who use such models must be aware of this limitation.

It can also be argued that the territorial model must address the legitimate objectives of the transport sector as set out by government. Such objectives may, in reality, be contrary to what constitute the main urban development issue, but the blame for this can hardly be assigned to the transport sector or to the modelling system used.

The territorial land-use/transport model used to assist in the TDS formulation process is far from ideal. While many of the socio-spatial issues which need to be addressed in this process exist in the model, accessing them is cumbersome. This is because the system was not designed to reveal such issues, but rather to produce traffic network economic performance indicators. While it may be tempting to use these model outputs, this should be avoided when testing development scenarios which vary significantly from that on which the model was calibrated, as they have no recognisable validity.

Attention should instead be focused on the locations of personal trip productions and attractions. The linkage of these trip ends represents the centrality pattern of land-use/transport interactions in the territory. The rationality of this pattern, and the extent to which this might reasonably be accommodated, is the initial purpose of the analysis leading to the possible rejection of some land-use development options. Ultimately, a probabilistic quasi-calibration is needed to adjust the centrality pattern, and if necessary, the travel characteristics. The objective of this exercise would be to restore some probabilistic validity into the network economic performance indicators.

The substance of Central Place Theory - which attempts to rationalise the natural formation of some optimality between land-use distribution and the subsequent interactions between the land uses - relies on opportunities being available for individuals to exercise self-determination over their residential location in accordance with their needs and means. The deterministic attempt to optimise land use and transport in LUTO was based on various cost objectives. However, the attempt to optimise residential locations around an unknown and presumed distribution of jobs, without any real control over the creation of such jobs, has limited practical value. Over the last two decades, the geographical expansion of urban development throughout the territory has been led by the public housing sector, while the creation of job opportunities has been left to market forces. In Hong Kong individuals in need of public housing, and lacking in means, have in fact been able to exercise very little self-determination over their residential location vis-à-vis the job market. As a result, there is no longer any reason to believe that the present territorial distribution of land use exhibits any natural optimality if compared with Pareto's concept.

This lack of optimality among the territory's land-use/transport interactions has become a major problem in Hong Kong which the various arms and sectors of government are unable to confront. Considered separately, the sectoral issues which contribute to this problem remain unrecognised so that development continues to diverge from, rather than converge towards, some natural balance between land-use distribution and interaction.

Notes

1 This represented the beginning of the 'love affair' of the USA with the motorcar, commencing in the 1950s with the rapid proliferation of low-density suburban development on the periphery of cities, made possible by the ownership of the motorcar and rising incomes which fuelled such growth.
2 A LUTO model was subsequently developed for Beijing upon the request of the Chinese authorities. In 1990, the Director of the Study (Dr. Choi) visited Beijing (in a private capacity) to present an equivalent model for use in mainland cities.
3 'Influence coefficients' represent traffic volumes generated by a unit of additional population or employment.
4 The utility of an orthogonal plan (i.e., the geometric division of usable land into private space and some geometric pattern of interconnecting mobility space) has

been understood in the Western world for the last two thousand years. Such socio-spatial regimentation of urban activities has been variously exploited over the centuries by authoritarian regimes to police the population effectively and by industrial regimes to transport both people and goods efficiently. A simple model of this kind allows decision-makers to control the incremental allocation of space, not only to meet demand, but to meet social, political and economic objectives of the civil authority. While these are abstract, and the perception of them might well vary between individuals in the disjointed sectors of government, space is an issue and spatial awareness is a characteristic which humans do not need to intellectualise. It is a genetic learned experience which man inherited from his ancestors. He will barter for, and if necessary, fight over spatial allocation rights until some equilibrium, or forced consensus is established. An orthogonal plan with space conveniently packaged into geometric shapes is therefore a powerful model over which a pluralistic society can debate its spatial needs in a civilised manner.

References

Black, J. (1981) *Urban Transport Planning: Theory and Practice*, Baltimore: John Hopkins University Press.

Choi, Y.L. (1982) *Transport Planning in Hong Kong* Paper presented to Highway Engineering Symposium, December, Guangzhou, China.

——— (1984) 'Land Use and Transport Optimisation Study' *Hong Kong Engineer* November, Hong Kong Institution of Engineers.

Christaller, W. (1966) *Central Places in Southern Germany*, Translated by C.W. Baskin, Prentice-Hall: Englewood Cliffs, N. J.

Creighton, R. L. (1970) *Urban Transportation Planning* Urbana: University of Illinois Press.

Dalby, E. (1967) *Hong Kong Passenger Transport Survey 1964-66* Hong Kong: Government Printer.

Dickey, J.W. (1975) *Metropolitan Transportation Planning* New York: McGraw-Hill.

Dimitriou, H.T. (1992) *Urban Transport Planning: A Developmental Approach* London: Routledge.

——— (ed.) (1994) *Moving Away from the Motor Vehicle* Centre of Urban Planning and Environmental Management, The University of Hong Kong and the Goethe Institute, Hong Kong.

Environment Branch (1979) *Keeping Hong Kong Moving: A White Paper on Internal Transport Policy*, Hong Kong: Government Printer.

Freeman, Fox, Wilbur Smith and Associates (1967) *Hong Kong Mass Transport Study* Hong Kong.

——— (1968) *Hong Kong Long-Term Road Study* Hong Kong.

Lösch, A. (1954) *The Economics of Location* Translated by W. H. Woglom, New Haven: Yale University Press.

Mitchell R. and C. Rapkin (1954) *Urban Traffic: A Function of Land Use*, New York: Columbia University Press.

Pareto, V. (1896-97) *Cours. d'économie politique* Lausanne.

——— (1909) *Manuel d'économie politique* Paris.

Planning, Environment and Lands Branch (1990) *Metroplan: The Foundations and Framework*, Hong Kong: Government Printer.

Ricardo, D. (1912) *The Principles of Political Economy and Taxation* London: Everyman's Library.

Stopher, P.R. and A. Meyburg (1975) *Urban Transportation Modelling and Planning* Lexington, Massachusetts: Lexington Books.

Transport Department (1983) *1981 Special Survey on Transport Characteristics* Technical Report No. 319, Traffic and Transport Survey Division, Hong Kong.

Transport Department and Wilbur Smith Associates (1989) *Hong Kong Second Comprehensive*

Transport Study, Hong Kong: Government Printer.
Von Thünen, J. H. (1875) *Der isolierte Staat in Beziehung auf Landwirthschaft und Nationalökonome* Berlin: Hempel and Parey.
Wilbur Smith and Associates (1976) *Hong Kong Comprehensive Transport Study* Hong Kong.
——— (1984) *Trucking Industry Study* Hong Kong.
Ziv, C. (1977) 'La Genèse des Modèles de Trafic aux Etats-Unis' Unpublished PhD Thesis, Cornell University, Ithaca.

8 Transport planning and Metroplan: progress or a missed opportunity?

Harry T. Dimitriou and Alison H.S. Cook

Introduction

Metroplan, a land-use/transport, environmental and institutional planning framework for the metropolitan area of Hong Kong, was released at the end of 1990[1]. Its planning horizon of 2011 is consistent with that of the Territorial Development Strategy (TDS) and the Port and Airport Development Strategy (PADS). Geographically the plan covers Hong Kong Island, Kowloon and New Kowloon, and Tsuen Wan/Kwai Tsing (see Figure 8.1).

Hong Kong's century-old entrepot function with China, temporarily displaced after the Korean War and the United Nations embargo in the export of goods to China, re-emerged during the 1970s and 1980s. Economic restructuring in this period resulted in a rapid decline of manufacturing as the staple of the economy with a concomitant expansion of the service sector (see Table 8.1).

Table 8.1 Percentage distribution of GDP by economic activity 1970, 1975 and 1980

Economic Activity	1970 %	1975 %	1980 %
Agriculture, fishing, mining, quarrying	2.2	1.5	1.0
Manufacturing	30.9	26.9	25.1
Electricity, gas, water	2.0	1.8	1.3
Construction	4.2	5.7	7.1
Wholesale, retail, import/export, restaurants, hotels	19.6	20.7	19.2
Transport, storage, communications	7.6	7.2	7.2
Financing, insurance, real estate, business services	14.9	17.0	25.9
Community, social and personal services	18.0	18.7	12.9
Other	0.6	0.5	0.3
Total	**100.0**	**100.0**	**100.0**

Note: GDP at Current Factor Cost *Source: Census and Statistics Department (1996)*

Figure 8.1 Geographic extent of Metroplan

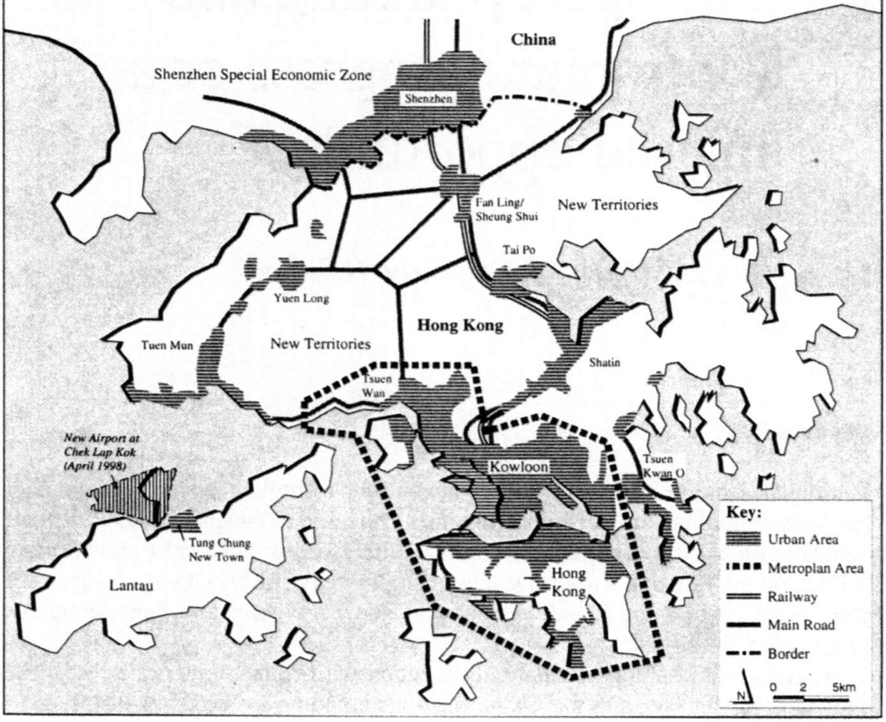

The main thrust of the urban development strategy in these two decades was the New Towns Programme which dispersed population from the metropolitan area to the New Territories. However, during the 1980s reappraisal of this urban expansion strategy took place, partly because the New Towns Programme was nearing completion, and partly because of the changed economic function of the city-state which brought new infrastructure and land-use demands.

The development focus turned from the New Territories back to the metropolitan area. The entrepot function, the rise of Hong Kong as an international city and prominent business and financial centre, and its role as a regional hub for the Pearl River Delta and Southern China, reinforced the need for new port and airport infrastructure. This in turn, increased the prosperity of the general population, and raised the expectations of the public on the quality of their living environment.

The pressure for the restructuring metropolitan Hong Kong also came from two other major sources, namely:

- The economic space opportunities offered by committed new infrastructure projects, particularly, the port, airport and associated infrastructure as proposed by PADS (now referred to as the Airport Core Projects).
- Concerns about the poor quality of living space expressed by low grade, high density public and private housing; the congested road network; obsolete and

congested industrial development; the lack of community facilities; and inadequate open space for recreational use.

Metroplan sought simultaneously to address existing problems, respond to new demands and identify courses of action within the context of the long term strategic plan (TDS). As an advisory document, its stated purpose is to provide:

> *...a framework within which both public and private sector agencies can formulate detailed plans and development programmes with the common aim of making our city a better place in which to live and work.* (Planning Department 1991d).

In 1986, the base year for the population statistics for Metroplan, the metropolitan area was home to 4.24 million (77 per cent) of Hong Kong's population, as well as 80 per cent of its workforce. In addition, the Metroplan area contained almost all office floorspace and 90 per cent of flatted factory floorspace and hotel rooms, most high-level government and administrative functions, and many of the territory's cultural and tertiary educational venues. It also handled all air traffic and most of the territory's seaborne trade.

On the basis of the above, the planners believed that the Metropolitan area's dominant economic, social, administrative and cultural position in the territory is unlikely to be seriously challenged in the future. This is so, with the exception of developments associated with the westward shift of the port, airport and related industrial developments, and the continued growth in the New Territories. Two major guiding principles of Metroplan were the presumed continued pre-eminence of the metropolitan area within the territory, and the structural changes resulting from the exodus of manufacturing industries to southern China and elsewhere in the Pacific Rim region.

Relationship with other plans and studies

As already noted (see Editors' Overview for Part One), planning studies in Hong Kong proliferated in the 1980s. Many were undertaken in parallel or with a large degree of temporal overlap. The most important of these for Metroplan were:

- the Territorial Development Strategy (TDS) (1984) and its reviews (1986 and 1988) and supporting studies (see Chapter 3);
- the Harbour Reclamations and Urban Growth Study (SHRUG) (1983);
- the Second Comprehensive Transport Study (CTS II), (1989) (see Chapter 7);
- the Port and Airport Development Strategy (PADS) (1989).

Because these studies were so critical to the metropolitan development study, their salient points are discussed below at some length. In so doing, it is hoped to provide the reader with a better idea of the origins of many of the assumptions and concepts employed by Metroplan.

TDS (1984)

The change of focus for strategic development from the new towns back to the harbour was initiated by TDS. The rationale for this was that the New Towns Programme was anticipated to be virtually complete by the early 1990s - the then current forecasts of land supply available in the New Territories indicated a fall from 655 hectares in 1983/84 to only 65 hectares in 1991/92. While development of land during this period could support a population up to 6.2 million, it was concluded that further growth would have to be accommodated elsewhere (Lands and Works Branch 1985). Other changes had also made their mark. Most significant, was the realisation that the self-containment premise of new towns was not working and that mass transit had the ability to move large numbers of the population, thereby creating a potential for easing traffic congestion and increasing the accessibility to, and the employment capacity of, the central area.

In searching for new urban areas, five sub-regional studies were carried out as part of the TDS process (see Chapter 3). Of these, the study of Hong Kong's urban area (SHRUG) indicated that it had the potential for accommodating a far greater population than had formerly been assumed. In choosing between possible land development alternatives for this study, the overall aim was 'to ensure that the resources of the community are used to produce the best value for money' (Lands and Works Branch 1985) and the alternatives were tested against a number of measurable costs and benefits.

Two options for the direction of urban growth were produced by this process, one which shared growth between the north-west New Territories and the western harbour area, and the other which concentrated development around the western harbour area. A 'common components approach' was adopted by the study (meaning that those elements common to both options formed the basis of the recommended strategy) with the result that the focus of new development shifted to the harbour area. Drawing on the findings of SHRUG, TDS indicated a number of areas for strategic growth within the Metropolitan area (see Figure 8.2), namely:

- West Kowloon reclamation (for 90,000 residents);
- Green Island reclamation (for 200,000 residents);
- the extension of the Wan Chai reclamation (for 60,000 residents),
- Hung Hom Bay reclamation (for 17,000 residents) and for office use;
- Wah Fu vicinity (for 16,000 residents);
- Central reclamation (for office and cultural uses);
- a Western Harbour road and rail crossing;
- an expansion of the KCR terminal;
- an expansion of the container port at Kwai Chung;
- a third Tsing Yi crossing and a number of new major road links.

SHRUG Final Report (1983)

The new towns, though successful in meeting Hong Kong's housing needs, fell far short of fulfilling the objectives of 'self-contained' and 'balanced' communities

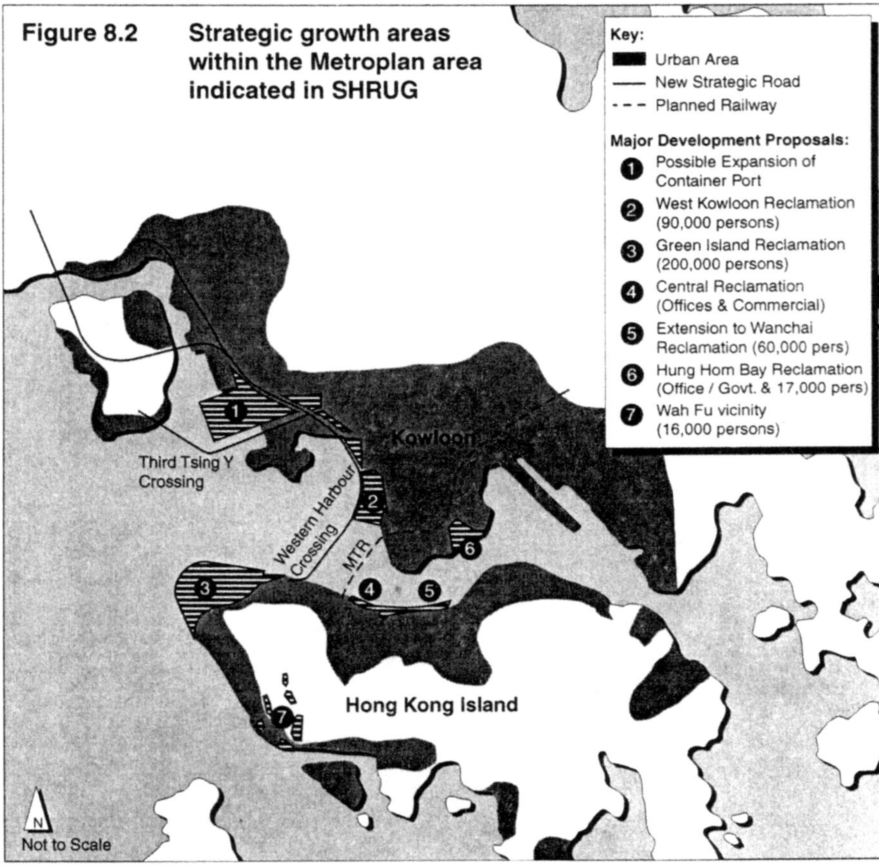

Figure 8.2 Strategic growth areas within the Metroplan area indicated in SHRUG

(Bristow 1989). They were, furthermore, public housing-led and jobs did not accompany the decentralisation of population with the result that commuting between these settlements and the metropolitan area became part of the Hong Kong daily work regime.

Industrial restructuring was also affecting the ability of the new towns to attract jobs, even though a major potential strength of the new towns was their ability to provide land for industry to enable the expansion of manufacturing from the crowded conditions of the existing urban area. However, the timing of this provision did not favour the internal territorial restructuring taking place. China's Open Door Policy had resulted in the relocation of a significant proportion of manufacturing industry, not to the New Territories, but across the border into southern China. This trend has been sustained to the present.

Hong Kong's economic base was thus rapidly changing from manufacturing-based to finance and services dominated, with the main generators of revenue centred on the harbour area. These developments provided the context in which SHRUG, one of the sub-regional studies of TDS, set out to investigate the development potential of areas, in terms of new sites, reclamation and redevelopment. The resultant proposals focused on the Kowloon Peninsula, Kowloon East, North-West Kowloon and part of Tsing-Yi, Hong Kong Island and Lamma Island (see Figure 8.3).

Although the site of Chek Lap Kok had been selected as the most appropriate location for a new airport, the decision on when (and indeed whether), the airport would move had not been made at this stage. Hence, two scenarios were addressed: one in which the airport remained at Kai Tak, the other with relocation to Chek Lap Kok.

SHRUG acknowledged there was scope for further development in the main urban area. However, as the introduction to the study points out, there was a certain polarisation of opinion on future directions of growth. Dissenters to the government view claimed that there was no scope for further reclamation and that the old urban areas were so congested that improved quality of life could only be attained by the

Figure 8.3 Land supply potential investigated by SHRUG

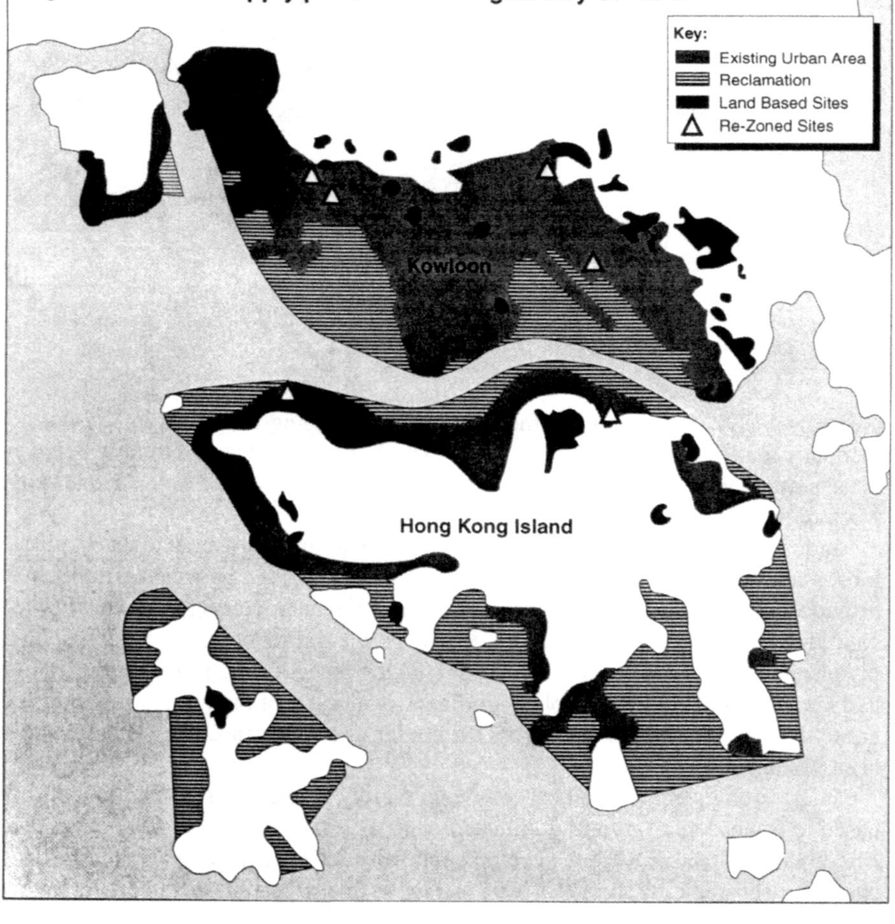

further decanting of population and jobs to the new towns. The main issues addressed by SHRUG were:

- population and housing, particularly overcrowding and inadequacies of the housing stock;
- the concentration of employment in the central area;
- transportation, in particular movement problems generated from the mismatch of jobs and residents, major concerns being the high demands for cross-harbour travel, the shortfall in capacity of public transport, and the overloading of the road network;
- declining 'quality of life', including a shortfall in community facilities and of open space; air, noise and water pollution;
- deterioration of landscape qualities, especially the hill areas of Hong Kong Island and Kowloon (particular concern was expressed regarding the importance of maintaining and enhancing the visual character of the harbour corridor);
- the capacity of the container port; and
- whether or not to relocate the airport.

In terms of the population capacity of the SHRUG area, the study concluded that 4.35 million was the practical maximum number that could be accommodated by the year 2001, assuming the relocation of the airport. Population to be accommodated on new land (provided by reclamation and the redevelopment of the Kai Tak airport site) was estimated at 1.15 million. To achieve this level of population absorption would also depend on the expansion of the MTR system to a network serving the SHRUG area, as well as linkage with the New Territories. Accordingly, a transport network of 146 kilometres of new roads (see Figure 8.4), 49 kilometres of new MTR lines and 39 new stations were recommended. Linkage of the south side of the island by MTR, and improved primary road access, were also proposed to integrate this area more closely with the central area.

A planning assumption of this study was that the new land (in terms of total area) would predominantly be used for public housing purposes, and that residential development by the private sector would be through redevelopment. The main revenue generating component of the proposals, for both the government and private sectors was assumed to be from office development on reclamation which would provide new land for extension of the CBD. It should be noted that at the time of drawing up the study, land revenues were estimated to exceed the proposal costs by some 30 per cent (including the costs of foundations, private utilities and MTR facilities). In recommending the extension of the MTR network, government policy was that the social needs for additional high speed/capacity urban travel must be matched by commercial viability, and that property developments could contribute in a major way to achieving such viability.

In respect to the ultimate population capacity, it was concluded that the study area had the potential to double its existing population to accommodate 11 million people, principally by means of a massive reclamation programme[2] which would provide land for accommodating 60 per cent of the population increase. However, in addition to development capability, there were a number of factors, predominantly environmental, which acted to reduce the area of lands considered suitable for urban development, and in turn, reduce the population capacity. These were:

- the airport (Kai Tak) noise exposure forecasts which restricted the amount of development on the north eastern side of the island as well as redevelopment in Kowloon;
- the harbour regime in terms of tidal flows and water quality which limited the amount of reclamation which could be carried out on the south side of the island;
- a landscape appraisal identifying adverse impacts on Country Parks if extensive reclamation were carried out on the south side of the island; and
- a geotechnical investigation which suggested that the use of Mount Davis as a borrow area would not be cost effective.

CTS 1 (1976) and CTS II (1989)

The First Comprehensive Transport Study (CTS I) investigated transport requirements from the base year of 1973 to 1991 (see preceding chapter). Recommendations for a mass transit rail supported previous proposals for a heavy rail system, comprising the Island and the Tsuen Wan Line, followed by the East Kowloon Line

Figure 8.4 Recommendations for the road transport networks

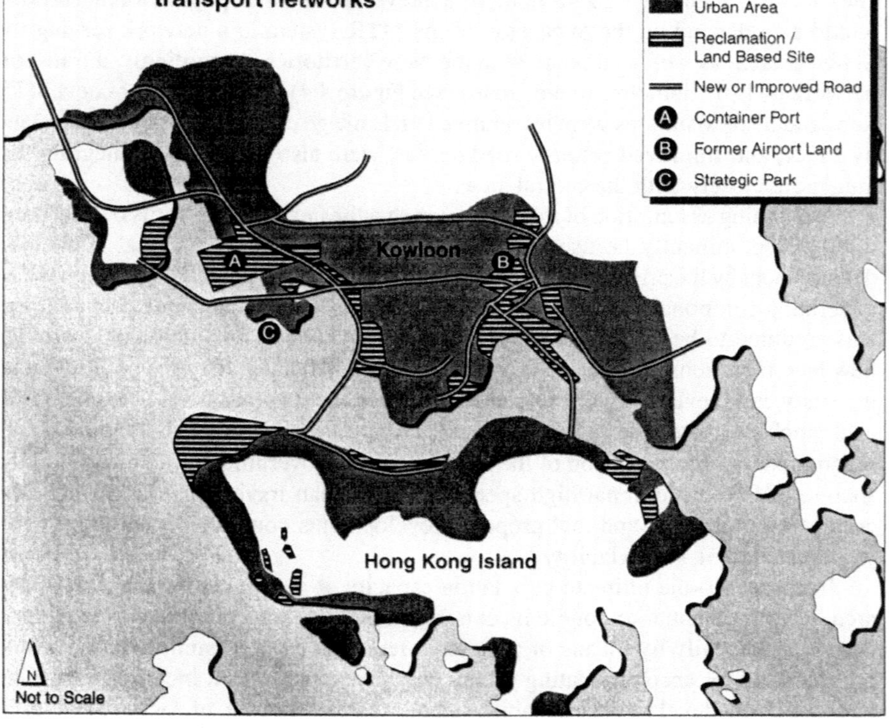

with provision for future extension further south on East Kowloon (see Figure 8.5). The study anticipated that the KCR would play a major role as a local transit service and recommended the double tracking and electrification of the KCR network.

Although CTS I examined transport needs in relation to three different land-use scenarios for the territory, namely, the dispersion of development to each of the Eastern New Territories and Western New Territories, and concentration in the existing urban area, in reality, its brief related to transport topics of concern rather than to the integration of transport and land use. Accordingly, although no specific recommendations concerning land use were made, the study concluded that 'the further concentration of population in the urban areas would lead to the greater utilisation of public transport, particularly the metro and the trams, and that policies to disperse population to the New Territories would result in overall higher levels of travel, particularly by private vehicles' (Wilbur Smith and Associates 1976).

The purpose of the Second Comprehensive Transport Study (CTS II) was to determine what had to be done to 'achieve and maintain an acceptable level of mobility for passengers and freight by road, rail and ferry up to the year 2001' (Transport Department and Wilbur Smith and Associates 1989). Essential to this study (as in the case of its predecessor) was a computerised model which was employed to

Figure 8.5 Recommendations for the rail network: Airport relocated

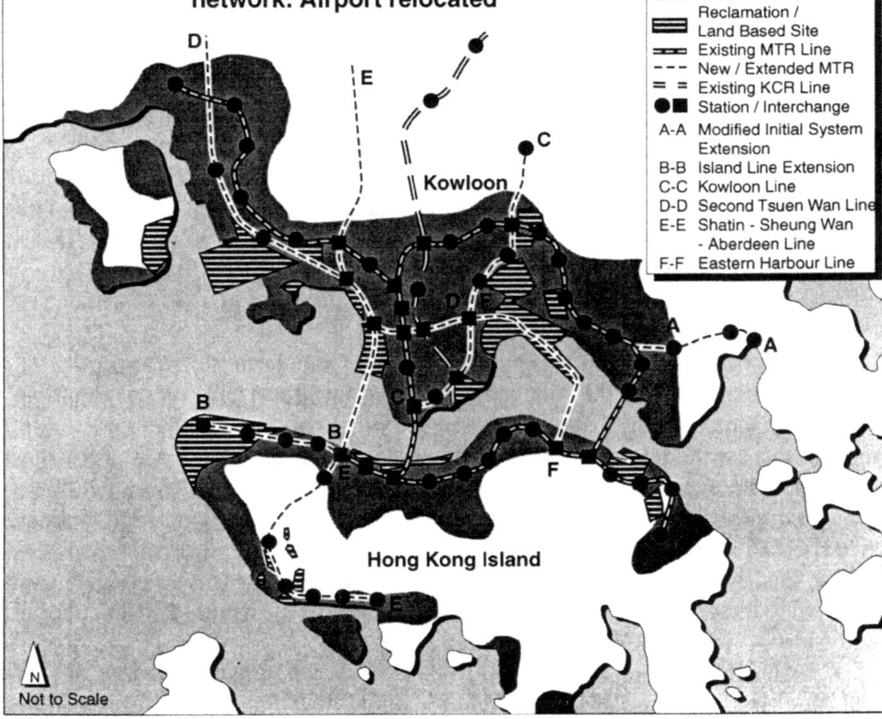

project traffic conditions using different assumptions in relation to population, employment, economic growth and international traffic (see Chapter 7). The study predated the decision to move the airport to Chek Lap Kok and only cursorily considered the implications of the relocation. In respect of road transport, recommendations were made for three new trunk highways (see Figure 8.4), namely:

- Route 3 (Hong Kong Island to Yuen Long via a Western Harbour Crossing);
- Route 7 (Causeway Bay to Aberdeen, utilising the proposed Central and Wan Chai reclamation); and
- an extension of the Hung Hom bypass (across the Hung Hom reclamation and under Kai Tak airport to the Kwun Tong bypass).

Recommendations in respect of rail transport were of two types: extensions to the new towns, and new links within the urban area. The former proposed rail links from Kowloon to Tseung Kwan O, and from Tsuen Wan to the northwest New Territories. The latter, from Diamond Hill in Kowloon to Sheung Wan (the East Kowloon Line) with further extensions from Diamond Hill to Sha Tin and from Sheung Wan to Aberdee (see Figure 8.5). The study also recommended a new rail harbour crossing, from Yau Ma Tei to Fortress Hill, and a new KCR/MTR interchange at Hung Hom. Virtually no attention was paid to the role of buses and prime bus routes. An implicit assumption was that the private sector would meet the demand for such services and that planning to accommodate or facilitate these services and routes was therefore not required!

PADS (1989)

Work on the formulation of the Port and Airport Development Strategy (PADS) commenced in 1987. The strategy was released and published in October 1989 and led to the ten PADS Core Projects currently under construction[3]. The PADS Study drew on a number of other studies, including those for the airport and port related projects, and various strategic planning studies (e.g. the Hong Kong Mass Transit Study of 1967, the Colony Outline Plan: Book 3 of 1972, the SHRUG Study of 1983, CTS I of 1976, Working Papers for CTS II prepared in 1987; and the TDS initial results of 1984).

The aim of PADS was to 'produce a medium to long term [infrastructure] strategy which would provide the context for decisions on airport and port development' (Mott MacDonald Hong Kong Limited et al 1990). This strategy had far reaching implications for the future structure of Hong Kong, not only of the port and airport but of the whole territory, and particularly the metropolitan area. These two major infrastructure developments, located immediately west of the hub of Hong Kong, gave rise to a number of linked and complementary projects concerned with accessibility, land formation, industrial and residential development. The projects were designed to integrate the port and airport with the rest of the territory, and in particular the airport with the CBD.

The decision to remove the airport from Kai Tak, left both a horizontal and vertical 'gap' in the urban fabric. Potential for future development was provided not only by the vacated site, but also through the release of the encumberment of a limitation on building heights set by the existing flight path. The removal of the airport would also result in many associated land uses and functions, from tourist hotels to freight related operations, being relocated over time.

Critical to understanding Metroplan is the fact that its recommendations were to a large extent dove-tailed with those of PADS as these studies were contemporaneous (PADS 1987 through to 1989, and Metroplan from 1988 through to 1990). Given that two scenarios were investigated by PADS, Metroplan initially also investigated the restructuring of the metropolitan area under two scenarios: one which assumed the retention of the airport at Kai Tak, and the other which assumed the relocation of the airport from Kai Tak to Chek Lap Kok.

Metroplan study process and framework

Role of the study

Metroplan was formulated as a citywide context for decision-making at the district level. Great pains were, however, taken by the Planning Department to explain that it was neither a 'blueprint' nor a programme of public works even though certain projects, particularly those related to PADS, are given a high profile by the study. The Planning Department also stressed that Metroplan was not a panacea for all Hong Kong's urban problems. The plan, in particular, sought to raise the environmental quality of the metropolitan area, meet and integrate all projects and programmes that had been agreed and committed earlier, and yet still offer flexibility to respond to changing circumstances through a proposed monitoring system and consultation exercises.

Study process

The broad study process employed was based on a 'continued refinement of land-use-transport options, using agreed objectives and evaluation criteria as performance yardsticks'. Constant interaction between the transport and land-use and economic streams of these options were claimed to be the hallmark of the process. There were four stages in this plan-making exercise (see Figure 8.6), namely:

- Initial Option Development - involving the generation, testing and evaluation of different 'solutions' for restructuring the metropolitan area.
- Hybrid Option Development - taking forward the most essential and desirable elements of the Initial Options for further testing and evaluation.
- Preferred Options Development and Strategy Selection - based on the findings of the preceding stage, the Preferred Options provided the opportunity for the refinement of the proposed land-use and transport framework which in turn led to the selection of a strategy.
- Planning and Development Guidelines Formulation - conducted in parallel with the previous stage, planning and development guidelines were formulated to

Figure 8.6 Stages of the Metroplan plan-making process

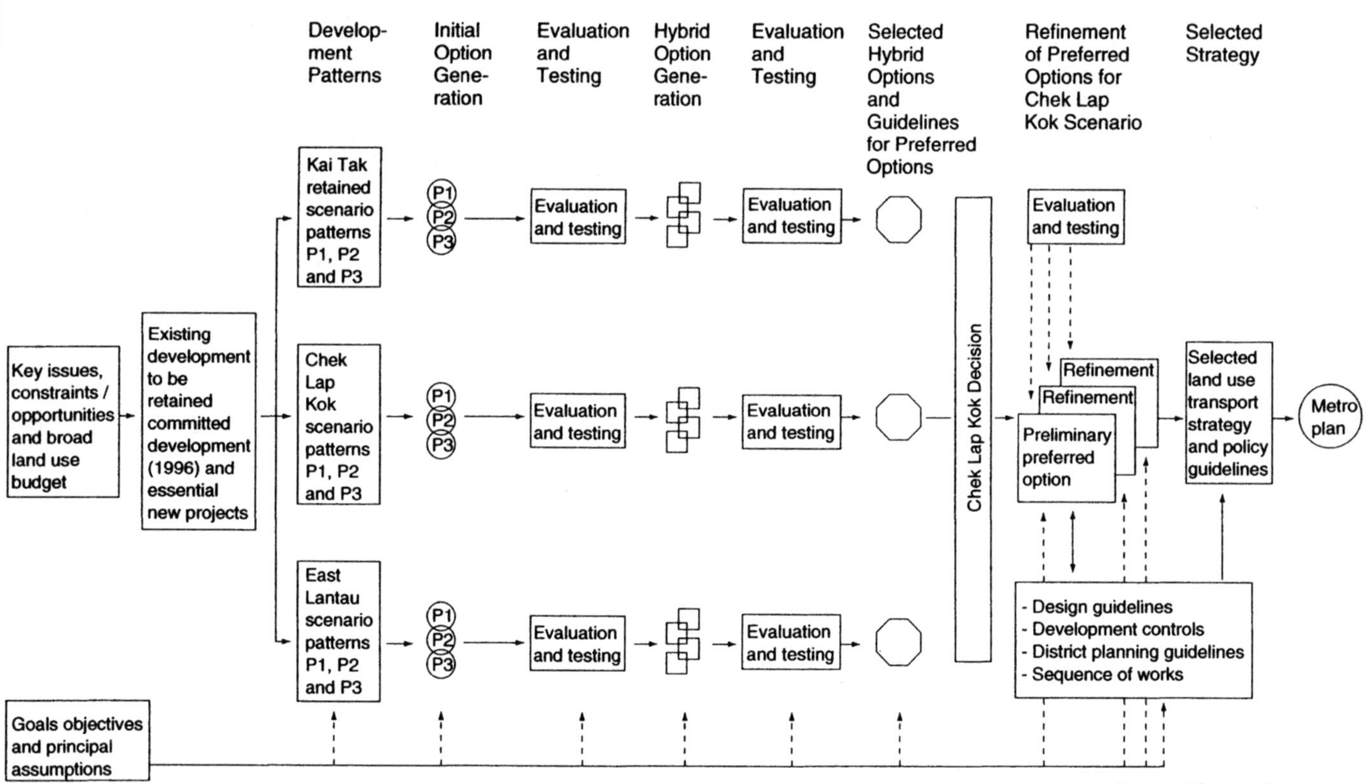

Source: Planning Department

facilitate the translation of the selected option and strategy into action at the district level.

Principal issues

The major issues Metroplan sought to address may be divided into three broad categories:

- Macro Key Issues - These include co-ordinating the development of old and new areas; limiting the overall metropolitan population to 4.2 million; responding adequately to forecasts of structural changes in the economy (and their locational needs); resolving the locational needs and impacts of the commercial office sector; developing a coherent and efficient transport network capable of meeting increased travel demands in an environmentally appropriate matter; accommodating changes in the level and pattern of housing demand; and responding to air, noise and water pollution problems.
- District Key Issues - These typically represent the physical manifestation of the macro key issues at the district level.
- Land-Use Budget Key Issues - These are closely related to sectoral land-use needs and requirements of the strategy arrived at to tackle the aforementioned key issues.

Development opportunities

The potential 'solution spaces' to address existing and forecast needs, and demands of the plan, were identified by Metroplan as being made available from:

- approximately 1250 ha of new reclamation areas;
- approximately 740 ha of new land based sites;
- approximately 300 ha of Housing Authority and Housing Society redevelopment schemes;
- approximately 380 ha of under-utilised/inappropriately located government uses;
- approximately 450 ha of obsolete private housing areas;
- sites and areas where a change or intensification of use might be permitted; and
- 4,400 ha of urban fringe areas with potential for recreation use.

Since the development opportunities cited above are all contained within the Preferred Options and Selected Strategy, which in turn are based on a detailed examination of the different land use and transport configurations generated by earlier studies and other principal components, the following discussion offers a more in-depth coverage of these fields to better appreciate the underlying assumptions and components of Metroplan.

Land-use/transport framework

It should be noted that the transport framework for the testing of initial options was based on proposals from CTS II (up to the year 2001) and from PADS, TDS and

SHRUG. As a starting point for the study, the network of highways and rail links up to 2001 were adopted from the CTS II proposal and those from PADS, TDS and SHRUG for the period to 2011.

The means of technically establishing the relationship between transport and land use for Metroplan, as in the case of TDS, was through a set of conventional transport modelling techniques, under a guiding framework of formally expressed objectives relating to transport (see Table 8.2).

Table 8.2 The transport objective

Transport Objective: The provision of an efficient, financially viable and integrated multi-modal transport system that also makes good use of human and energy resources.	
Sub-objective	**Measurement criteria**
• To minimise the need for road travel	Total road travel distance (in PCU-kms)
• To maximise the operational efficiency of the road network	% of total road travel distance on roads with $v/c \geq 1$
• To promote public transport	% of total person trips by public transport
• To encourage rail travel	% of public transport person trips by rail
• To minimise the land-take from the community for transport infrastructure consistent with serving the social and economic demands of the community	not quantifiable
• To minimise the need for multi-mode changes	"
• To maximise the efficiency of mode changes	"
• To promote off-street parking and servicing to meet demand	"
• To promote walking as a mode of transport	"

Source: Planning Department, Transport Studies Unit (1990)

The main transport objective which was adopted for Metroplan is 'to provide a multi-choice, high capacity transport system that is financially viable, energy efficient and makes provision for the safe and convenient movement of pedestrians' (Planning Department 1990b). This is supported by the sub-objectives listed in Table 8.2.

The plan generation process of Metroplan used a series of iterations for determining and quantifying land-use distributions, within a scenario-building framework. Initial Options were postulated in a manner intended to represent a series of distinctly different directions which the socio-economic development of Hong Kong might take (see Figure 8.7).

The main purpose of these options was to identify advantageous and disadvantageous components from the point of view of transport dynamics and from these to develop hybrid options, which again would be subjected to an evaluation process. The process then reduced these options to three (based on their likelihood and desirability) from which a final selection was made after further evaluation. The main differences, in relation to land uses, between Options 1, 2 and 3 are indicated in Table 8.3. Common components of the three final options, on a sectoral basis, were as follows:

- Industry: New land for export-related industries to be provided in proximity to the Kwai Chung container port and expanded West Harbour facilities.
- Housing: Obsolete public housing estates to be redeveloped in order to produce a more diverse mix of public and private housing. Private obsolete housing to be

Table 8.3 The Metroplan options for final evaluation

Option 1	Option 2	Option 3
Industry		
Redevelopment and relocation of industrial uses within the metropolitan area	Limited relocation of factories to new towns	Major relocation of factories to the new towns
Dispersed pattern of new low-medium density estates for light industry	Limited number of major new light industry estates	
Business and commercial activities		
Expansion mainly around established business centres on HK Island and E Kowloon	Limited expansion around existing centres; major expansion of secondary office centres on MTR axis of Kowloon Peninsula.	Limited expansion around existing centres; major development of satellite business centres at transport nodes.
Decentralise major government office functions within the metropolitan area at key transport nodes	Centralise major government office functions.	
Housing - no difference between options		
Port-related activities - no difference between options		
Transport related development - no difference between options		
Open space and landscape		
Provide large number of small interlinked open spaces.	Provide small number of medium-large interlinked open spaces.	Compromise between Options 1 and 2.
Use of Kai Tak site		
Develop part for low-medium density light industrial and related facilities.	Develop most for low-medium density light industry and related facilities.	Develop most for mixed residential, community facilities and district open space.
Removal of Airport Height Controls		
Height restrictions lifted to fullest extent possible.	Height restrictions lifted mainly around transport nodes.	Height restrictions retained.

Source: Compiled from Planning Department, 1990d

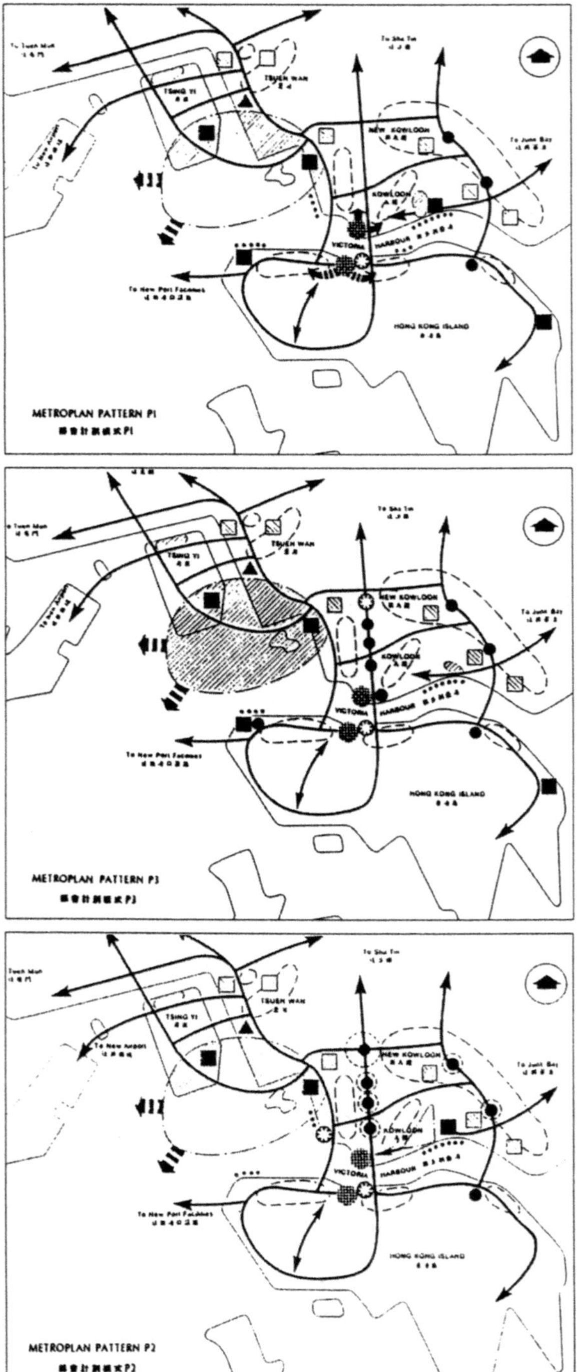

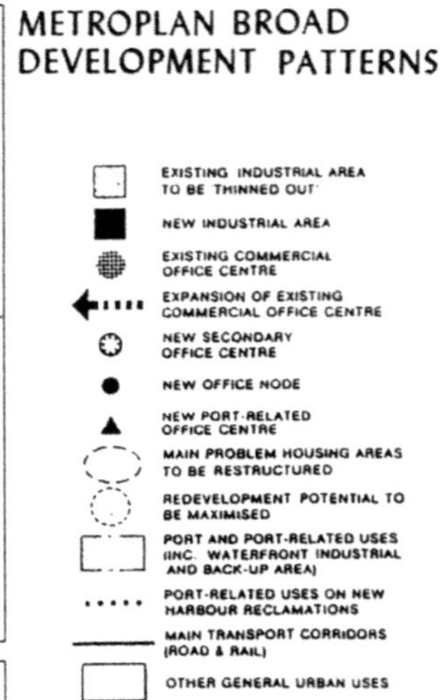

Figure 8.7
Metroplan broad development patterns
Source: Planning Department

up-graded as far as possible through LDC, and new housing sites to be provided to create a more diverse housing mix within each broad district.
- Port-related activities: To comply with the PADS proposals.
- Mass transit transport-related activities: Development to be concentrated within walking catchment of mass transit stations; nodes of major economic activities to be developed along Route 3, and nodes of new activities along Route 6.
- Open space and landscape: Efforts to be made to rehabilitate, develop and manage suitable urban fringe areas for public recreation. Hong Kong Island South and the coastline of Tsuen Wan West to be conserved for low intensity recreation.

Selected strategy and recommendations

Strategy components

The principal components of the selected Metroplan strategy are a broad land-use/ transport plan (see Table 8.4), a written statement covering the land-use distribution and transport framework, an open space framework for recreation and a landscape strategy for urban fringe and coastal areas. The following discussion concentrates on the land-use/transport framework.

Broad aims

The broad aims of Metroplan are shown in Table 8.5 (these are further subdivided into objectives but are not shown in the table for reasons of brevity).

Proposals

In summary, the basic proposals of Metroplan are:
- Area: The metropolitan area is to grow in size from 6,500 ha to about 8,600 ha, excluding port facilities, cemeteries and urban fringe parks.
- Time Frame: Plans up to 1996 are to remain unchanged, with medium and long term strategic development programmes up to 2001 and 2011 taking over the direction of future development thereafter.
- Population: The population of the metropolitan area is to be kept at approximately 4.2 million inhabitants; population growth is to be accommodated by the New Towns Programme and further development in the New Territories.
- Land-Use and Planning Standards: The predominant land use is to be housing, planned and developed increasingly to new towns standards 'wherever practical' and in accordance with the Hong Kong Planning Standards and Guidelines.
- Housing: New life will be injected to old areas through the redevelopment of obsolete housing, jointly pursued and financed on commercially prudent principles by both quasi public agencies and the private sector.
- Growth Areas: Principal areas of growth are to be on major reclamations, and the redevelopment of the Kai Tak site.

Table 8.4 Principal components of the Metroplan strategy

<table>
<tr><th colspan="2">Land-Use Framework</th><th>Transport Framework</th></tr>
<tr><td>Housing</td><td>approx. 530 ha gross</td><td rowspan="2">2001 Scenario
• new cross Kowloon route and improvements
• extension of Route 7 to Kennedy Town
• new rail link between Sha Tin and Diamond Hill</td></tr>
<tr><td>Commercial Office
• existing centres
• secondary centres
• office nodes</td><td>approx. 87 ha. gross</td></tr>
<tr><td>Hotels</td><td>approx. 58 ha. gross</td><td rowspan="3">2006 Scenario
• cross Kowloon route to Tseung Kwan O access
• Route 16 between Sha Tin and West Kowloon
• rail link from Diamond Hill to West Kowloon Reclamation
• extension Tseung Kwan O rail to Lam Tin</td></tr>
<tr><td>Industry</td><td>approx. 93 ha. gross</td></tr>
<tr><td>Port and Port-related Facilities
• container terminals
• back-up areas
• multi-purpose berths</td><td></td></tr>
<tr><td>Other Uses</td><td>approx. 149 ha</td><td rowspan="2">2011 Scenario
• extension of Tseung Kwan O rail from Lam Tin to Kowloon Bay Reclamation
• extension of MTR from Sheung Wan to Green Island</td></tr>
<tr><td>Recreational open space</td><td>total provision about 17 ha per 100,000 pers.</td></tr>
</table>

Source: *Compiled from Planning Department (1991b)*

Table 8.5 Principal aims of Metroplan

Overall Goal:

To bring about a better organised, more efficient and more desirable place in which to live and work.

Broad Aims:

- Enhance Hong Kong's role as an international port and airport; as an international business, finance and tourist centre; as a centre for a diverse range of light manufacturing industry and as the centre of Government for the territory.
- Provide opportunities to satisfy, as far as practicable, housing needs according to what people can afford and where they would like to live.
- Achieve a more balanced distribution of jobs relative to population concentrations, the location preferences of new enterprises and the ease of travel.
- Sort out land-use patterns to create a more acceptable urban environment
- Create an urban form that will foster a sense of community identity.
- Conserve and enhance major landscape attributes and important heritage features.
- Provide a multi-choice, high capacity transport system that is financially and economically viable, energy efficient and makes provision for the safe and convenient movement of pedestrians.
- Produce a strategy that can be carried out by both the public and private sectors under variable circumstances, particularly with respect to the availability of resources and significant changes of demand.

Source: *Planning Department, 1990d.*

- Industry and Employment: Manufacturing industry is to be relocated to the new towns and areas close to the new airport and West Harbour port facilities. This is to take place against a backcloth of a continued decline of manufacturing industry and a rapid growth of tertiary services.
- Transport: The recommendations of PADS and CTS II are to provide the main basis for the transport component of the strategy employed.
- Finance: Public expenditures generated by the strategy (as in the case of PADS) are not to increase at a greater rate than GDP growth, while revenues from reclamation projects and the redevelopment of Kai Tak are assumed to more than cover the costs of the planned growth for these areas. All new rail investments are to be provided on a commercially viable basis.
- Institutional: Current institutional mechanisms/frameworks are to be used for implementing Metroplan, with new means being introduced where appropriate.

Basic principles

Eight basic principles were claimed to be employed in the formulation of the selected strategy, namely:

- The 'thinning out' of existing congested urban areas in tandem with the planning and development of new urban areas.
- The recognition of practical limits to urban growth in the metropolitan area.
- The location, scale and timing of new commercial office developments in a manner that recognises the demands created by market forces and the needs of financially and economically viable efficient transport systems.
- The movement of the industrial base of the territory towards automated, capital intensive activities that will decentralise manufacturing activities.
- The joint planning of land use and transport.
- The increased dependence on public transport.
- The rectification of current environmental problems and avoidance of the generation of new.
- The fullest use of current agencies and institutional mechanisms.

Based on the premise that there are 'good opportunities in the Metro area to create new city forms which can at least equal those achieved in other leading world cities' (Planning Department 1990b), the above principles were complemented by the adoption of the following design principles:

- The provision of a good mix of compatible land uses.
- The provision of selected public spaces at focal points.
- The concentration of relatively high density developments around high capacity transport interchanges, mainly MTR stations.
- Within each neighbourhood, provision for high capacity pedestrian system.
- The creation and design of certain building and landscape landmarks.
- The design of building forms and spaces in a manner that offers a 'coherent and visually exciting environment'.

Urban restructuring

On the basis that new reclamations are seen by Metroplan as providing important 'solution spaces' for the additional land needed to help restructure metropolitan Hong Kong, four strategies were advanced for restructuring land use (see Figure 8.8):

- Strategy 1: In the case of a congested area (A) adjacent to a new development area (B), to provide community facilities and open space together with residential use to new towns standards.
- Strategy 2: In the case of a congested area (A) adjacent to a new development area (B), to provide community facilities and open space together with limited residential use so that the total density (in A+B) is of an acceptable level.
- Strategy 3: To upgrade area (A) by providing community facilities and placing density limits on redevelopment schemes.
- Strategy 4: To comprehensively redevelop area (A) to new town standards (as in the case of new development area (B)).

All the above responses require the existence of adjacent new development areas, which Metroplan refers to as 'solution spaces' which are principally to be provided through reclamation.

Land-use/transport configuration

The conceptual framework of the selected transport structure for Metroplan is shown in Figure 8.9. The latter illustrates the progressive development of the principal road and rail networks proposed for the years 2001 and 2011. The transport network for 2001 is based on the TDS Review of 1987 and recommendations from CTS II and PADS, plus the findings of the harbour reclamation studies.

The most significant features of the selected Metroplan land-use/transport configuration are perhaps best summarised by a direct quote from the pertinent government report where it is explained that the metropolitan area:

> ..... *will be served by a high capacity, multi-mode transport system with high density development concentrated around major interchanges. The highway system will be expanded considerably. There will be three tunnels across the harbour ..... forming part of an East-West, North-South framework of major roads that join up with the wider territorial system providing links to/from the new airport facilities ..... and the wider hinterland of the Pearl River Delta. Within that framework will be a metropolitan network of local roads and more extensive, grade separated pedestrian systems linked to major transport interchanges. Passenger rail systems will penetrate areas characterised by high concentrations of housing and jobs with supplementary public transport services being provided by buses.* (Planning Department 1991a)

Light rail passenger systems may be provided in certain areas, including schemes on some reclaimed areas. Provision will also be made for commercially viable cross-harbour and outlying district ferry services, as well as ferry services to China. The

Figure 8.8 Approaches to restructuring the city

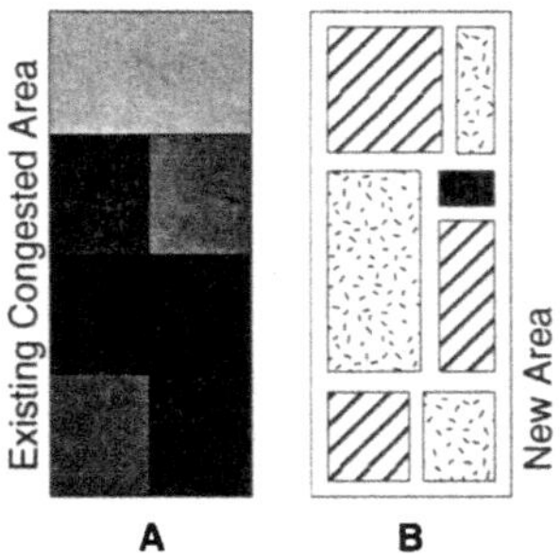

Scenario 1

- Needs of area A for additional community uses provided mainly in area B
- Density in area A stays high
- Overall density for areas A & B is moderate
- Public sector and social costs are high
- General urban environment and layout are poor

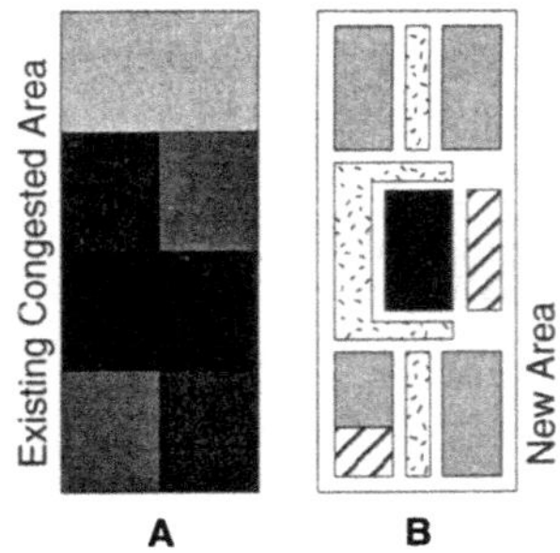

Scenario 2

- Minimal improvements made in area A where densities stay high
- Area B developed for new housing and associated community facilities to new town standards
- Overall density for areas A & B increases
- Public sector revenues from land sales are high
- Social costs stay high for area A
- General urban environment better in area B

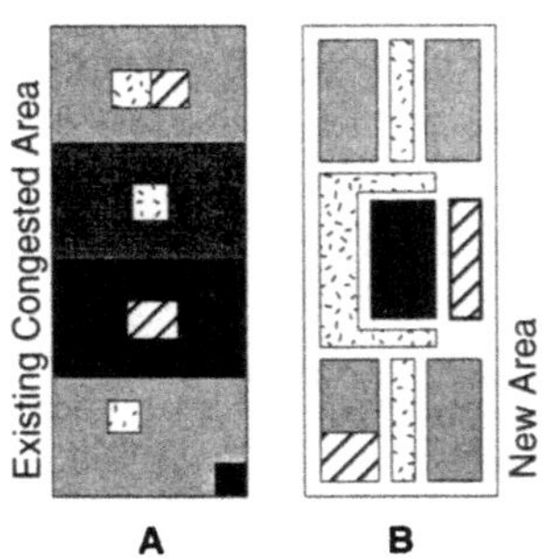

Scenario 3

- Area A upgraded by providing community facilities
- Area B developed as for Scenario 2
- Overall densities are moderate
- Public sector revenues and costs in better balance
- Overall urban environment and layout improved

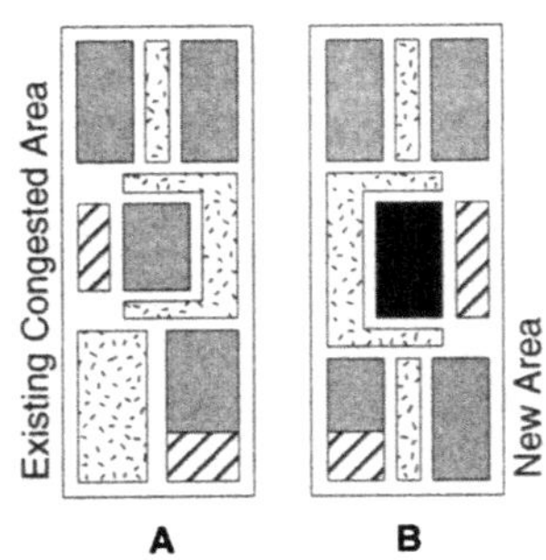

Scenario 4

- Area A redeveloped to new town standards
- Area B developed as for Scenario 2
- Overall densities are moderate
- Public sector costs are high and public revenues are moderate
- Overall urban environment and layout greatly improved

Source: Compiled from Planning Department

Figure 8.9 Conceptual framework for the selected transport structure for Metroplan: progressive development of the principal road and rail networks for 2001 and 2011

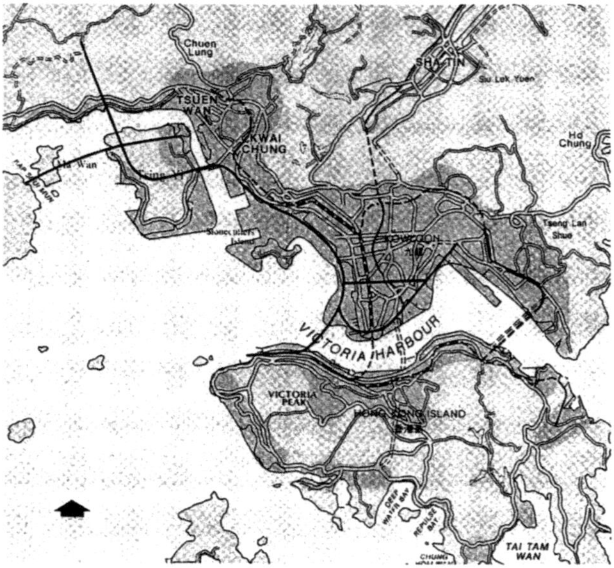

Road development, 2001

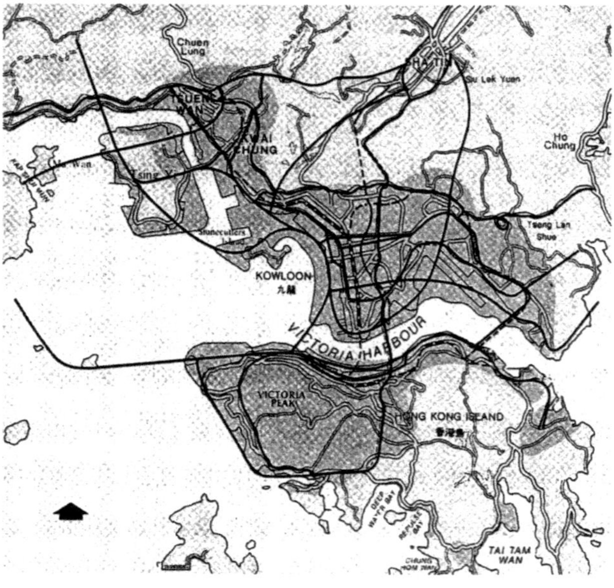

Road development, 2011

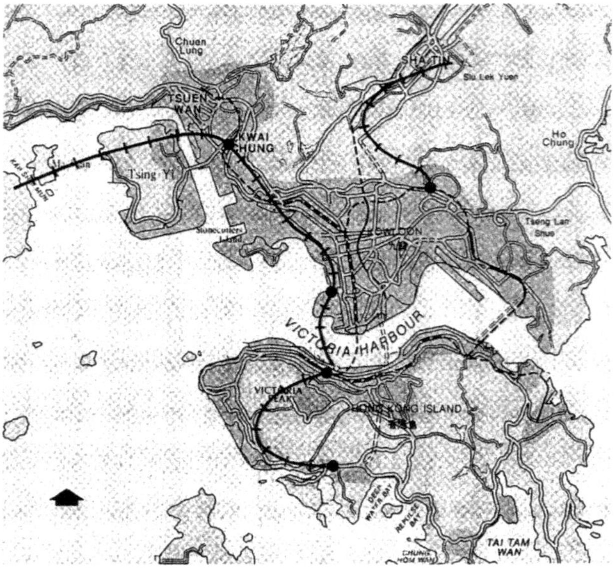

Rail development, 2001

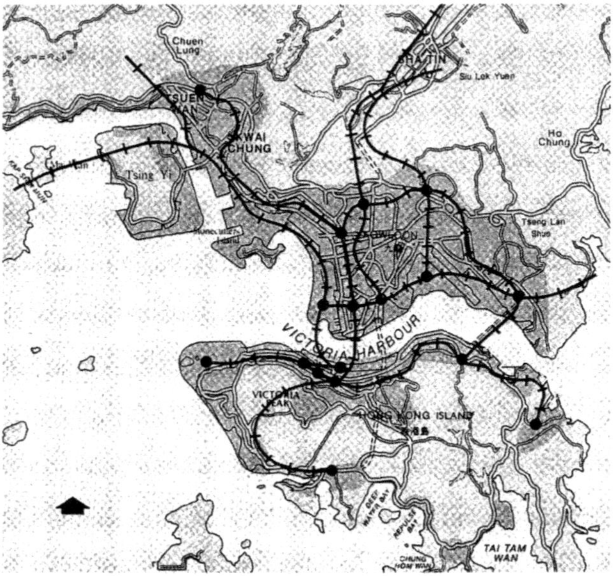

Rail development, 2011

Source: Planning Department

working assumptions of the above include:

- GDP in the year 2001 will have grown to 2.38 times that of 1986, and in 2006 by 3.03 that of 1986, representing in both cases an approximate growth rate of five per cent per annum[4];
- private car numbers will also have grown by five per cent per annum, compound[5]; and
- goods vehicles will grow in line with GDP but restrictions on vehicular activity within the metropolitan area will limit it to the 1986 level[6].

Implementation

In terms of Metroplan implementation, the programme of core projects for PADS virtually directs the sequence of proposed works. The two major elements, transport infrastructure and reclamation, govern the restructuring process.

The institutional support for the execution of Metroplan relies very much on already introduced policies, the development and co-operation of the private sector (particularly construction, real estate, transport and banking interests), and the government's resource allocation system. The institutional framework within which Metroplan is to be implemented is shown in Figure 8.10. While a number of unresolved issues remained at the time of the strategy's completion (e.g. what to do about obsolete industrial areas and how to deal with urban renewal and industrial development problems without an agreed policy for each), it was considered that the Land Development Policy Committee (LDPC) of the Hong Kong government (supported by the Town Planning Board at the district level) could provide the necessary executive leadership required.

In the Hong Kong hierarchy of plans, Metroplan is a second level plan (under TDS). The translation of the Metroplan strategic concepts into Development Statements for providing more detailed guidance on how geographic components or districts within the framework of Metroplan will be developed is critical. In theory, these statements are in turn further elaborated on by Outline Development Plans and Outline Zoning Plans (OZP). The only plan which has statutory power is the OZP which is a development control plan, rather than an implementation plan to achieve a vision. OZPs tend to reflect and perpetuate current land uses rather than being forward looking or enabling change.

Consultation and review

A public consultation exercise was embarked upon in early 1990 to seek responses on the preliminary proposals of Metroplan. Two documents for public reference were published: one on the Initial Study Options (Planning, Environment and Lands Branch 1990b), the other on the Study Foundations and Framework (Planning Department, 1990d). In addition to these, various committees and professional, as well as interested bodies, were briefed. A number of seminars for selected interest groups were also conducted.

According to government records, 'strong support for Metroplan as a framework for restructuring the city' was expressed in the written and verbal returns received, with a number of concerns expressed, including, among other things:

- the impact of relaxing airport height controls at Kai Tak after the removal of the existing airport;
- the long term needs for marine facilities and re-provisioning of affected shipyards;
- the integration of different rail projects and financing of rail extensions to new development areas; and
- the integration of the Metroplan strategy for office growth and redistribution of population and manufacturing jobs.

Recognising that the formulation and development of a metropolitan development strategy is a complex dynamic exercise, and inevitably subject to a host of uncertainties and changes, the Planning Department saw a need for monitoring the plan's progress and implementation. The Planning Department therefore set up a Metroplan Monitoring Unit to monitor trends and make periodic adjustment (especially to the land-use/transport components), and to also incorporate the outcomes of decisions at the district planning level. These observations and findings are to be presented in quinquennial reviews.

Critical review of Metroplan

The following comments are drawn from a variety of sources. In particular, these include the authors' critical analysis of Metroplan and its planning process, as well as other government documentation observations and impressions from one of the two author's participation in the Metroplan Steering Committee[7]. An article written by Victor Sit of the University of Hong Kong (Sit 1989) is also drawn upon, as are notes prepared for this chapter by a Planning Officer of the Strategic Planning Unit in the Planning Department (Robb 1996). Due acknowledgements have been cited.

Ultimately, the outcome of all planning exercises needs to be judged against what they set out to achieve, the principles and assumptions employed, the opportunities and constraints encountered, and the resources available with which to pursue the aims. There is also the added factor of whether the plan has experienced sufficient passage of time to be fairly judged.

Innovative moves

Whatever the limitations of Metroplan, even its strongest critics acknowledge that it marked the beginning of a more open planning procedure in Hong Kong. This is the case not withstanding the fact that many, indeed most, of the studies upon which it is based, were formulated behind 'locked doors' (Sit 1989). Metroplan also made the first real attempt to bring to the forefront a number of qualitative parameters that were so badly ignored in earlier planning studies. These particularly concern land-

Figure 8.10 Institutional framework for implementing Metroplan

A Main Targets	**Target 1** The level of population in the Metro Area should not exceed 4.2 million	**Target 2** The type, scale, location and timing of land uses should be coordinated with the provision of transport and other services to minimise travel/ service demands	**Target 3** The type, capacity, routing and timing of transport infrastructure and other services should be p coordinated with development projects to minimise the need for land and optimise investment opportunities	**Target 4** The Metro Area should be replanned to achieve, wherever practical, new town standards	**Target 5** Future planning and development in the Metro Area should aim to redistribute and regulate land uses and infrastructure systems to help create a better place in which to live and work.
B Policy Foundation	a. Metroplan aims & objectives b. Territorial Development Strategy	Metroplan aims & objectives	White Paper on Transport Policy	a Metroplan aims & objectives b. Hong Kong Planning Standards & Guidelines	a. White Paper on Pollution b. Metroplan Landscape Strategy for the Urban Fringe and Coastal Areas
C Principal Policy Agents	a. Executive council b. Land Development Policy Commitee c. Housing Authority d. Development Progress Committee	a. Land Development Policy Committee b. Transport Policy Coordination Committee c. Development Progress Committee	a. Executive Council b. Land Development Policy Committee c. Transport Policy Coordination Committee d. Transport Advisory Committee	Land Development Policy Committee	a. Land Development Policy Committee b. Country Parks Board c. Environmental Protection Advisory Committee d. Antiquities Advisory Board
D Tactical Plans and Programmes	a. New Town Development Programmes b. Housing Authority's Long Term Housing Programme c. Land Authority's Land Sales Programme	As in D1	a. Transport Infrastructure Development Programme b. Review of Comprehensive Transport Study c. Railway Development Study	a. District level plans & programmes b. Redevelop-ment & new housing programmes of Housing Authority, Housing Society, LDC c. Major private developers	a. Sewage Strategy b. Waste Disposal Plan c. Country Parks Development Plans and Programmes
E Principal Resource Allocation Agents	a. Government's resources allocation system b. Budgetary processes of Housing Authority, Housing Society, Land Development Corporation (LDC)	As in E1	a. Government's resources allocation system b. Budgetary process and investment programmes of transport operators and statutory undertakers	a. As in E1 b. Budgetary processes of Urban and Regional Councils	Government's resource allocation system

F Principal Executive Agents	a. Planning Department b. Building & Lands Department c. Housing Authority & Housing Society d. Territory Development Department (TDD) e. Land Development Corporation f. Private developers g. Town Planning Board	As in F1	a. Highways Department & TDD b. Transport Department c. Railways - MTRC / KCRC d. Other transport operators e. Water Supplies Department f. Drainage Services Department g. Statutory undertakers for gas & electricity, telephone & other tele-communication services	a. As in F1 b. Urban Council c. Regional Council	a. Environmental Protection Department b. Country Parks Authority (Director of Agriculture & Fisheries) c. Marine Department d. Civil Engineering Services Department (Geotechnical Control Office) e. Electrical & Mechanical Services Department (Gas Standards Office) f. Drainage Services Department g. Antiquities Authority (Secretary
G Main Mechanisms /Powers Available to Executive Agents	a. Territorial, sub-regional and district level development plans; density zoning controls b. Land resumptions; land sales/ allocations; lease enforcement; building demolitions/repairs c. Design, construction, management of new estates; redevelopment of obsolete estates; squatter control and clearance; housing allocation and affordability rules d. Public Works Programme; New Town Development Programmes; e. Redevelopment of obsolete private properties under Land Development Ordinance backed up by Crown Land Resumption Ordinance f. Development of new sites made available via (b) g. Preparation of Outline Zoning Plans and coordination of development proposals under the Town Planning Ordinance	Generally as in G1	a. Planning, design, construct-ion, maintenance of highways & road related structures b. Planning & implementation of new transport services & systems; formulation & implementation of traffic management schemes; regulation of public transport services c. Planning, design, financing, construction and operation of rail systems d. Provision and operation of franchised public transport services e. Design, construction, maintenance of water services f. Design, construction, maintenance of drainage services g. Design, construction, maintenance of other utilities services	a. Generally as in G1 b. Works Programmes of Urban Council c. Works Programmes of Regional Council	a. Air Pollution Control Ordinance; Water Pollution Ctrl. Ordinance; Waste Disposal Ordinance; Noise Control Ordinance; promulgation of Environmental Impact Assessment in context of development studies b. Country Parks Ordinance; control of mariculture c. Regulation of shipping and other marine activities d. Regulation of quarrying and rehabilitation / stabilisation of slopes; control of public dumping and extraction of marine fill; promotion of use of underground space e. Control / abatement of potentially hazardous gas storage installations f. Provision and operation of stormwater and foul water drainage and treatment systems g. Listing of places and buildings under the Antiquities and Monuments Ordinance

Source: Planning Department

scape and townscape qualities, open space and urban design considerations, the role of railways and the poor quality of life generated by high density living and fast economic growth. In its own way, it contributed to the foundations for the current belated concern in government and some private sector circles about sustainability[8]. These innovative moves in part account for the plan's limited adoption to date, in that it presented proposals that conflicted with certain interests, for example Metroplan's density guidelines have met with opposition from a number of quarters, including the real estate industry.

Especially impressive in the Metroplan process was the management of the exercise. Bringing together numerous different government departments, often clearly with very different interests and prejudices, and arriving at a negotiated strategy is an achievement by any measure. It was made possible not only by the skilled direction of the then Secretary for Planning, Environment and Lands (Graham Barnes), but also by the highly competent leadership provided by the first Director of Planning (Robert Upton), the determination of a Principal Government Town Planner (Ted Pryor), and a highly committed team of government planners and consultants.

Reinforcing existing trends and integrating infrastructure commitments

While the Planning Department went to great lengths to claim that Metroplan was not a panacea for all Hong Kong's urban problems, nor a blueprint for action, its critical dependence on the port and airport projects, which were given priority in the schedule of works, makes it difficult to distinguish the difference between the Metroplan Strategy and a blueprint for public works action around which Metroplan was based[9].

In retrospect, it may be concluded that the political price the Planning Department had to pay for its innovative plan to be accepted was an unchallenged acceptance of all previously agreed and committed programmes, particularly PADS. The following discussion suggests that this price has indeed been high and perhaps will become costlier as sustainability considerations are increasingly taken into account (see Chapter 12). Whether this cost will be any higher than any alternative that could have been considered is conjecture as no real alternative to an economic growth based strategy was ever seriously considered. Much of Metroplan was driven by the economic aim of reinforcing existing trends and integrating committed major infrastructure proposals (these two being interdependent) to what, in essence, was an already decided infrastructure planning framework for the metropolitan area determined by PADS.

This Metroplan strategy not only reinforced the shift of focus of development from the new towns to the central and western harbour, but in responding to the relocation of the airport and the large scale expansion of the container port, incorporated a vast programme of associated land-based transport infrastructure and building construction. This programme was dominated by economic interests, in particular the needs of trade and transport industry on the presumption that what was good for economic interests was also good for the inhabitants of Hong Kong. Social issues took a secondary position.

The question of affordability is extremely important in a society where almost half the inhabitants live in public housing, and where 72 per cent of daily trips are made by public transport (excluding taxis) (see Lau 1996). Leaving the private sector (only loosely regulated) to decide the pricing of such public goods at a time when the metropolitan strategy itself makes them more dependent upon such interests (e.g. by focusing developments on MTR access, by re-housing people on reclamation sites and in new towns, and by the displacement of persons through redevelopment schemes) can have serious repercussions where the affordability levels change. Already the numbers of unemployed are growing, while those in poverty are assessed to represent six per cent of the population[10].

Furthermore, planning practice in Hong Kong in the last few decades has presumed the continuation of stable growth trends in population, industrialisation and the export of goods - all within the limited territorial boundaries of Hong Kong. External circumstances however, must be expected to have a marked effect on the socio-economic structure of the city over the next twenty years, particularly the socio-economic implications of integrating the enclave with its natural regional hinterland. This aspect of Metroplan was grossly under-investigated although it has subsequently, and belatedly, been addressed in the Review of TDS (Planning Department 1996).

Influence of earlier studies

TDS

The change of focus for strategic development from new towns back to the harbour presented in the TDS deprived Metroplan of considering how it would respond in the context of a more decentralised strategy that sought a greater component of growth accommodated by the New Territories.

In addition to this, the failure of TDS to address the problems of isolation experienced by the residents of the new towns resulted in these problems being ignored by Metroplan. The further concentration of employment in the central areas envisaged by Metroplan will exacerbate the mismatch between residence and workplace. Long commuting journeys are reinforced by the plan's recommendations. The consequences are social and economic hardship for those who cannot afford to finance such trip-making. The government insistence on commercially viable public transport links, furthermore, has prevented (or delayed) new rail links being introduced, while the belief that bus services 'do not need planning' has reinforced the isolation of many of the new town inhabitants.

The emphasis of TDS on a financial viability ('best value for money') approach and transport efficiency optimisation over and above other socio-environmental development concerns pre-empted the consideration of more socially sensitive Metroplan options. This viability pre-requirement, interestingly, was overlooked when it came to financing the airport railway, thus suggesting that the travel needs of visitors to the territory and the business community have priority over local inhabitants![11]

What is surprising about TDS is that even though its modelling procedure (the LUTO Model - see Choi, 1986) had been widely discredited (even ridiculed) as a basis for planning the future of the territory (the most persuasive of the criticisms being provided by Taylor 1987), the Planning Department based much of Metroplan on this methodology (see Chapter 3). This led Sit (1989) to accuse the Planning Department of using Metroplan as a vehicle for re-introducing TDS under another name.

The LUTO Model assigned no importance to non-quantifiable aspects of land-use/transport interaction and paid no heed to the fast-changing internal and neighbouring environments (including population changes), the effects of the Sino Joint Declaration on how Hong Kong's future land supply and land-derived revenues would be generated, and the rapidly changing circumstances of the Pearl River Delta and its interaction/impact on Hong Kong (Sit 1993).

The Hong Kong planning horizon in the 1980s was narrowly focused within the territorial boundaries. This is understandable as Hong Kong had no mandate to take into consideration changes across the border even though economic and social ties with China were strengthening.

The 1990s, however, saw a change of thinking in the planning sphere. The much belated Railway Development Study (Transport Branch and Highways Department 1993), the Study of Development Trends in Guangdong Province (Shankland Cox et al 1993), and the TDS Review (1995 and 1996) have all taken more positive steps in considering the implications of the relationship with Southern China for land-use planning.

This change of direction is welcomed on all fronts. It is, nevertheless, too late to affect the fundamentals of Metroplan.

SHRUG

The widely held views at the time of SHRUG, that the old urban areas were 'so overcrowded that only by decanting population and jobs to the New Towns could the quality of life in the main urban areas be significantly improved' (Town Planning Department and Lands Department 1983) have returned to haunt the present administration. The resultant environmental degradation, sewage pollution and traffic congestion the metropolitan area is now experiencing as a result of its harbour-centred strategy is a vindication of the critics. One wonders whether the traffic generation and pollution implications of the reclamation areas were sufficiently investigated, particularly in the case where light rail and tram options were not selected, as in the case of Aldrich Bay.

Port expansion/PADS

The implementation of PADS 'sealed the fate' of the territory as a large scale international transport hub so long as the region, especially China, is unable to provide competing facilities. Others have claimed that PADS has in effect held to ransom the future of Metroplan, whatever its qualities and however good its intentions.

Today, the container port handles more than 11 million twenty-foot equivalent units (TEUs) a year at a rate of approximately one such unit every three seconds (Planning, Environment and Lands Branch 1995), with a further three million TEUs being handled mid-stream. Some 70 per cent of the port's throughput is bound for, or originates in, China. The bulk of this is being handled by the eight existing container terminals located at Kwai Chung (all financed, built and operated by the private sector). Agreement for constructing a ninth terminal on Tsing Yi Island has now been reached.

On the basis of an 18 per cent average annual growth rate over the last decade, it is forecast that by the year 2011 there will be a demand for Hong Kong 32 million TEUs a year. This provided the justification for planning a completely new port on Lantau Island consisting of between 17 and 24 berths to be operated by private companies. The implications of continuing to bring more and more freight to Hong Kong harbour, by road and by rail are by no means fully understood. The question of how much growth can, and should, Hong Kong absorb is a very important one, yet few people in decision-making positions are asking it or appear to be concerned (see Chapter 12). Similar points of concern can be raised regarding the implications of growth of air traffic, although these are likely to have far less impact than the port-generated traffic and activities. The decision to expand the port and add new terminals at the location chosen ensures that so long as Hong Kong offers the most well managed port facilities and affordable rates there will always be large volumes of freight traffic traversing the territory with all the environmental impacts and traffic disruption this entails.

The choice of the location of Hong Kong's new airport raises the question of whether it was politically motivated. The new airport is seen by some to have been purposely situated as far away as possible from the territory's soon to be defunct international border with China. The decision, by design or otherwise, has cancelled out the option of an airport that could be shared with Shenzhen along an enhanced capacity KCR route, which would no doubt have accelerated the inevitable joint future of the twin cities of Hong Kong and Shenzhen. Some critics of the Hong Kong Administration have claimed this 'back towards China' development approach of Hong Kong was intentional and locked the territory into a strategy that did more to separate Hong Kong from its hinterland than to join it.

CTS studies

It is surprising that in the late 1980s, Hong Kong employed a type of comprehensive urban transportation planning process which had long been discredited in industrialised countries. The mathematical models and assumptions used in this process were based on trend and road-based traffic analysis which paid little (if any) heed to water transport, pedestrian movement, environmental issues and public transport.

The heavy emphasis given to transportation evaluation criteria in choosing among Metroplan options thus generated 'desirable options' more for the transport industry than for Hong Kong's citizens in general, on the basis that the 'trickle down benefits' of such an approach would produce more employment, increases in land values (and thus government revenues) and higher levels of accessibility along certain routes. The long term 'trickle down costs' in terms of social and environmental impacts were not seriously considered.

The CTS studies (and its reviews) played down the role of railways, at least up to the publication of the Railway Development Study (which took place after Metroplan had been completed). Bus and ferry services, as well as pedestrian movement, were given scant attention. Provision for these was left to the mercy of market forces despite an attempt by Metroplan to place greater emphasis on these modes.

Transport planning

The CTS derived transport component of Metroplan - which ultimately determined much of the feasibility (and sustainability) of its alternative urban development options - relied on transport planning techniques which were dependent on an empirical understanding of existing travel behavioural characteristics for a range of given income categories (see preceding chapter) and observed and estimated goods vehicle generation of major employment types. As has already been noted, such techniques are based on acknowledged questionable assumptions that growth of urban economic activity will inevitably give rise to an increase in private motor vehicular travel, and that space within the built-up urban environment can/should be found to accommodate the growth of moving and parked vehicles (Dimitriou 1992).

Robb (1996), argues that there is evidence to suggest that growth in economic activity in Hong Kong is adapting to the constrained mobility opportunities available. He believes that in many of the territory's metropolitan districts, transport networks are operating at full practical capacity, suggesting that the city would face an intractable transport problem if further growth were to follow the conventional pattern of increased economic development, leading to an additional increase in demand for motorised transport. This warning was echoed by one of the authors of this chapter in a recent local publication (Dimitriou 1994).

Robb also claims that the evidence that the MTR system in Hong Kong is proactive to responses to travel demand is circumstantial. He argues that this supposition largely rests on the basis that current MTR route alignments serve districts and desire-line corridors which were already heavily developed. He believes this masks the fact that the majority of urban redevelopment investment over the last five years or so has taken place within walking distance of an MTR station. This suggests that the coincidence of interest the real estate developer, the public transport operator and the urban resident can be a very strong formula, especially where employment opportunities and welfare facilities can be easily and affordably accessed, and where road space and motor vehicle ownership remains constrained.

Finally, there is the unknown impact on urban transport of the current telecommunications revolution, so important today to commercial ventures in Hong Kong, and potentially useful to the transport industry in the enhancement of its operations management and control. Unlike Singapore, which has formulated a specific strategy to encompass new telecommunication technologies in its urban development proposals, Metroplan has ignored this very important facet of development.

Spatial structure and urban renewal

Robb (1996) makes the very important point that Metroplan deals not only with committed and possible new developments and restructuring but also with the bulk of the existing metropolitan form and structure. These, he emphasises, will not remain unchanged but will be subject to both renewal and redevelopment[12]. Apart from the creation of opportunity spaces for new development, and some change and increased intensity of land use, the bulk of metropolitan urban form and structure is already established as described by the CTS model (see Chapters 3, 4 and 7).

Given the absence of an Urban Renewal Strategy and insufficient powers conferred on the Land Development Corporation (LDC) to make a significant contribution in this field (Cuthbert and Dimitriou 1992), urban renewal is primarily determined by private sector real estate interests. These interests have displayed no real regard to the traffic congestion impacts they generate, or the environmental problems they contribute to, given the perceived high opportunity costs these might entail. A similar lack of regard is also shown in making available more open space.

Central to the restructuring of the metropolitan area, and a major thrust of Metroplan itself, is the intention to reduce the residential population densities in old overcrowded areas, referred to as 'thinning out'. There are two interdependent facets of this problem - the high density of population per se and the poor quality of the living conditions. Both stem from the fundamental problem of the typical low-income levels of the residents living in such areas (with incomes in the region of US$4,600 per annum), and their inability to obtain access to better accommodation[13]. The physical restructuring of such areas through land-use planning policies alone will not, however, offer the solutions to problems whose root cause is low income levels.

Furthermore, the improvement of the living conditions of the residents in areas identified as requiring 'thinning out' will also not necessarily or automatically occur through market forces, despite the claims of some. Many of these same areas also support high levels of local employment with the presence of many small business uses at street level. Redevelopment can, in certain circumstances, therefore, result in a loss rather than a gain to the existing residents, as new developments tend to be less flexible in terms of their capability to create provision for a mix of activities.

Population densities and distribution

Given that the key premise of Metroplan was that there would be no increase in the population of the metropolitan area within its time horizon, and that population growth would be accommodated in the New Territories, one could conclude that the approach adopted represents much more of an 'improvement scheme' rather than a strategic plan.

Metroplan concluded that 4.2 million should become the upper limit for the population of the metropolitan area. This envisaged a decrease in the average metropolitan population density from approximately 600 persons per hectare in 1986 to 480 per hectare in 2011. In effect, virtually all population growth up to 2011 would, by implication, be accommodated in the New Territories, with negligible growth within the Metroplan area. At the time Metroplan was formulated, the 2011 projected population for the territory was 6,627,000 of which 4,142,000 was allocated to the metropolitan area (see Table 8.6).

This suggests a projected increase in population of 1.14 million would need to be accommodated in the New Territories. The estimate, however, seems to contradict the findings of SHRUG, based on which TDS had indicated that the focus of accommodating strategic growth would shift from the New Territories back to the central area.

Table 8.6 Metroplan projected population to the year 2011

Year	Metropolitan Area	Rest of Hong Kong	Total
1986	4,239,900	1,247,800	5,487,700
2001	4,084,000	2,251,000	6,335,000
2011	4,142,000	2,485,000	6,627,000

Source: *Planning Department, 1990a*

Planning guidelines and control

Many of the issues raised and proposals made by Metroplan resulted in principles and recommendations that would ultimately prove very difficult to enforce. For example, the concept of limiting the metropolitan population to approximately 4.2 million requires a set of workable enforcement procedures that were never spelt out. Similarly, the intention that travel demands should be accommodated in an 'environmentally appropriate manner' leaves much open to conjecture when environmental standards are already difficult to enforce and at a time when the private sector is increasingly looked to provide transport infrastructure.

Robb (1996), furthermore, makes the important point that the imposition of statutory control over urban development is perceived by finance capital in Hong Kong (and elsewhere in Asia) as restrictive and consequently an inhibitor of investment. This of course may partially explain why Metroplan is not a statutory document and why it is not taken too seriously in certain quarters. On the other hand, it is being increasingly recognised internationally that structure planning guidelines can be perceived (if so desired) as identifying opportunities for private investment and public-private sector investment, thus promoting rather than inhibiting capital investment in locations with good accessibility opportunities. This lesson has not been fully appreciated by the private sector interests of Hong Kong who take a particularly short-term view on development investment[14].

Notwithstanding the fact that Metroplan attempted to introduce New Towns standards within the metropolitan area, the Hong Kong Planning Standards and Guidelines (HKPSG) remained socially and politically acceptable at the time Metroplan was formulated and thus govern many of the characteristics and configurations of Metroplan's proposed land use. This is despite their questionable origins[15].

Economic foundations and employment

In 1986, 86 per cent of manufacturing employment was in the metropolitan area, as were 96 per cent of office jobs and 75 per cent of other employment. Manufacturing comprised 36 per cent and office jobs 10 per cent of total employment. According to earlier cited population assumptions for 2011, this would result in 62 per cent of the Hong Kong population, and 61 per cent of its labour force living in the metropolitan area by the year 2011.

This contrasts with 77 per cent and 80 per cent respectively in 1986. Whereas jobs in the metropolitan area were anticipated to fall from 81 per cent in 1986 to 75 per cent in 2011, estimates of the population to be housed in the New Territories far outpaced the job increases, the net result being a further widening of the gap between residence and workplace (Planning Department 1991a). The growth of the service sector (associated particularly with finance and import/export activities), as Hong Kong's role as an international centre and entrepot for China strengthens, favours the creation of jobs in the CBD and harbour locations, rather than in the new towns.

Industrial growth and restructuring

Metroplan's extrapolation of industrial growth trends of the past twenty years has, according to Robb, produced alarming spatial consequences. These trends are contrary to current evidence that employment in the local manufacturing sector is falling as a consequence of the development of the special economic zone (SEZ) of Shenzhen in the PRC, and the economic activities generated by the Hong Kong-Guangzhou axis and by the coastal periphery of the Pearl River Delta which have also been loci of industrial capital investment (Shankland Cox Ltd. et al 1993).

Bearing in mind that Metroplan was intended to be a structure plan for the metropolitan area, rather than for the territory of Hong Kong, it was reasoned that the traditional industrial zones of the metropolitan area could not be expected to continue to make a significant contribution to the growth of the territory as a whole. From other planning points of view, however, particularly the desire to mitigate environmental problems, the prospect of reducing the manufacturing industry in the metropolitan area is regarded as favourable for Hong Kong by virtue of the transfer of an intractable industrial and pollution problem to a neighbour. However, rather than resolving a problem it is in fact pushing it out of sight. Quite possibly this will not be for very long.

The above concerns raise the questions of whether the Hong Kong economy is still changing and of whether the spatial effects of these changes have yet fully worked through the land-use/transportation system (Robb 1996). A study of the restructuring of employment spatially in Hong Kong between 1981 and 1991 found that there had been a drastic reduction in jobs in the manufacturing sector and a strong compensatory growth in the import/export sector (Cook and Wang 1996). The question of the succession of uses on account of economic restructuring in the Hong Kong urban area is not well understood. Many questions remain unanswered:

- is there under-utilisation of industrial buildings?
- have uses such as storage, warehousing and office functions taken up the space vacated by manufacturing industry?
- is industrial floorspace simply less intensively used?
- does industrial floorspace need to be redefined to meet high-tech space requirements?
- is the very nature of industry changing and are people's perceptions of industry also changing?
- are new industries emerging, and will old industrial areas be suitable for redevelopment for these new industries, or will they require different site attributes, such as an 'industrial park/business park' or 'science-or technology-city' environment?
- are obsolete industrial areas suitable for redevelopment for residential use, as has occurred in many post-industrial Western cities?

Reclamation areas

A major concern raised several years before Metroplan, but still of critical importance today, is whether the extensive reclamation in central areas with planned large scale commercial floorspace will undermine the viability of any major non-CBD commercial nodes, as well as dampen demand for commercial redevelopment and upgrading.

These and other concerns re-emerged in 1995 when the harbour reclamation schemes became a major issue in the public arena. Reclamation as a means of increasing land supply is not new to Hong Kong. It has been a response to the lack of space for urban development as long ago as the 1840s. However, the recent magnitude of reclamation works resulting from Metroplan and earlier studies has evoked considerable criticism from a number of quarters, and on a number of accounts, including environmental, economic, aesthetics and marine safety.

The current reclamation in the central harbour area of some 400 hectares is now highly visible to the public (see Figure 8.11). It is partly the sheer scale of the reclamations that led to questions re-examining their necessity (Ng and Cook 1997). Proposed reclamations up until 2003 will add a further 630 hectares bringing the total reclamation in the harbour area to over 1000 hectares, reducing the central harbour to what in effect will become a narrow river.

In addition Container Terminals 9 and 10 will bring total new harbour reclamation since 1994 to over 2000 hectares (Planning, Environment and Lands Branch 1995).

Environment

Metroplan has often been described as having a land use, transport and environmental framework. While certain environmental concerns were indeed addressed by the strategy, particularly with regard to open space, and landscape and townscape attributes, not to mention the laudable attention given to urban design, there is no

Figure 8.11 Central Harbour reclamation
Source: Courtesy of Hong Kong Tourist Association

doubt that the environmental impacts of the transportation infrastructure proposals and the urban redevelopment and reclamation components were far from satisfactory.

During the Metroplan Steering Group meetings it was patently clear to all involved that the more established and powerful agencies, such as the Public Works Department, Building and Lands Branch, and the Transport Department, as well as (unfortunately) the Planning Department, considered the then newly established Environmental Protection Department too demanding and unrealistic in its expectations of the other sectors. This, and subsequent experiences may account for the outgoing Director of EPD in 1996 acknowledging in his resignation speech that

> *... perhaps more could have been achieved by the EPD on the environmental front than had been.*

As well as environmental concerns about poor living conditions (in terms of congested living quarters)[16] and the deterioration of the existing housing stock, concerns were expressed in the Metroplan Study about the inappropriate mix of land use activities, such as polluting industries in, or impacting on, residential areas. More generally other prevailing conditions were also recognised as requiring attention. These included air (vehicle and industry emissions) (see Chapter 10), water (sewage and industry wastes) and noise (industry and roads) pollution.

While Metroplan highlighted the major environmental changes arising from the proposed restructuring and new proposals, it did not incorporate any real enforceable measures to prevent the identified problems from becoming worse. Quite the contrary, an impression is emerging that the plan is actually contributing to these problems through its reclamation programme and acceptance of additional large scale transport infrastructure expansion (both of which were inherited from earlier study proposals incorporated in Metroplan).

The main source of air pollution - the motor vehicle - emits carbon monoxide, hydrocarbons, nitrogen oxides and particulates. The principal origin of particulates and nitrogen oxides in Hong Kong is goods vehicles, and to a lesser extent buses. Taxis and cars contribute only a small fraction of these pollutants even in the densely populated areas of canyon-type urban development (Rusco and Walls 1995). By accepting and accommodating high levels of goods vehicle projections, Metroplan (unwittingly perhaps), has in fact made itself in compliance with this worsening trend of air pollution[17].

Notwithstanding these criticisms, Metroplan did put the urban environment more firmly on the government's planning agenda, leading among other things to White Paper reviews (Planning, Environment and Lands Branch 1993), the initiation of new legislation (Planning, Environment and Lands Branch 1996), as well as a more serious coverage of environmental matters in the TDS Review (Planning Department 1993c).

Institutional arrangements

The principle of 'thinning out' of population densities, referred to above, very much depends on establishing a partnership between the 'overcrowded area' and the 'solution' space. It also rests on nurturing a close working relationship between the private and public sectors, among the different agencies and between the communities involved and affected.

This, in turn, implies a high degree of government management of the partnership process. The planning process is further complicated by the fact that although the 'solution space' (generally provided through reclaimed land) may be in public ownership, the overcrowded area is likely to comprise a number of different buildings, each with multiple ownership, which without compulsory purchase powers can make the concept of thinning out unfeasible.

Public consultation

It could be argued that selecting a single scenario from the Metroplan plan-making process was unrealistic. However, it should be appreciated that the motive of producing a single 'end product' (plan) was partly political. It was considered important to be seen to have produced the plan from an immense amount of technical analysis and, whatever may be claimed to the contrary, there were also powerful interest groups that wanted to see a blueprint and Metropolitan programme of public works. While denying that Metroplan was such a document, the Planning Department did provide the strategy in such a form that (in conjunction with the use of Development Statements) one could have a fairly accurate picture of what the future would be like and what kind of urban development would be supported.

While Metroplan clearly offered a milestone in the 'opening-up' of plan-making to the public in Hong Kong, this belated gesture was really largely about already committed works and projects, and the scope for change was somewhat limited (Sit, 1989). Having made this point, the question of the extent to which the public at large can appreciate a macro-scale planning effort of this kind is clearly a problem, irrespective of whether earlier studies of a more closed nature limited the scope of public input.

Metroplan signalled a new direction of urban development and a new planning process for Hong Kong well into the next century. It was, however, criticised by Sit for offering an 'over-planned planning framework' during and beyond the transition to 1997. He also claimed the plan exaggerated the powers of the Hong Kong planner to unrealistic levels and that this commenced with the setting up of the Strategic Planning Unit within the Planning Department which was charged, among other things, with the preparation of the Territorial Development Strategy and Metroplan. To loosen the grip of the planners on plan making, Sit recommend the setting up of a Commission that would 'enhance public participation and accountability in the (metropolitan) planning process'. It would be interesting to see whether such a proposition would be welcomed by the post-1997 administration of Hong Kong.

Post-1997 implications

The main concern of Sit and others about strategic planning for Hong Kong is the extent to which plans will impose a development strategy on the future Special Administrative Region (SAR) based on the underlying misconstrued premise that Hong Kong is a separate entity from China. He points out that 'the initial years in China's opening-up clearly had benefited Hong Kong and helped its expansion in the re-export trade'. Yet Sit argues that, with China extending trading relations with other countries throughout both the region and the world, Hong Kong's growth cannot be taken for granted. Added to this, competing infrastructure port and airport projects, both elsewhere in the region and in China itself, suggest that Metroplan should incorporate these other developments into its plan rather than seek to absorb all the traffic that Hong Kong can attract.

Conclusions

Of the many conclusions one can draw from the preceding discussion, one of the most significant is that although Metroplan was publicised as having a strong vision of its own, it relied heavily on the proposals of preceding studies for this vision. This is particularly the case with regard to its transportation component, which placed the economic and efficient transportation of labour, and goods and economically beneficial use of land above all other objectives.

This dependency of its vision on other studies has led some to argue that Metroplan represents a lost opportunity to sweep aside the 'ill-consequences of some bad planning in the past' (Sit 1989). Instead, it continues with a scenario that perpetuates misconceptions and undesirable trends into the future. Others, however, see Metroplan as a pioneering step in Hong Kong's planning efforts, to be built upon further, and monitored. Such parties see the plan as an important attempt to involve more stakeholders in the plan making and vision development exercise. On the basis of current trends, what appears most likely is that 'business will continue as usual', whereby the captains of industry and commerce will 'hand in glove' with government principally decide the future of the metropolitan area. Indications are that while there will be different personalities in power in government, and some new faces in industry and commerce, public participation, as promoted by Metroplan, will have had its day.

Looking back, rather than forward, the achievements of Metroplan are perhaps best judged when examined against its professed planning principles. Firstly, one can only conclude that the claim that Metroplan 'recognises the practical limits to urban growth in the metropolitan area' is clearly untrue. The scale of its PADS and reclamation based proposals present the most vivid contradiction to this stated principle. It is clear that the growth levels and trends cited by the plan appear insatiable, yet Metroplan offers no response to this other than to provide more infrastructure. Secondly, the claim that Metroplan pursues a 'joint planning of land and transport' also is not matched by reality; an observation already acknowledged by other contributors to this book (see Chapters 3, 4 and 7). Thirdly, while Metroplan places much greater emphasis on public transport, increasingly the government policy is to abandon the public to the mercy of the private sector bus operators and ferry operators. These will decide what services should be provided - when, where and at what fare level - which not only exposes the public to the insecurity of not having their travel needs adequately met, but may undermine the Metroplan Strategy itself. Fourthly, the claim that Metroplan rectifies environmental problems and avoids the generation of new ones is also somewhat far-fetched. The environmental implications of the reclamation schemes have already been cited as a case where more problems may be generated than they solve. Similarly, the environmental impacts of new transport infrastructure generating both freight and passenger traffic in already heavily congested urban corridors, is also attracting critical public attention - and these are early days in terms of Metroplan implementation!

Finally, and paradoxically, one of the principles that Metroplan is promulgating, namely that Metroplan is best implemented by the 'fullest use of current agencies and mechanisms' is, in the view of the authors, contributing to the plan's ineffectiveness. To cut across existing government departmental interests, to free the plan from the existing inflexible financial mechanisms, and to provide strong executive direction that enables a more equitable public/private partnership to materialise, requires a powerful Metropolitan Development Agency headed by a strong political personality who sees the need for change rather than a perpetuation of past ways. An agency of this kind needs to further elaborate on the vision of Metroplan and offer a balanced mix of visionary, enabling and control policies (see Chapter 12).

There is still time to steer metropolitan area energies and economic growth in a direction that does not suffocate itself. There is also a dire need for the city to protect itself and its inhabitants from the worst impacts of further unsustainable fast growth. Those currently directing Metroplan (ultimately the members of the LDPC) have failed to do this.

This is not so much because of a lack of a vision provided by the plan, but because the implementers of the plan do not have the power (provided by legislation, government and the community), the commitment of their leaders, and the co-operation of the private sector to make Metroplan work where it is in conflict with the interests of industry and commerce.

Notes

1 Metroplan is currently (1997) under review.
2 This programme has since become increasingly controversial and has attracted growing opposition as the real and feared disbenefits of reclamation become apparent (see Ng and Cook 1997).
3 These comprise of the new airport at Chek Lap Kok, Tung Chung New Town Phase 1, North Lantau Expressway, the Airport Railway, the Lantau Fixed Crossing (Tsing Ma and Kap Shui Mun bridges), Route 3 (part), West Kowloon Reclamation, the Western Harbour Crossing, the West Kowloon Expressway and the Central Reclamation (phase 1).
4 The fluctuations in GDP growth since 1986 have been substantial and current indications are that the assumed 5 per cent rate may be unsustainable. The rate of growth of GDP has been declining in the 1990s from a high of an annual increase of 6.3 per cent in 1992 to a 4.6 per cent increase in 1995 (Census and Statistics Department 1996).
5 Private car numbers grew by an average of 8 per cent per annum between 1986 and 1995, but with a slowing of the rate from 1994.
6 Goods vehicle numbers grew by an average of 5 per cent per annum between 1986 and 1995 but there has been a distinct slowing in more recent years.
7 Professor Dimitriou was the only non-government official member of the Steering Committee, with the exception of Mr. Razack who, as Director of the LDC, was a semi government official.
8 The Planning Department commissioned a consultancy study on Sustainable Development in Hong Kong (SUSDEV) in 1997.
9 On several occasions at Steering Group meetings on Metroplan, it was felt necessary by the Planning Department to re-emphasise to a key spokesman for the Lands Branch (and for the Secretary for Planning) that Metroplan was not a public works programme (Dimitriou 1996). This was done in order to counter the apparent confusion in the minds of key Lands Branch staff as to the difference between a strategic plan and large projects that in essence made up the plan.
10 Hong Kong Council of Social Service Survey, reported by the South China Morning Post 13/10/96.
11 The funding of the airport railway link on non-commercially prudent principles was rationalised by government on the basis that it would call on revenues generated for related land use developments - something government had refused to do earlier.
12 In the view of the authors there has been a constant confusion of terminology

between the concept of urban redevelopment and urban renewal in Hong Kong government and LDC circles. The former is the demolition of existing buildings and the resultant dislocation of local populations to make way for new construction typically yielding financial returns, whereas the latter also includes the economic and social upgrading and renovation of existing buildings and areas.

13 Hong Kong Council of Social Service Survey reported in the South China Morning Post 13/10/96.

14 Until recently, most developers and investors did not consider ventures or projects viable unless they made a healthy profit within five years.

15 The authors have found it impossible to trace the exact origins and logic of the various standards and guidelines employed beyond the fact that, being a colonial administration, there are some clear indications of reliance on British planning standards in some instances, but not in most.

16 Hong Kong has some of the highest residential densities and overcrowded housing conditions in the world (over 2,000 persons per gross ha in some parts of Kowloon) despite the territory being one of the top ten trading economies of the world.

17 The concept to be incorporated by Metroplan in its proposals of introducing environmental capacity ceilings on major transportation corridors resulting from the PADS Study, was rejected by a technical committee on transport for Metroplan with no convincing reasons. Post-construction environmental impact studies of such corridors were proposed instead. This, it was argued, would then lead to the introduction of more appropriate amelioration measures. Given that the traffic forecasts already exceeded the planned infrastructure provision, one can only conclude from this decision that it was feared that the use of environmental capacity ceilings would prove detrimental to the maximisation of economic growth.

References

Barnes, G. (1990) 'The Shape of Hong Kong After PADS' *Planning and Development*, Vol. 6 No. 1.

Bristow, R. (1989) *Hong Kong's New Towns. A Selective Review.* Hong Kong: Oxford University Press.

Census and Statistics Department (1996) *Estimates of Gross Domestic Product 1961-1995.* Hong Kong Government Printer.

Choi, Y. (1986) 'Land Use Transport Optimisation Model (LUTO). Model on Strategic Planning in Hong Kong' *Asian Geographer* Vol.5 No.2.

Cook, A.H.S. and J.J. Wang (1996) 'Employment Concentrations in Hong Kong: Changing Patterns, Changing Structure' *Occasional Paper* No. 133 Geography Department, The Chinese University of Hong Kong.

Cuthbert A.C. and H.T. Dimitriou (1992) 'Redeveloping the Fifth Quarter: Practices and Prospects of Redevelopment in Hong Kong', *Cities Journal*, Vol. 9. No.3.

Dimitriou, H.T. (1992) *Urban Transport Planning: A Developmental Approach*, London: Routledge.

——————— (ed) (1994) 'Visions of a More Motor Vehicle Dependent Free Strategy' in *Moving Away from the Motor Vehicle: The German and Hong Kong Experience*, CUPEM, University of Hong Kong.

——————— (1997) *Toward a Sustainable Urban Transport Strategy for Medium-Sized Cities in Asia and the Pacific Region*, Unpublished Draft Consultant Report to World Bank for UNDP and ESCAP, Bangkok.

Eason, A.E. (1985) 'Territorial Development Strategy Studies: A View of the Process' *Planning and Development* Vol. 1.
Government Information Services (1990) *Metroplan. The Selected Strategy: An Overview*, Hong Kong.
———— (1991) *Hong Kong's Port and Airport Development Strategy. A Foundation For Growth.*
Lands and Works Branch (1985) *Planning For Growth*, Information Services, Hong Kong Government.
Lau, J. (1996) *Evaluating the Performance of Public Transport Operations in Hong Kong from the Point of View of the Disadvantaged,* MPhil. Dissertation, CUPEM, The University of Hong Kong (unpublished).
Morris, K. (1990) 'PADS: An Overview' *Planning and Development*, Journal of the Hong Kong Institute of Planners Vol. 6 No. 1.
Mott MacDonald Hong Kong Limited et al (1989) *Port and Airport Development Strategy*, Final Report. Prepared for the Government Secretariat and the Lands and Works Department, December 1989.
———— (1990) *Port and Airport Development Strategy,* Executive Summary. Prepared for the Government Secretariat and the Lands and Works Department, January.
Ng, M.K. and A.H.S. Cook (1997) 'Reclamation: an urban development strategy under fire' *Land Use Policy* Vol.14, No. 1.
Osborn, R. (1985) 'Harbour Reclamations and Urban Growth Study', *Planning and Development* Vol. 1 No. 2
Planning Department (1990a) 'Technical Report on Transport', Metroplan Studies, Hong Kong Government, Unpublished.
———— (1990b) *Metroplan. The Foundations and Framework.* Planning Department, Hong Kong Government, January.
———— (1990c) *Metroplan. The Foundations and Framework.* Hong Kong Government.
———— (1990d) *Metroplan. The Foundations and Framework.* Planning Department, Hong Kong Government.
———— (1991a) *Metroplan. The Selected Strategy.* Executive Summary. Hong Kong Government.
———— (1991b) *Metroplan. The Selected Strategy* Volume 1: Technical Report. Hong Kong Government.
———— (1991c) *Metroplan. The Selected Strategy* Volume 2: Appendices. Hong Kong Government.
———— (1991d) *Metroplan. The Selected Strategy. An Overview.* Hong Kong Government.
———— (1993a) *Territorial Development Strategy Review Foundation Report*, Hong Kong Government.
———— (1993b) *Territorial Development Strategy Review: Development Options,* a Consultation Digest, Hong Kong Government.
———— (1993c) *Territorial Development Strategy Review: Environmental Baseline Conditions*, Hong Kong Government.
Planning, Environment and Lands Branch (1990a) *Metroplan: Full Seminar Report*, Government Secretariat, Hong Kong.
———— (1990b) *Metroplan: Initial Options* Government Secretariat, Hong Kong.
———— (1993) *The Hong Kong Environment: A Green Challenge for the Community*, Government Printer, Hong Kong.
———— (1995) *The Shape of Things to Come: An Overview of the Role of Harbour Reclamations in the Future Development of Hong Kong*, Government Printer, Hong Kong.
———— (1996) *Consultation Paper on Town Planning Bill* Government Printer, Hong Kong.
Pryor, E.G. (1985) 'An Overview of Territorial Development Strategy Studies', *Planning and Development* Vol. 1, Hong Kong.
———— (1992) 'Metroplan: A Framework for Restructuring Hong Kong's Main Urban Areas' in *Proceedings of the Conference on Planning for High Concentration Development*, Hong Kong Institute of Planners and School of Professional and Continuing Education, The University of Hong Kong, Dec.

Pun, K.S. (1980) 'Urban Planning in Hong Kong - Concepts and Approaches' *Asian Architect and Builder* 9 (12, 1980), Hong Kong.

Robb, B. (1996) 'Transportation Planning and Metroplan', Unpublished Paper.

Rusco, F.W. and W.D. Walls, (1995) *Clearing the Air: Vehicular Emissions Policy for Hong Kong*, HKCER Paperback, The Chinese University Press, Hong Kong.

Sit, V.F.S. (1989) 'Is Metroplan Setting the Right Strategy for Urban Hong Kong?' *Planning and Development* Vol. 5.

Shankland Cox et al (1993) *Study of Development Trends in Guangdong Province*, Executive Summary Report for Planning Department, Hong Kong Government.

Taylor, B. (1987) 'Rethinking the Territorial Development Strategy Planning Process in Hong Kong' *Asian Journal of Public Administration* Vol. 9 No. 1 .

Town Planning Division, Lands Department (1983) *Harbour Reclamations and Urban Growth Study* (SHRUG), Final Report, Hong Kong Government.

Transport Branch and Highways Department (1993) *The Railway Development Study*: Final Report, in association with MVA Asia Ltd., Acer Consultants (Far East) Ltd. and Maunsell Consultants Asia Ltd. Government Printer, Hong Kong.

Transport Department and Wilbur Smith and Associates (1989) *The Second Comprehensive Transport Study*, Hong Kong Government Printer.

Wilbur Smith and Associates (1976) *Hong Kong Comprehensive Transport Study* Hong Kong Government Printer.

9 Traffic management and environmental protection

Kelvin C.W. Leung and John Spiers

Introduction

There is an increasing global awareness that the preservation of a satisfactory living environment for today and the future is the responsibility of everyone, not just of those few professionals who specialise in environmental protection. Traffic and transport engineers, and transport and land-use planners alike, can all make a significant contribution to this area of concern as they deal on a daily basis with road traffic - one of the key factors that has direct impacts (both beneficial and detrimental) on the living environment.

The purpose of this chapter is to discuss how traffic management techniques in Hong Kong are being adjusted to try to incorporate environmental concerns in urban transport planning, with a view to minimising the adverse environmental effects of traffic.

It begins by discussing the general relationships that exist between traffic and the environment. The chapter continues with an overview of traffic engineering and management practices in the territory, and then concentrates on the ways in which these are being modified to incorporate environmental concerns.

Traffic and environmental management

Aims

Broadly, the aim of traffic management is to utilise the existing street and road patterns to increase the efficiency of traffic movement, using traffic regulations (such as one-way systems) and traffic control equipment (such as traffic signals) to make the best use of the road system to the benefit of all road users, and only by extension, to the benefit of the wider community. In comparing alternative traffic management approaches, it is becoming increasingly appreciated that it is also important to consider the need for pedestrian access and circulation.

Environmental management, on the other hand, aims to preserve the quality of the environment for the benefit of the community as a whole. In so doing, it seeks to modify or support development proposals in line with its objectives. To do this, government, on behalf of the community, typically sets specific quantitative and qualitative standards as the main environmental measures to be used in environmental appraisal and evaluation studies. When these measures are applied to the management of traffic, they typically address the adverse impacts of traffic on humans and the surrounding environment (including air pollution, noise pollution and safety), using measures designed to prevent the entry of extraneous traffic to certain areas and re-organise traffic flow patterns so that the damaging effects are minimised.

Although there are some common aims in traffic and environmental management, inevitably there are conflicts too, particularly between the needs of the transport planners and engineers, and the environmental specialists and planners (see Chapters 3 and 7).

Common and conflicting aims

Both forms of management aim to maintain and improve the living standards of society, and are seen to be undertaken for the ultimate benefit of the community as a whole. Traffic management in urban areas is intended to provide adequate and satisfactory transport access and circulation ideally for road users and pedestrians alike. Road safety is a common concern of both traffic and environmental management, although the latter concentrates more on people, and the former more on vehicles.

The conflicting aims of traffic and environmental management are of increasing concern to the community. Introducing traffic into any area immediately gives rise to adverse effects on the living environment, especially with respect to air pollution, noise pollution, and road safety. Even with careful planning, it is impossible to remove totally the adverse effects of traffic. It is more a matter of ensuring that they are maintained within acceptable limits.

Modern cities are dependent on good transport and the ease of movement. It is, therefore, increasingly acknowledged that there must be some kind of balance between the mobility needs of the people and the preservation of the environment. This is especially true in Hong Kong where urban land is at such a premium and urban population densities are among the highest in the world.

A principal objective of traffic management is to provide conditions that help vehicles maintain higher speeds so as to maximise road capacity. However, higher speeds also create greater environmental concerns. Thus, this is one instance where the aims of traffic and environmental management are in direct conflict with each other. Traffic engineering practice, in general, attempts to avoid conflicts between people and vehicles, and this is best done by reducing traffic speeds, irrespective of traffic capacity.

Perversely, however, a significant improvement to the environment of a particular location may have the opposite effect to that intended. For example, the environmental enhancement of a particular site may attract additional people and traffic, in turn, giving rise to the need for additional traffic and environmental management.

This vicious circle (see Figure 9.1) is in evidence in many fast growing urban areas, especially in the rapidly developing economies of Asia. It in turn highlights the special need for co-ordinated land-use and transport planning, advocated elsewhere in this book, and the collaboration of traffic and environmental management recommended here, if this vicious circle is not to be perpetuated. However, the increasing use and ownership of the private motorcar in Hong Kong indicates that the vicious circle is not being (nor about to be) broken. There is, thus, a particular case to be made for pursuing traffic and environmental management jointly in Hong Kong.

Figure 9.1 Vicious circle of environmental improvement / impairment

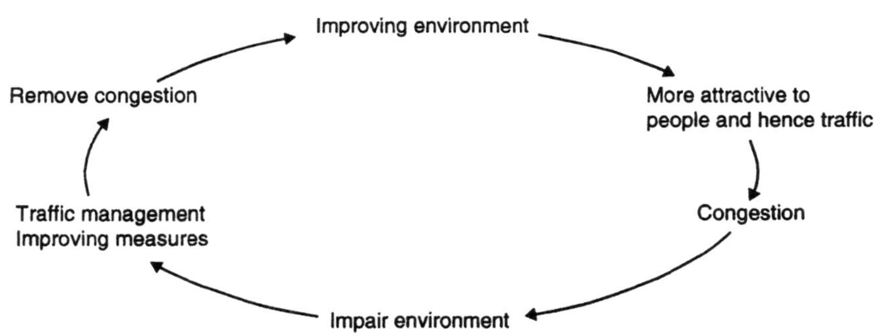

Causes of environmental impact by traffic

As indicated above, at least three main components constitute the environmental impact caused by motor vehicles in an urban area. These are noise pollution, air pollution and safety concerns.

Noise pollution

Noise pollution caused by motor vehicles mainly arises from the combustion engine and tyres. There can also be wind noises from vehicles which have a poor aerodynamic body designs. In general, however, this is not a very significant source. The use of vehicle horns, especially at night or close to hospitals and other sensitive locations, also contributes to noise pollution.

By and large, the bigger and heavier the vehicle, the greater the noise levels incurred. Also, the greater the traffic flow the higher the related noise level. Besides the size of vehicle and volume of traffic flow, other factors which affect traffic-related noise levels include:

- road gradient characteristics;
- road surfacing characteristics;
- the separation of the noise source from the receiver;

- the nature of the topography of the surrounding environment through which the road runs; and
- whether the vehicles are free-flowing or frequently stopping/starting at junctions.

Interestingly, in Hong Kong, road traffic noise is not included in the Noise Control Ordinance which covers general noise level standards for the territory from general commercial, industrial and building activities. Recommended acceptable limits for residential and other sensitive developments are instead laid down in the Hong Kong Planning Standards and Guidelines (HKPSG)[1]. This latter document recommends that sensitive receivers should not be subject to L10 road traffic noise levels of more than 70dB(A) for dwellings, 65dB(A) for schools, and 50-60dB(A) for hospitals. The 'L10 noise level' is the peak hour noise level exceeded for 10 per cent of the time. In effect this is the 'average maximum noise level'.

Air pollution

Air pollution arises from exhaust gases from motor vehicles (see following chapter). The main forms of engine used world-wide have two main oil-derived fuel types: petrol (gasoline) and diesel. In Hong Kong, as in most countries, the larger and heavier (goods) vehicles and buses run on diesel fuel, the lighter vehicles (motorcars and motorcycles) mainly run on petrol. Taxis in Hong Kong are a mixture of both, some run on diesel, and some on petrol.

The main air pollutants that are emitted from motor vehicles are carbon monoxide, particulates, sulphur dioxide, oxides of nitrogen, and lead[2]. In general, goods vehicles emit more particulates (both respirable and non-respirable) than cars, but less carbon monoxide and sulphur dioxide. Cars emit more sulphur dioxide at high speed, and more carbon monoxide at low speed. After much debate (see Chapter 10), there has been a widespread move towards lead-free petrol in Hong Kong. Price incentives, coupled with the shift by motorcar manufacturers to catalytic converters, suggest that problems of lead pollution related to vehicles emissions in Hong Kong should decline.

The Hong Kong Road Traffic (Traffic Control) Regulations provide for the control of polluting emissions from motor vehicles in two ways: firstly, vehicles must be designed to meet certain pre-specified emission standards; secondly, vehicles must be constructed and maintained so as not to emit excessive smoke. Current Hong Kong air quality objectives stipulate the concentrations for the main air pollutants as shown in Table 9.1.

Due to the high development densities in Hong Kong which contribute to the creation of canyon-type urban form conditions (see Figure 9.2), the combination of traffic generated pollution and development generated pollution means that both noise and air quality objectives and standards are often exceeded in many parts of the urban area where such kinds of development may be found. Figure 9.3 indicates the annual respirable suspended particulate concentrations attributable to vehicle emissions. The greatest problem facing the urban planner and traffic engineer, therefore, is how to control pollution to manageable levels without stifling the very development that keeps Hong Kong so prosperous.

Table 9.1 Hong Kong air quality objectives

	Average Concentration ugm^{-3}				
Parameter	**1-Hour**	**8-Hour**	**24-Hour***	**3 Month**	**Annual**
Carbon Monoxide	30,000	10,000			
Nitrous Dioxide	300		150		80
Total Suspended Particulates	500**		260		80
Respirable Suspended Particulates			180		55
Lead				1.5***	

* *Not to be exceeded more than once per year.*
** *In addition to the above established legislative controls, it is generally accepted that an hourly average TSP concentration of 500 ugm-3 should not be exceeded. Such a control limit is particularly relevant to construction work and has been imposed on a number of construction projects in Hong Kong in the form of contract clauses.*
*** *If it is assumed that 10% of daily traffic flows in the peak hour, the worst case 24 hour average will be approximately half the 1-hour worst case average. The 24-hour average then can be used as an estimate of a 'very pessimistic' worst case 3-monthly average.*

Figure 9.2
Canyon-type urban form
Source: O. Trueb

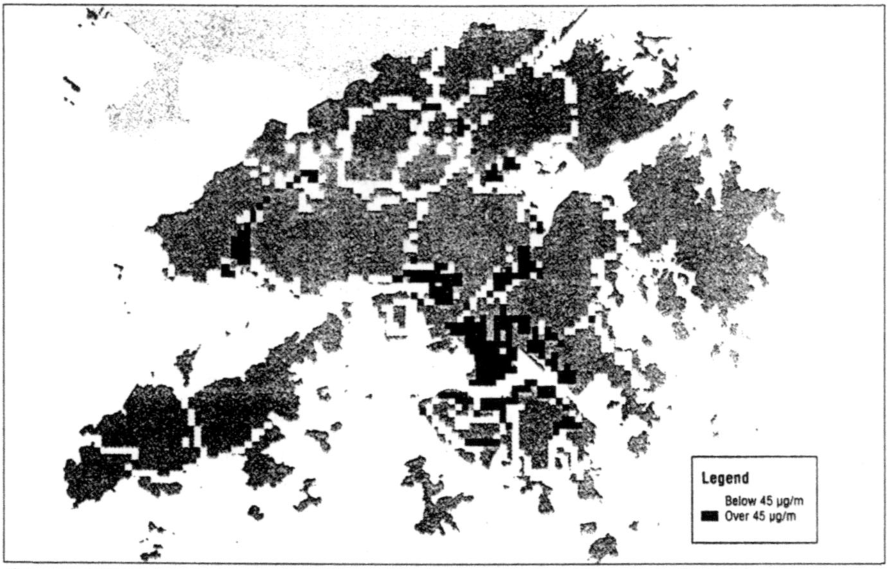

Figure 9.3 Annual respirable particulate concentrations attributable to vehicles

Source: Environmental Protection Department.

Safety

The third main environmental concern with regard to traffic is that of road safety. There are two concerns here: 'vehicle-vehicle' conflicts and 'vehicle-pedestrian' conflicts. The faster vehicles travel, the greater the likelihood of serious injury in the event of an accident. The aim of traffic management, to increase the general flow and speed of traffic, is thus in conflict with reducing speed for safety reasons. A large body of research has been undertaken to understand the relationships between the design of traffic systems and the propensity for accidents[3]. Reflecting progress made in this research, road and traffic engineering design standards are constantly being enhanced world-wide to try and increase traffic performance, whilst maintaining an acceptable level of safety.

Interaction between traffic management and environmental protection expertise

Conflict of interests

It is clear from the preceding discussion that there are many conflicts between the interests and objectives of traffic management, and those of environmental management. Inevitably, therefore, numerous compromises need to be made. This is often reflected in both the derivation and application of the standards and guidelines employed. Until the early 1980s, only lip-service was paid to environmental concerns of new traffic and transport schemes in Hong Kong. This is despite the abundant literature regarding such aspects available to the professions in the UK dating back as early as 1963 (Buchanan Report 1963, Report of Independent Commission on Transport 1974, and Institution of Highways and Transportation with Department of Transport 1987)[4]. It has to be appreciated, furthermore, that any successful traffic management measures which facilitate the introduction of additional vehicles on the road in urban areas will inevitably increase traffic-related noise levels and air pollution, and possibly also generate additional safety concerns.

In the recent past, it has not been considered the role of the traffic engineer or transport planner in Hong Kong to consider the environmental consequences of traffic management schemes. Instead, it was typically assumed that an increase in traffic levels was inevitable as economic growth increased, and that therefore, there was little point in worrying too much about such matters. It was, furthermore, often believed that provided any negative impacts were restricted to the 'less influential' parts of the community, these impacts were 'unavoidable', i.e., a price to be paid for 'progress'. By default then, these became 'acceptable' (Chau and Dimitriou 1994).

The overriding concerns in Hong Kong of traffic engineers and transport planners up until the mid-1980s in their examination of traffic management measures, was orientated toward improving traffic conditions and reducing the costs of achieving this. Through the 1960s and 1970s, however, the government's main concern had been to provide a rapidly expanding population, much of whom lived in temporary low-quality accommodation, with the basic material necessities. In this context, traffic and transportation improvements concentrated on providing for the essential mobility needs of the people and industries of Hong Kong, especially of new public housing estates (see Chapter 2).

Legacies of past

Looked at from today's perspective, a large proportion past building developments and associated transport and traffic facilities in the territory, are very basic indeed with very little concern for the environment. Indeed, prior to the 1980s, a number of 'environmentally unfriendly' road schemes were constructed in Hong Kong. A well known example of such a scheme is the Hill Road Flyover in Western District. As may be seen from Figure 9.4, the edge of this flyover is built so close to adjacent buildings that it is possible not only to peer through the windows into the rooms of adjacent buildings, but, from the roadside, to shake hands with the residents of these buildings.

During the 1980s, Hong Kong's increasing affluence, and the government's slow recognition of the world-wide concern for the effects of global warming and other environmental issues led to the creation of an Environmental Protection Department (EPD) in 1986. Prior to this, there had been a small Environmental Protection Agency within the Hong Kong Administration which had had very limited powers.

The EPD is a much more powerful agency, whose responsibilities are clearly defined under the relevant ordinances[5]. It is responsible for all environmental pro-

Figure 9.4
Hill Road flyover in Western District
Source: MVA Asia Limited

tection issues in Hong Kong, including water, sewage, air pollution and noise pollution, and has a current staff of level of about 1,200 professional and administrative personnel.

The Hong Kong government now insists on Environmental Impact Assessments (EIAs) for each major traffic and transport scheme in Hong Kong. These assessments must not only cover the environmental effects of the completed scheme, but also consider pollution effects during construction. The steps of a typical EIA process in Hong Kong are outlined in Figure 9.5. The introduction of EIAs has led to much more direct contact between traffic and transport planners and environmental specialists - both within government and outside.

The Hong Kong government (through the EPD) is now very concerned that the environmental effects of traffic schemes are fully understood before a implementation decisions are made. However, the scale and density of Hong Kong, coupled with the land costs and constraints that have been inherited from history, mean that there is a practical limit to what can be achieved in existing urban areas. All professionals involved in development, including developers, architects, urban designers, land-use planners, civil engineers, transport and traffic engineers, and environmental specialists, need to work together to produce acceptable compromises to enable Hong Kong to continue to prosper, while maintaining an acceptable and sustainable environment (see Chapter 12). This will entail a process of compromise and evolution, both as standards change and as society's expectations rise.

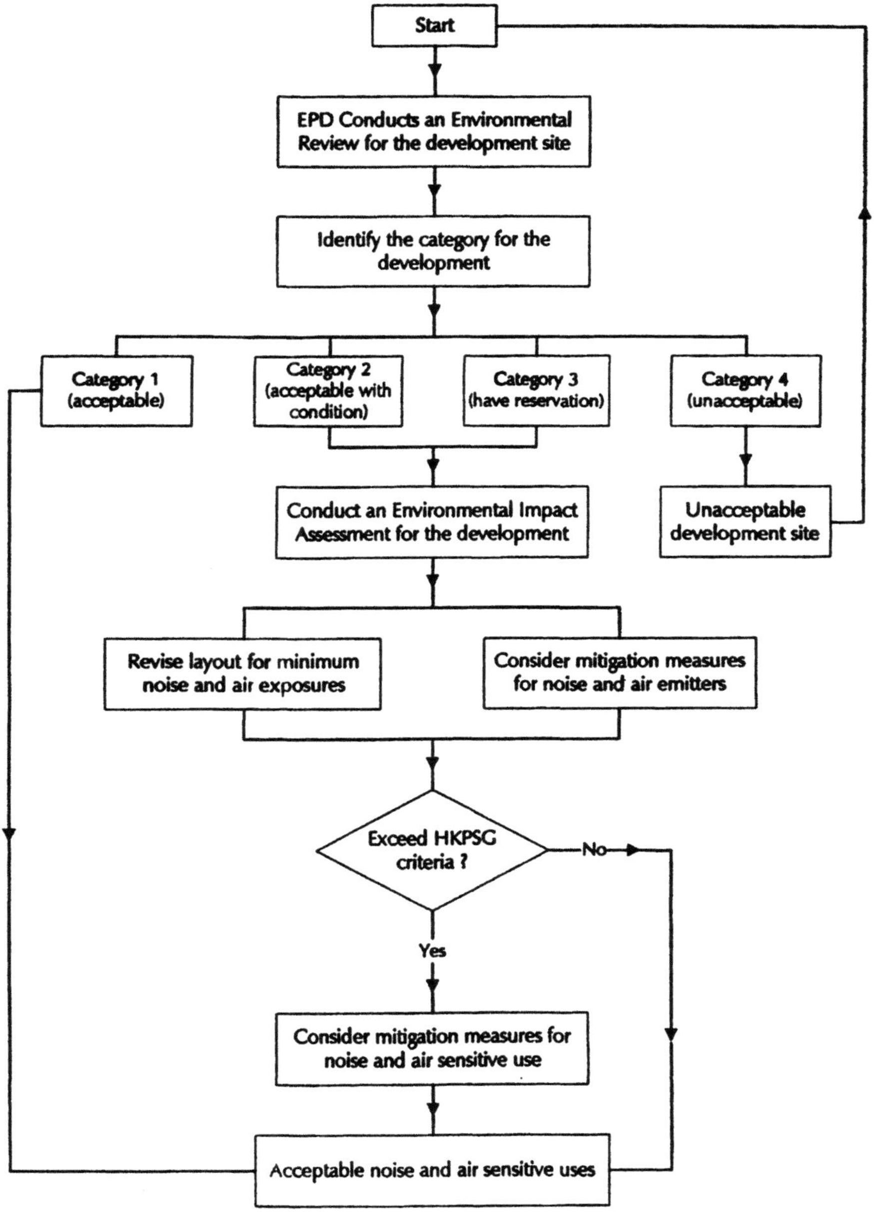

Figure 9.5 *Steps of a typical EIA process in Hong Kong*

Inevitably, there has been a keen debate over the applicability and interpretation of international and local environmental protection standards and guidelines. The often conflicting aims of traffic engineering and management, and the aspirations and aims of the environmental specialists, require project designs and solutions that are acceptable to both parties. The two disciplines need to interact at an earlier stage than has traditionally been the case, both in scheme development and evaluation, rather than waiting until a scheme has been taken to the detailed design stage, ready for construction. More recently in Hong Kong, traffic engineers and planners, together with environmental specialists and land-use planners, have begun to work more closely during the feasibility stages of schemes, thereby helping to mitigate against the adverse effects of traffic. This close collaboration has involved:

- examination and adoption of more wide-ranging techniques such as new traffic management measures;
- greater use of noise barriers and open-textured road surfacing materials to reduce noise;
- more care over the planning, siting and design of buildings; and
- greater use of grade-separation for pedestrians.

The way in which traffic management techniques have been modified to incorporate environmental concerns is discussed below in general terms and, in the closing sections of the chapter, more specifically in the context of a case study.

Traffic management techniques and the environment

Traffic management techniques

Traffic management in urban areas covers any measure introduced to control the movement of vehicles. The main forms of such management are:

- junction control (signals, roundabouts, priority and more recently Area Traffic Control [ATC])[6];
- one-way streets and gyratory systems;
- speed restrictions;
- crossing facilities (for pedestrians etc.);
- access control;
- controls on parking and stopping;
- public transport priority; and
- traffic-calming[7].

Early traffic management measures were aimed at controlling conflicts between vehicles and less concerned about optimising capacity. As a result, basic junction priority rules were introduced to define rights of way for conflicting traffic streams. However, as traffic flows increased, simple priority junctions were not sufficient, and more sophisticated controls were subsequently introduced in the form of traffic signals and roundabouts.

Implications of traffic growth

The inexorable rise in city traffic volumes world-wide over the last two decades has brought widespread problems due to the shortfall of infrastructure and transport systems capacity.

These have manifested themselves in the form of long traffic queues contributing to motor vehicle user delays, an increase in road accidents, and environmental degradation to the areas through which traffic passes.

Confronting motor vehicle traffic growth in Hong Kong over the last ten years (at five per cent per annum average growth), traffic engineers have looked at several alternative ways to improve the territory's traffic capacity, and to enhance traffic circulation and improve road safety[8].

Less attention, however, has been paid to environmental issues. Interestingly, certain manufacturers of Hong Kong's imported motor vehicles (such as BMW [Monheim, 1994a]) have recently shown more concern about this aspect by attempting to produce vehicles that are much less noisy and polluting than their previous models.

Selective speed restrictions, one-way streets and gyratory road systems, have also been applied in Hong Kong as ways to improve road safety and traffic capacity simultaneously. This was done by:

- simplifying junction operations;
- reducing conflicts between opposing streams of traffic;
- providing more pedestrian crossing facilities (especially 'zebra crossings') to allow people to cross busy roads more safely; and
- constructing more grade separation pedestrian routes across major roads, using either under-passes or footbridges.

Other measures include the introduction of selective stopping restrictions, controls on parking, and bus priority measures. All such measures, however, were principally introduced for transport planning and traffic engineering purposes, rather than environmental reasons.

Emerging environmental concerns

It is only recently that strong 'environmental concerns' have forced traffic engineers and planners in Hong Kong to rethink their rationale for traffic management, and encompass a broader remit than just to maximise traffic flows and speeds. This contrasts rather starkly with the European experience, where a number of traffic management measures have received widespread acceptance, and are seen as necessary for preserving (protecting) high quality living environments (Monheim 1994b and 1994c). These measures include:

- banning certain types of vehicle from specific areas;
- imposing punitive speed restrictions to discourage particular types of vehicles;
- introducing large-scale pedestrianisation schemes in environmentally sensitive areas (such as historic centres, shopping areas etc.).

Such concepts have been very slow to catch-on in Hong Kong for a variety of reasons (Dimitriou 1994). Notwithstanding this slow progress, traffic engineers do now consider the environmental effects of traffic schemes in Hong Kong. They use whatever techniques are available to them to minimise any negative impacts caused. Given a careful consideration of the particular circumstances of the territory, the basic package of urban traffic management techniques advocated include:

- speed control restrictions;
- vehicle access and use restrictions;
- vehicle re-routing measures;
- the location of car parks on strategic sites; and
- traffic-calming measures.

All the above have the common aim of reducing traffic in areas where environmental concerns are important. In reducing traffic, motor vehicle-related noise and air pollution are automatically reduced, as are road safety hazards[9]. The ultimate intention of vehicle access restrictions and re-routing measures is to keep extraneous traffic away from local streets that have been designed to serve local traffic circulation only. As levels of road congestion rise on major urban routes, the use of local back-streets as 'rat-runs' increases. This places an intolerable burden on these streets as they are not designed to sustain heavy traffic. This in turn, not only causes environmental problems for residents living in the buildings adjacent to these streets but also incurs high maintenance costs to the highway authorities.

Speed control

Traffic speed restrictions in urban areas may be introduced either through the imposition of speed limit regulations, or through road design. In the case of the latter, the road geometry can be deliberately made tortuous so that low speeds are inevitable. Artificial impediments to vehicles, such as speed-humps, may also be placed in the road, or localised street narrowing can also be used. Both these measures are more common in the new towns and housing estates than in the existing built-up areas of Hong Kong. As already indicated, slower speeds help to reduce traffic noise, especially of heavy goods vehicles. However, steep gradients tend to increase traffic noise levels as gear changes take place. This is of particular significance in the hilly parts of Hong Kong.

Air pollution is much less dependent on speed, although the constituents of the pollutants generated will vary by speed (as well as gradients encountered). The worst air pollution is generated when vehicles frequently stop and start, and where high rise buildings are very close to each other and trap the pollution emitted by the congested traffic. Such conditions have been reported in high-density areas of Central and Kowloon, as confirmed by EPD air quality readings. Under these circumstances, it is desirable to maintain as smooth a traffic flow of vehicles as possible. The introduction of ATC whereby traffic signals are linked and timed to allow a steady flow of traffic between junctions at the designated speed will help as these encourage traffic to travel at set speed limits.

Vehicle access and use restrictions

Access restrictions can normally be applied by vehicle type and/or by time of day. Most commonly, they are applied to restrict the movement of goods vehicles - either during peak periods to minimise traffic congestion, or at night in residential areas to minimise environmental impacts (i.e., noise, air and safety problems).

Restrictions on goods vehicles in Hong Kong are normally applied by weight or by length. In either case, there is a need for enforcement to ensure that the regulations are obeyed, especially around seaport and airport areas which both generate and attract large volumes of goods traffic. In Hong Kong, the enforcement is provided by the Hong Kong Police Force.

By means of careful traffic engineering design of junctions, it is often possible to ensure that vehicular length restrictions of goods vehicles are self-enforcing. For example, it can be made almost impossible (through geometric design) for a vehicle to enter certain roads. The resultant reduction of traffic along these routes has the effect of improving road capacity and traffic circulation for those vehicles permitted to use the road. It also improved the quality of the surrounding environment. Access restrictions are now commonly applied to many city centres in Europe, not only for goods vehicles but also in certain cases (and times), for private motorcars (Fan et al 1992).

City centre traffic congestion levels have now become so acute in many countries that numerous governments are again examining the feasibility of road pricing as an option for restructuring vehicular access to city centres. One such scheme is Electronic Road Pricing (ERP). This scheme was examined and rejected in Hong Kong in 1985 (Transpotech 1985), primarily for three reasons:

- there were fears that the system constituted an 'invasion of privacy' in that it would enable specific vehicles to be located as they passed the ERP toll points;
- the government's introduction of large increases to the First Registration Tax and Annual Vehicle Licence Tax coincided with a period of economic downturn. Vehicle ownership actually fell, and it was felt that further vehicle restraint was not necessary;
- there were fears that ERP was just another means of raising general tax revenue, and that there would be no real increase in spending on alternative transport systems (bus and rail) to compensate for displacing people from their own vehicles.

Despite its rejection at this time, many transport and traffic specialists still believe that some form of road pricing is inevitable in Hong Kong (and elsewhere) if motor vehicle use is to be kept to manageable levels (Hau, 1992a and 1992b). A new investigation of ERP was begun in 1997.

Road pricing advocates often draw upon Singapore's experience of vehicle access restrictions to support their views. Singapore has applied a selective scheme since 1975, using an Area Licensing System[10] (Fan et al 1992). The system was extended in January 1994 to all vehicles entering the Central Business District (CBD) between 07.30 and 18.30.

Kuala Lumpur also investigated (in the late 1970s) the possible application of vehicular access restrictions to its city centre but this was turned down because of the lack of adequate public transport. It is currently re-examining the introduction of an ERP-type system now that it has started to build a Light Rail Transit system to serve the downtown area. Jakarta introduced another form of restricted access. This scheme is based on the occupancy levels of cars entering the CBD, whereby only cars with three occupants or more are permitted to enter the CBD by certain major routes during peak hour travel periods.

In addition to restricting the number of vehicles on the roads on any given day by limiting the total fleet (as in Singapore), an alternative option is to restrict particular vehicles on certain days of the week (as in Athens), whereby vehicles with number plates ending in odd numbers and even numbers are prevented from entering the city centre on alternate days.

Other measures

Other vehicle access and use restrictions include:

- Vehicle Re-routing - This entails the imposition of selective bans on turning movements at junctions and can force motor vehicles to avoid sensitive areas of a city centre, and to keep to the main routes preferred by the transport and traffic engineers. Measures of this kind have similar effects and benefits to access restrictions, but achieve similar objectives in a more subtle way.
- Strategic Provision of Car Parks - This involves the creation of car parks on the fringe of urban areas at strategic locations. These are believed to be a major determinant of traffic routings and to reduce the number of vehicles that have to drive through the centre. Many cities in North America and Europe (Retzko 1994, and Monheim 1994b) have adopted fringe car parks and 'Park-and-Ride' schemes with a view to greatly reducing city traffic and significantly improving the environment.
- Traffic-calming - The calming of traffic is normally applied to a specific urban area and incorporates the full range of traffic engineering measures available, including speed restrictions, access restrictions and turning restrictions (Hass-Klau 1992). Routings through these areas are deliberately made indirect, so as to discourage through-traffic 'rat-runs'. Road width restrictions, and humps in the carriageway are also used. In some cases roads are turned into 'cul-de-sacs' to physically prevent through routing.

All of the above can be seen as tools to be used by traffic engineers and transport planners in consultation with environmental specialists and land-use planners at the scheme development stages of a study. The following section of this chapter discusses how developments in the use of these traffic management options have influenced, or are likely to influence, practices in Hong Kong. The chapter concludes with an outline of traffic management issues in the territory.

Trends and practices in Hong Kong

Limited progress

Because it has not been common practice in Hong Kong to be overly concerned with the environmental effects of transport projects, new road schemes have been 'squeezed' into available land spaces, while related noise and air pollution, and increased levels of road accidents have all been seen as an inevitable part of 'progress'. In other words, the needs of mobility and commerce have to date far outweighed the adverse environmental effects of these new projects. As a result, much of the urban area of Hong Kong does not meet noise and air pollution standards laid down either by the EPD or the Planning Department's HKPSG. Lessons from other Asian cities, especially Bangkok, Jakarta, and Seoul, however, show what can happen to city air quality and noise levels if vehicular growth and use continue unconstrained, with no regard for environmental consequences (Pendakur 1993).

Notwithstanding the above, some progress has been made in Hong Kong on the transport/environment interface. The network of overhead walkways in Central is an example of how government engineers and developers have worked together to provide a safer and more 'environmentally friendly' pedestrian network for the community. Furthermore, the opening of the hillside escalator in Central (see Figure 9.6) in 1993 represents a bold move to provide a more 'environmentally acceptable' network for pedestrian journeys to work that would otherwise require motorised transport.

Recent road improvement schemes in Hong Kong are also being designed with more concern for the environment. A large section of the Kwun Tong By-pass, for example, has been covered by a noise barrier to protect adjacent residential blocks (see Figure 9.7). However, more traditional roadside noise barriers, such as light-

Figure 9.6 Central Mid-levels hillside escalator, Hong Kong Island
Source: MVA Asia Limited

weight panels or earth berms, have not found widespread acceptance in Hong Kong. This is because:

The barrier should ensure that there is no line of sight between vehicles and the noise sensitive windows in buildings. With the very tall buildings located close to busy highways this is often not practicable.

An earth-berm barrier usually requires more land than a barrier constructed from panels. The value of land in Hong Kong often makes the use of berms economically non viable. The exception is in Hong Kong Housing Authority estates where it is now becoming standard practice to build noise barriers around the site boundary of the developments. This helps to provide some noise reduction to households on the lower levels.

A more popular approach to traffic noise reduction in Hong Kong is the location of transport interchanges underneath major developments, as in New Town Plaza in Sha Tin and Exchange Square on Hong Kong Island. In fact, for most of the new public housing estates, locating the estate roads and transport interchanges underneath the podium of the development is now common practice. This raises the pedestrian circulation and access level to the first storey of the developments, thereby enhancing the environment and safety.

Where possible, on more free-flowing roads, the use of an open-textured macadam road surface above an asphabol cushion course can further reduce traffic noise

Figure 9.7 Kwun Tong By-pass noise barrier
MVA Asia Limited

by up to 5dB(A). However, this increases maintenance costs, and is inappropriate on stretches of road where traffic frequently stops and starts. New roads on reclamations (such as Route 7 on the Green Island Reclamation are being planned with adequate set-backs for development so that noise impacts are minimised. However, these extensive set-back requirements for trunk roads makes land-use planning more difficult and costly).

Case study - The Green Island Reclamation

A recent study illustrates the new procedures that are now built in to all significant traffic and transportation studies undertaken in Hong Kong. The Green Island Reclamation study (Ove Arup and Pypun 1993) was a Hong Kong government commissioned feasibility study for approximately 190 hectares of reclamation off the north-west corner of Hong Kong Island. The study involved a multi-disciplinary team of civil and infrastructure engineers, land-use planners, traffic engineers and transport planners, landscape architects, and environmental scientists/engineers.

The objective of the feasibility study was to produce an outline development plan incorporating housing (both public and private) and associated recreation and government/institutional facilities for up to 120,000 people, and employment (in a variety of commercial, industrial and marine activities) for about 30,000 people. In addition, the reclamation had to make provision for two major expressway routes and an MTR line: the Route 7 Expressway between the Western Harbour Crossing and Aberdeen; the Green Island Expressway Link which will connect from Route 7 at Green Island to the proposed Lantau Port Peninsula; and the MTR extension from Sheung Wan, via Kennedy Town to Green Island, and with an optional extension to the Lantau Port Peninsula.

One of the major concerns was the feasibility of incorporating these two major expressway links, and the MTR link into the reclamation without causing widespread environmental problems for the proposed housing, and institutional and recreational facilities. In addition, it was an objective to make the design of the housing areas as 'environmentally-friendly' as possible, by reducing extraneous traffic.

At the outset the study involved extensive interaction between each of the technical disciplines. The aim was to optimise the design of the reclamation and to ensure that the requirements of the standards and guidelines were met, with respect both to the 'hard' engineering and traffic design aspects, and to the 'softer' planning, landscape and environmental aspects. Typical environmental traffic and engineering measures proposed are:

- the use of underground roads to minimise noise and air pollution, and also to reduce the 'segregation' effect of major roads.
- the use of larger housing development estates, and of cul-de-sac arrangements to minimise 'rat-running' and extraneous traffic. Within these estates, the full range of traffic calming measures would be implemented, including speed humps, road narrowing and design of car-park and access road locations to maximise pedestrian and open-space areas.
- extensive set-backs of up to 150 metres from the major roads and expressways.

- the use of earth berms and cut-slopes to reduce traffic noise.
- the provision of adequate public transport facilities to discourage the use of private vehicles.

These traffic and engineering measures will be implemented in conjunction with planning and urban landscaping measures (such as single-aspect housing and tree planting) to improve the overall environment and reduce air, noise, and visual pollution and improve safety.

Future action needed

It is becoming increasingly apparent that the needs of environmental management in Hong Kong will have to be much more carefully balanced against the needs of mobility, especially along the ever busier major routes. If, however, motor vehicle growth and usage are allowed to continue at the current rate of about five per cent per annum, then the likelihood of any significant improvements in the environment throughout the territory is indeed very small, as has already been pointed out by several parties (see Chapter 12).

The traffic engineer in such circumstances will merely assign maximum priority to keeping traffic moving with all the environmental implications this implies. While promises have been made by the Hong Kong government that improvements to the environment will be included in transport projects wherever possible, it is difficult to see where significant gains can be made in the existing densely developed urban areas without considerable disruption and cost. To ensure, therefore, that the maximum amount possible is done to enhance environmental qualities, the following traffic management steps have been followed in all major transport studies in Hong Kong since 1989:

- identifying transport and traffic problems in the short, medium and long term;
- proposing transport and traffic solutions to those problems;
- carrying out an initial screening of the proposed solutions to identify those with unacceptable environmental consequences;
- undertaking more detailed evaluation of the remainder of the schemes, including an environmental impact assessment (EIA) and the identification of the scope for any mitigating measures; and
- producing a set of recommended transport and traffic measures with an associated set of environmental improvement measures.

The crucial step above is the penultimate one, where the possible mitigating measures must be identified. At this point, the temptation will be to argue that measures are not available or applicable on the grounds that they are too costly, or the land provision required is too difficult, or that the measures are 'not necessary in Hong Kong'. However, common sense must be used to strike a balance between the extremes of total environmental protection (e.g., advocacy of noise covers on minor local roads) and no protection (as on the Hill Road Flyover).

Conclusions

Perhaps understandably, the priorities for Hong Kong throughout the 1960s and 1970s were to cater for the basic needs of a rapidly expanding population. The achievement of a stable economy with adequate housing, sanitation, power, and water supply overrode the secondary considerations of air and noise pollution and of road safety. The concerns of traffic engineers and transport planners were the satisfaction of mobility demands quickly and at as low a cost as possible. However, increasing affluence and the growth of trade through the 1980s has led to two conflicting effects:

- an increase in both the ownership and usage of the private car and the usage of goods vehicles. This has increased vehicle-generated pollution.
- a more affluent, educated population who are more concerned about the environment and the adverse effects of vehicle-generated pollution.

The government has reacted by creating the EPD and ensuring that all significant traffic and transport schemes are subject to an EIA. A set of statutory standards and of guidelines has been compiled from a variety of international sources (the USA, Europe and Japan). The job of the EPD will be to mould these standards and guidelines to form a workable set for Hong Kong.

Transport and traffic engineers, urban planners and environmental specialists must work together within the ordinances and guidelines to produce realistic proposals that take into consideration the unique nature of Hong Kong. This should not, however, provide grounds for flouting international standards. Striking the correct balance between 'laissez-faire' and 'over-regulation', as in other aspects of urban transport, will be the principal challenge for the future. Since the methodology and tools are now in place, the territory needs the practitioners to use them wisely for the overall benefit of Hong Kong.

Notes

1 The HKPSGL recommend levels of provision for all aspects of a development including parking, servicing and lay-by provision for residential, industrial, commercial, retail, hotel, and government/institutional uses.
2 To illustrate the impact of motor vehicles on the environment it has been reported between 1986-87 that world-wide these vehicles emitted 2,860 Mt of CO2, 160 Mt of CO, 22 Mt of NOx, 27 Mt of hydrocarbons and 0.28 Mt of CFCs (Faiz et al 1992).
3 The UK Road Research Laboratory, formerly the Transport and Road Research Laboratory (TRRL), has undertaken extensive studies on the relationship between vehicle design and accidents.
4 The Buchanan Report is a particularly significant landmark in traffic engineering and transport planning literature in that it brought to the forefront for the first time in UK the concept of environmental capacities of roads to complement the

more traditional concepts of vehicular capacities. Buchanan's environmental capacity concept may be described as 'horses for courses' i.e., that the capacity of a road should be governed by its function and use. A local road serving small residential developments should have the traffic restricted accordingly - both in terms of numbers and speed of vehicles.

5 The most important of the regulations governing the EPD standards are the Noise Control Ordinance (1988), the Air Pollution Control Ordinance (1983 revised in 1987 and 1993) and the HKPSGL.

6 ATC is the co-ordinated control of the signalised junctions within a given, normally urban, area. The traffic signal controllers are linked to a central computer which controls the timing and operation of each of the traffic signals. The two most well-known ATC systems are SCOOT (a UK system) and SCATS (an Australian system).

7 Traffic-calming is a relatively new generic term which encompasses a range of measures (some old, some new) to reduce the impact of traffic on sensitive areas. The primary aim of traffic-calming is to rid an area of unwanted extraneous traffic. It attempts to limit vehicles in an area to access-only traffic (Hass-Klau et al 1992).

8 The Hong Kong Government White Paper on Internal Transport Policy 1990 sets out policy on the development of road and rail transport and includes a discussion of such measures as vehicle restraint, ownership controls and user-related charging such as ERP.

9 Lower speeds will also reduce the severity and risk of vehicle-vehicle and vehicle-pedestrian accidents. In general, however, high speed roads have fewer accidents overall, but the percentage and absolute number of fatal and serious accidents is much higher.

10 An Area Licensing System is a means of controlling the entry of vehicles into a specific area. To enter the area a licence must be purchased and, normally, prominently displayed. Non-display of the licence can result in a heavy fine.

References

Buchanan C. (1963) *Traffic in Towns: A Study of Long Term Problems of Traffic in Urban Areas*, (Buchanan Report), Reports of the Steering Group and Working Group appointed by the Minister of Transport, London: HMSO.

Chau, S. and H.T. Dimitriou (1994) 'Introduction' in Dimitriou, H.T. (ed.) (1994) Centre of Urban Planning and Environmental Management, University of Hong Kong, Hong Kong.

Dimitriou, H.T. (1994) 'The Urban Transport Policy Agenda Revisited' in Dimitriou, H.T. (ed.) (1994) Centre of Urban Planning and Environmental Management, University of Hong Kong, Hong Kong.

——————— (ed.) (1994) *Moving Away from the Motor Vehicle* Centre of Urban Planning and Environmental Management, University of Hong Kong, Hong Kong.

Faiz, A., C. Weaver, K. Sinha, M. Walsh and J. Carojo (1992) *Air Pollution from Motor Vehicles: Issues and Options for Developing Countries* Washington D.C.: World Bank.

Fan, H.S.L., A.P.G. Menon and P. Olszewski (1992) 'Travel Demand Management in Singapore' *ITE Journal* Vol. 62 No. 12.

Hass-Klau, C., I. Nold, G. Bocker, and G. Crampton (1992) *Civilised Streets: A Guide to Traffic Calming* Brighton, England: Environment and Transport Planning.
Hau, T. (1992a) 'Congestion Charging Mechanisms for Roads: An Evaluation of Current Practice' *World Bank Staff Report* Transport and Urban Development Department, Washington D.C.: World Bank.
——— (1992b) 'Economic Fundamentals of Road Pricing: A Diagrammatic Analysis' *World Bank Staff Report*, Transport and Urban Development Department, Washington D.C.: World Bank.
Independent Commission on Transport (1974) *Changing Directions* London: Coronet Books.
Institution of Highways and Transportation with Department of Transport (1987) *Roads and Traffic in Urban Areas* London: HMSO.
Monheim, R. (1994a) 'From Pedestrian Zones to Traffic Calmed Cities' in Dimitriou, H.T. (ed.) (1994) Centre of Urban Planning and Environmental Management, University of Hong Kong, Hong Kong.
——— (1994b) 'Problems of Traffic Growth and Strategies for Traffic Calming' in Dimitriou, H.T. (ed.) (1994) Centre of Urban Planning and Environmental Management, University of Hong Kong, Hong Kong.
——— (1994c) 'Cooperative Urban Traffic Management' in Dimitriou, H.T. (ed.) (1994) Centre of Urban Planning and Environmental Management, University of Hong Kong, Hong Kong.
Ove Arup and Pypun Consultants (1993) *Green Island Reclamation* Final Report to Hong Kong Government, December Hong Kong.
Pendakur, V.S. (1993) 'Congestion Management and Air Quality: Lessons from Bangkok and Mexico City' in *Asian Journal of Environmental Management*, Vol. 1, No. 2, Hong Kong.
Retzko, G. (1994) 'New Developments in Urban and Suburban Public Transport' in Dimitriou, H.T. (ed.) (1994) Centre of Urban Planning and Environmental Management, University of Hong Kong, Hong Kong.
Transpotech (1985) *Electronic Road Pricing Pilot Scheme: Main Report*, Report to Hong Kong Government, May, Hong Kong.

Part Four
NEW DIRECTIONS

New directions: Editors' Overview

Harry T. Dimitriou and Alison H.S. Cook

Following Part 3 which focused on particular plans and policies for land use and transportation in Hong Kong, this final group of chapters reflects on current issues and new directions that the territory might take.

With the motor age now having already begun in China[1] and with Hong Kong's return to the mainland, it is timely to consider whether the motor vehicle restraint experience of the territory can assist China to circumvent traffic congestion problems rapidly building up in its major urban areas. While these pressures have been encountered elsewhere in the developed and developing world, with the latter being even more critically affected as experiences in many cities of the Asia Pacific Region suggest, China's predicament is even more special given the scale of growth anticipated. The government's publicly announced policies supporting motor vehicle ownership, and the use of the emerging car industry as an economic agent of modernisation, clearly indicate the path the country is aiming to follow[2].

Two questions immediately come to mind in this context, as far as Hong Kong is concerned. One is whether it is realistic to expect Hong Kong to keep its traffic restraint and public transport-based model into the next millennium, when it is about to become the richest city in China by an extraordinary margin, and the other principal cities of China see motorisation as a symbol of economic virility. The other, is whether it is more realistic to presume that the territory will ultimately also succumb to the forces of motorisation as its land-use and transport policies become free of strict colonially administered controls and more exposed to the forces of the market place, the aspirations of its inhabitants, not to mention the influence exerted by national government policies in favour of the automobile industry.

Motorisation trends in the region and government responses to these, particularly in urban areas, suggest a bleak future for Hong Kong. These trends indicate that Hong Kong's administration quickly needs to exult and clearly disseminate the advantages of the traffic restraint measures and public transport policies it has pursued despite fast its rising per capita incomes[3]. The administration would also benefit from publicly articulating the perceived advantages of the compact city development model it has been obliged to follow. (It perhaps could be assisted in doing this by international development agencies such as the World Bank, the Asian Development Bank and UNDP). Recent developments in the region clearly demonstrate that other traffic management, transport planning and public transport investment responses are typically introduced too late, on too limited a scale and/or with inadequate enforcement and funding.

The experience of the more affluent cities of the so-called 'developed world' suggests that accommodative approaches to urban vehicle growth are just not sustainable on a long-term basis. These same developments also demonstrate that the mobility needs of the poor are typically ignored as traffic congestion levels rise and as the cost of motor vehicle use increases. It is, furthermore, argued that these trends lead to the inevitable outcome of a privileged minority of motor vehicle owners 'ultimately kicking away the ladder behind them' after they have improved their lot. Introducing measures to ensure that additional vehicles being brought onto the roads are increasingly restricted as roadspace becomes scarce (Adams 1996) does this. With the above considerations in mind, this final part of the book examines:

- The policy and planning responses to transport-related air pollution problems in Hong Kong in Chapter 10.
- The institutional mechanisms for effective land-use and transport planning in the territory in Chapter 11.
- The case for the need of a new lead in transport and urban development for Hong Kong that adheres to the principles of sustainable development in Chapter 12.

Pollution and congestion are two inevitable outcomes of unabated motor vehicle use. Today they are interlinked critical factors affecting Hong Kong's environment and (Figure 1) its population's health. William Barron, (an environmental economist and Associate Professor at the Centre of Urban Planning and Environmental Management) addresses questions concerning the deteriorating air quality from road transport pollution in the territory in Chapter 10. He places emphasis on remedial measures for air pollution from road transport and points out that although pollution from

Figure 1 *Pollution haze over Hong Kong Island*
Source: A.H.S. Cook and H.T. Dimitriou

industry is decreasing, overall exposure to air pollution health risks is increasing on account of the rising numbers of vehicles on the roads. This rise is not only of private cars but of goods vehicles - ranging from container lorries to small vans.

Barron points to the increasing levels of emissions from diesel vehicles as being of particular concern since these account for two thirds of the vehicle kilometers travelled in the territory. He also points out that Hong Kong's bus services enjoy a very high patronage despite the popular image of the MTR and KCR being the main providers of mass transit. Rail ridership remains secondary to bus patronage. Bus trips account for more than 50 per cent of all public transport journeys. The use of diesel by buses, as well as by freight vehicles, is a major contributor to air pollution in Hong Kong. Barron canvasses the options available to the territory for reducing air pollution from such vehicles.

The role of rail, not simply as a means of bringing people to the central area but in facilitating a more dispersed population strategy, has in Barron's view been paid too little attention in Hong Kong and contrasts sharply with the rail-based system of Singapore. Furthermore, the emphasis on the private sector and market forces to resolve location issues in Hong Kong has resulted in an extreme situation whereby residential expansion is now primarily taking place in the New Territories and employment expansion in the central area. This is creating a mounting journey-to-work commuting problem for public transport and leading to more air pollution as diesel bus services expand.

The role of government and the institutional mechanisms used for land-use and transport infrastructure planning are examined in Chapter 11 by Colin Arnott, (a Director of the London-based consulting firm of urban economists - Roger Tym and Partners). He presents the important contribution of public sector investment from two perspectives, namely the negative effect of failing to resolve the land development problems of urban restructuring in the old urban areas; and the positive impact of investment in mass transit rail in enabling land development.

Arnott comments on how the recent Territorial Development Strategy Review seeks to link land-use and infrastructure planning with the process of economic development and management and relates this to the central role of public sector decision-making in the land development process under a leasehold system. He sees the key function and issues of strategic planning in Hong Kong as being the facilitation of economic restructuring by the redevelopment of old industrial areas; the establishment of a job balance necessitating the decentralisation of development and investment in public transport infrastructure; and the creation of new patterns of land use and development for promoting the financial viability of new rail-based passenger transport systems.

Arnott traces the history of institutional arrangements for planning, environmental protection, and transport, as well as the emergence of the Planning, Environmental Protection, and Transport Departments, the former two under the Secretary for Planning, Environment and Lands (SPEL). He also refers to the roles of various government corporations having an impact on the activities of these government departments, namely, the Hong Kong Housing Authority; the Land Development Corporation (LDC); the KCRC; and the MTRC.

In addressing the issue of urban restructuring, Arnott points out that the scale of redevelopment that can take place simply through linked sites is very limited. He further claims that the principles of subsidisation, which have been strongly resisted by the Hong Kong government, will result in far less renewal than social and environmental needs demand (Figure 2).

In contrast to urban land assembly, public investment in transport infrastructure, particularly the MTR, is shown to have a profound impact (both direct and indirect), on the economic development of the city and stimulation of urban development to meet wider social and economic objectives. In respect of the latter, Arnott believes it is significant that the institutional arrangements of MTRC enable it to run at a profit. A final important area addressed by Chapter 11 is the fact that there is today an emergence of regional planning in the Pearl River Delta (of which Hong Kong is an important part) for which there is as yet no institutional framework for land-use environmental and transportation planning. Arnott presents this as a major challenge for the future.

Figure 2 Redevelopment of profitable sites: Midlevels, Hong Kong Island
Source: A.H.S. Cook

The final chapter to this book written by one of the editors (Dimitriou) discusses the planned vision for the territory in the context of the sustainability debate. It highlights the uncertainty in the transition period of Hong Kong and poses a number of critical questions facing territory as it enters a new socio-political arena and era.

Dimitriou argues that Hong Kong's current urban development course is one of a set of 'enablement strategies' based on technocratic vision serving the 'captains of industry' expressed in the form of major plans in infrastructure and housing investments, with little democratic input from the public at large. Such development has addressed short-term goals and focused on short-term economic gain - both to government and to the private sector. He points out that the international debate on long term sustainability has not been embraced until very recently by the territory's leaders, and only now, in response to public pressure, is it on the planning agenda.

Ideas from work undertaken elsewhere in the world concerning efforts to link urban transport and sustainable development are applied to the Hong Kong context. Dimitriou examines the concepts of sustainable economic development; sustainable environmental development; sustainable social development; sustainable institutional development; and sustainable urban development. In so doing, he seeks answers on what mechanisms can be adopted to encourage or ensure that the sustainability factor is integrated into policy-making and planning practice for the territory, particularly in relation to transport.

Finally, Dimitriou makes recourse to developments in his own research on the 'developmental' approach to urban transport planning (comprehensively outlined in Dimitriou 1992). This is an approach which embodies the principles of planning urban development within the capabilities of the local transport infrastructure, the economic context and the institutional framework and which sees transport as an agent of development and not as an end in itself. He poses the question as to how best the developmental approach can be transformed to a 'sustainable development approach' and speculates what the implications of such an amended approach might be for Hong Kong if it incorporates a wider interpretation of 'stakeholder consultation and participation' into decision-making by the authorities.

A clear and important message from these three chapters is that Hong Kong's focus on short-term economic expediency needs to be urgently broadened to encompass a much wider based set of goals concerned with land-use planning and environmental protection which are in the best interests of the whole community. This is particularly important, since in the near future the bounds of this community will not be confined to the Hong Kong territory but will extend to its Chinese hinterland and beyond.

Notes

1 The Economist Intelligence Unit (EIU) estimated that in 1994 there were between 700,000 and 1.3 million passenger cars in China, depending upon the data source. The same source two years later estimated that passenger car production in China increased approximately tenfold between 1980-1990, and then doubled in each of the following years until 1996 (Dimitriou 1997).

2 In its Seventh Five-year Plan (1986-90) the Government of China designated the motor industry as a priority for the promotion of investment and a 'pillar of the national economy'. In March 1988 further plans were drafted to increase the country's vehicular production fourfold by the year 2000, with priority given to passenger car production (EIU 1994).

3 It is important to do this sooner rather than later since the huge road infrastructure investments in the territory and current rising motor vehicle growth trends suggest that Hong Kong's future is ultimately not too different from its Asian neighbours. This will make its successful public transport and traffic restraint policies more of historical note than a model for the future (see Chapter 12).

References

Adams, J. (1996) 'Can Technology Save Us?', *World Transport Policy and Practice*, MCB University Press, Vol. 2/3.

Dimitriou, H.T. (1992) *Urban Transport Planning: A Developmental Approach*, London: Routledge.

——— (1997) Toward a Sustainable Urban Transport Strategy for Medium-sized Cities in the Asia and the Pacific Region, Unpublished Draft Consultant Report to World Bank for UNDP and ESCAP, Bangkok.

Economist Intelligence Unit (1994) 'China's Automotive Industry and Market: Powerhouse of the 21st Century' *International Motor Business Research Export Examining Worldwide Automobile Industry*, 3rd Quarter, London: EIU.

Maddison, D., D. Pearce, O. Johansson, T. Litman and E. Verhoef (1996) *The True Costs of Road Transport*, London: Earthscan Publications, Ltd.

World Bank (1996) *Sustainable Transport: Priorities for Policy Reform*, Environmentally Sustainable Development, Transportation, Water and Urban Development, Washington D.C.: World Bank.

10 Policy and planning responses to transport-related air pollution problems

William F. Barron

Introduction

Hong Kong, as indicated in previous chapters, has already achieved much in terms of improving its transport services by making public transport widely available, affordable, and greatly utilised. Yet additional steps are needed for the transport system to remain effective in the face of growing demands and to lessen the threat to environmental quality this growth poses.

This chapter looks at air pollution issues associated with Hong Kong's transport and assesses options for expanding services and making the transport system itself less damaging to the environment. While a larger context is also considered, the major focus of the chapter is on remedial measures for air pollution from road transport, since this is where one of the most serious environmental problems occurs. Because problems of this kind will increase if projections for growing traffic congestion are borne out, what is needed is a more effective dampening of the growth in road traffic, coupled with steps to make those vehicles which are on the road far less polluting, particularly in terms of on-road emissions.

Air quality environmental considerations

Air quality

For the most part, Hong Kong does not at present experience prolonged periods of extremely poor air quality which characterise a number of other major cities in the region (e.g., Bangkok, Taipei, Shanghai)[1]. In part this is due to its geography (e.g, a marine location with ocean breezes and a mountainous terrain which precludes vast urban sprawl development). Another important factor limiting air quality problems is the territory's transport system. Hong Kong enjoys an extensive and relatively inexpensive system of public transport which provides most of the journeys in the territory (see Chapters 4 and 5).

Nevertheless, air quality is far from good, especially in terms of respirable particulates, and the transport sector is the major source of these and other types of air pollution (e.g., nitrous oxides (NO_x) carbon monoxide (CO)). Ambient air quality in much of Hong Kong fails to meet the minimum air quality objectives (AQO) for particulates, while levels of sulphur dioxide (SO_2) and nitrogen dioxide (NO_2) often come close to failing to meet the minimum standards represented in the AQOs (Environmental Protection Department, Environment Hong Kong 1995).

Monitoring

Measured concentrations for various air pollutants in Hong Kong for 1995 are shown in Figure 10.1. With one exception (Mong Kok), the ambient air quality indicators are taken at roof-top level some twenty to fifty metres above street level and often much higher above heavy traffic spots in hilly areas. It is important to note that the measured levels of most pollutants are much higher in the commercial-residential area of Mongkok where air quality is measured at street level. Occasional street-level measurements by the Environmental Protection Department (EPD) show that the street-level air pollution in Hong Kong's areas of congested traffic is much higher than indicated in Figure 10.1. For example, a temporary roadside monitoring programme in 1992 showed average readings of TSP levels of 350 μm³ compared with

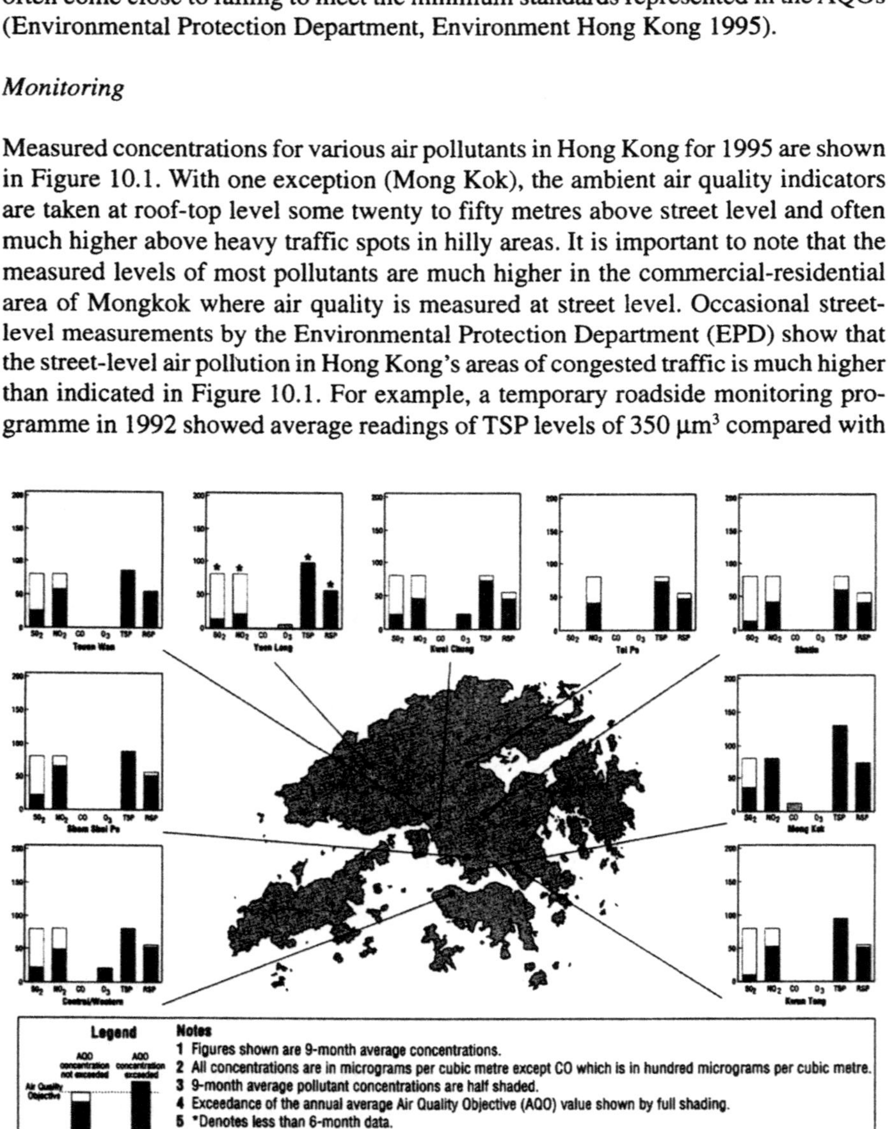

Figure 10.1 Annual average concentrations of air pollutants, 1995
Source: Environment Protection Department

AQO average annual concentrations of 260 μm^3 for maximum daily concentrations (EPD, Environment Hong Kong 1993).

While the one street-level monitoring station in Mongkok and the occasional short-term measurements provide valuable information on pollution exposure risks, the fact is that no one knows precisely the quality of the air people breathe on the streets of Hong Kong. However, there is obvious cause for concern.

Transport as a source of air pollution

Air quality in Hong Kong deteriorated significantly during the 1980s especially with regard to sulphur dioxide (SO_2), particulates, and nitrous oxides (NO_x). In 1988, the EPD estimated that more than half of the territory's population was exposed to unacceptably high levels of SO_2, and that particulates were higher than in previous years, with high concentrations over virtually the entire territory (EPD, Environment Hong Kong 1989). Sulphur dioxide levels dropped substantially following the mid-1990 ban on high sulphur fuel, which had been extensively used by industry. However, levels of other pollutants continued to be worrying (EPD, Environment Hong Kong 1991).

Motor vehicles contribute about 75 per cent of the total urban respirable suspended particulates (RSP) and NO_x in the territory. About two thirds of vehicle kilometres of travel in Hong Kong are by diesel-fuelled vehicles whose major pollutants are particulates and NO_x. Of the vehicular-related NO_2 emissions (NO_2 being one form of NO_x), an estimated 78 per cent are from diesel vehicles, while an even greater proportion of vehicular RSP comes from diesels (EPD, Environment Hong Kong 1993).

SO_2 is linked to respiratory health because it aggravates heart and lung problems. In Hong Kong as a whole, motor vehicles contribute a modest share of SO_2. However, in congested areas with slow-moving traffic and high numbers of pedestrians (such as the commercial areas of Central, Causeway Bay, and Tsim Sha Tsui), local concentrations of vehicle-related SO_2 could pose a significant health risk.

Probably more important, however, is the fact that motor vehicles in the territory contribute an estimated 45 per cent of the fine particles emitted to the air (EPD, Environment Hong Kong 1991). Since RSP pass relatively deeply into the lungs, where they provide a mechanism for presenting mutagenic material to cells over a long period of time, they constitute a health problem (Royal Commission 1991). They are also largely invisible and can be a problem even when there is no visible smoke from a vehicle.

Larger-sized suspended particles, TSP, are a combination of uncombusted carbon and miscellaneous other material which are seen as black smoke. These cause eye, throat and lung irritations. NO_x (especially NO_2) are another health hazard and are believed to contribute to the greenhouse effect, added to which pollutant emissions in typical diesel vehicle exhausts are thought to be cancer inducing (Royal Commission 1991).

In mid-1990, industrial sources of SO_2 were greatly reduced in Hong Kong through regulations forcing industry to switch from high sulphur (2.5 per cent sulphur content) fuel to low sulphur (0.5 per cent) diesel (EPD, Environment Hong Kong 1991). This also had the effect of moderately reducing particle levels in industrial areas. However, even these limited improvements had a significant measurable effect on the respiratory health of children in a heavily industrial area of Hong Kong (Hedley et al 1993). With these reductions in industrial air pollutant emissions, the transport sector of Hong Kong has emerged as the most prominent source of local air pollution, especially particulates (Barron 1993).

Despite the reduction in industrial air pollution in Hong Kong, minimum air quality objectives (AQO) continue to fail to be met with regard to TSP, and to a lesser extent NO_2. In a number of cases, air pollution levels in the territory have exceeded the AQO in areas with relatively little or no industry (e.g., Mongkok, Sham Shui Po, Tsim Sha Tsui) (see Table 10.1). Even in Central/Western (an area completely without industry), average annual concentrations of TSP and RSP exceeded permitted standards in 1994. At EPD's only street-level monitoring station in Mongkok concentrations of TSP and RSP are, as already indicated, well above the AQO for annual average pollutant concentrations, being double for TSP and about 25 per cent over the RSP objectives (EPD, Environment Hong Kong 1995).

Congestion

Figure 10.2 shows projected road usage for the year 2001 compared with the base year of 1986. Person trips are projected to increase by about 40 per cent by 2001 compared with the base level of 1986 (Transport Branch 1990). This would lead to an increase of about 38 per cent in public transport boardings (including a doubling of light rail travel demand), and an increase of about 152 per cent in the number of private cars and motorcycles.

In addition, the number of goods vehicles is projected to more than double by 2001 from the 1986 level. Underlying such projections is the steady growth in real income (which leads to increases in both private car ownership and increased freight movements to meet higher consumption levels) and the decentralisation of population as housing units are developed outside the urban core area. Although new road projects are underway or planned, it is unlikely that Hong Kong can avoid increasing congestion, and as average travel speeds fall and serious bottlenecks become more common.

Table 10.1 Air pollution and motor vehicles in Hong Kong

	TSP	RSP	NO_2
Mongkok street-level pollution in 1993 as fraction of minimum acceptable average annual AQO[1]	200%	125%	95%
Motor vehicle contribution to emissions[2]		45%	75%
Diesels as a fraction of vehicle emission source in Hong Kong[3]		98%	82%

Source: *Environmental Protection Department 1. Environment Hong Kong 1993, 2. Environment Hong Kong 1991, 3. Environment Hong Kong 1990*

Figure 10.2 Composition of road traffic 1986 and 2001

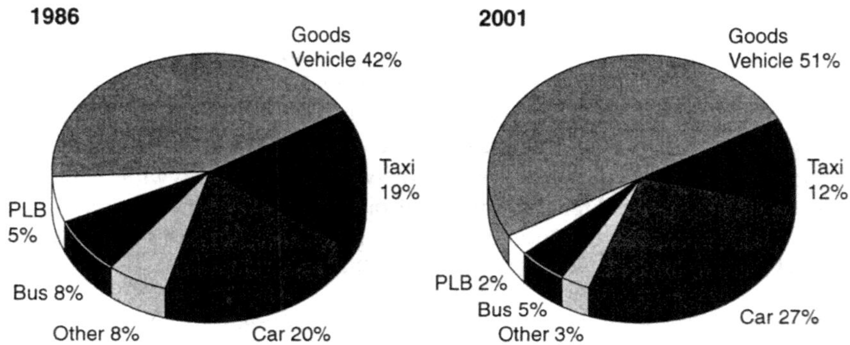

Source: Transport Branch

Air pollution and motor vehicles

Vehicle composition and pollution

Unlike many other high income economies, the private motor car is not the major source of transport-related air pollution in Hong Kong. Private cars are overwhelmingly petrol fuelled and since 1992, new petrol cars effectively must have catalytic converters and run on unleaded petrol in order to meet emission standards. Petrol cars account for just over one third of vehicle kilometres travelled by road vehicles with diesels accounting for the rest (EPD, Environment Hong Kong 1990).

Probably the most important contribution of private motor cars to air pollution in Hong Kong is somewhat indirect: it is the private car's contribution to traffic congestion which tends to lead to greater emissions from diesel vehicles and the resulting risk of people being exposed to them in congested areas. Despite dips in vehicle ownership in 1973-76 and 1982-85, private car registrations generally have shown a strong long-term upward trend. By the 2001, road traffic is projected to increase by 130 per cent from the 1986 level, with projected increases in the number of private cars being a significant contributor (Transport Department 1995). During this same period the road network is projected to increase by 27 per cent (Transport Branch 1990).

Average peak hour travel time in Hong Kong - 22 kilometres per hour (kph) in 1986 in the urban core area - is projected to fall to about 15-17 kph by 2001 (only about three times the speed of a brisk walk). With much slower traffic, the density of vehicles and the level of emissions over a given period of time in a given area tend to increase. Hence, unless such congestion can be reduced or the vehicles themselves be made much cleaner, Hong Kong faces worsening (rather than stable) air quality in the coming years.

An overview of basic options

The existing passenger transport system

With Hong Kong's very high population densities, naturally heavy reliance is placed on public transport. According to the latest Travel Characteristics Survey (1992) about 80 per cent of all journeys are by public transport, which is up from 78 per cent in 1981 and from the 73 per cent share for 1973 (MVA Asia 1993) (see Table 10.2). Of all public transport journeys about one-third are by rail while another one-half are by bus. One noteworthy feature of public transport in Hong Kong is the extensive use of taxis which account for about 12 per cent of all public transport journeys (Information Services Department 1997).

Table 10.2 Features of Hong Kong transport systems for 1992

	Length of route (km)	Av. daily pax journeys (million)	Number of vehicles (thousand)	Kilometres travelled (million)
Rail	121	6.6		
Road	1,559	16.3		
All buses				642
- Large bus		3.5	8.4	
- Light bus		1.8	6.7	
Taxi		1.3	17.7	1,811
Private car		0.7	265.8	3,566
Motor cycles		<0.1	26.1	
All goods vehicles				3,959
- heavy/medium goods vehicles			34.9	
- light goods vehicles			105.6	
Other vehicles			5.6	
Ferries		0.2	0.08	

Source: MVA Asia Limited, 1993

The territory has several extensive electric rail systems, including the world's busiest underground mass transit railway (MTR) serving the main business area and many of the residential areas. Electric-powered rail systems are, of course, not pollution-free. These systems simply produce most of their emissions at the power plants where the electricity is generated.

However, from an air pollution control standpoint large stationary sources of air emissions are far easier to control than numerous small mobile sources. Another advantage is that even when emissions do escape they tend to be released through tall stacks which carry them away. While certain important negative effects remain (e.g., CO_2 and its greenhouse implications, SO_2 and acid rain impacts), in general air quality for the local population is significantly improved where much of the transport is electric-powered. In 1990, transport accounted for about 26 per cent of all primary energy consumed in Hong Kong, with electricity generated for railway systems representing about two thirds of total energy for transport[2].

Yet despite the high use made of rail travel in the territory, in Hong Kong as in virtually every city in the world, most movements of people and goods are by road-based transport vehicles (see Chapter 6). In 1996, large buses carried 3.8 million passengers daily on a network of 464 scheduled routes. Mini-buses carried about 1.7 million passengers daily. Taxis carried a daily average of about 1.3 million passengers (Information Services Department 1997). As already highlighted in several other chapters in this book, the private motor car plays a rather modest role in Hong Kong's transport on a day-to-day basis. Probably between 0.4 and 0.8 million passenger journeys are made daily by private cars (Barron and Hills 1994).

What tends to separate Hong Kong from many of the more established higher income economies is that its road-based public transport is dominated by diesel vehicles. Unfortunately, the basic transport-related environmental problems in Hong Kong stem to a great extent from these road transport modes, of which the diesel vehicle is the most problematic.

Basic options for passenger transport

In deciding how to tackle these problems, one can consider several basic strategic planning options. These are shown in Table 10.3 and may be described somewhat more fully as follows:

- lowering the need for transport services (Option 1);
- increasing the reliance on transport modes with higher efficiency in moving people and goods (Option 2);
- improving the efficiency of existing road-based vehicles (Option 3);
- making changes in the fuel system to those vehicles which are either inherently cleaner or which do not leave the pollutant emissions along their routes (Option 4);
- cleaning up exhaust streams from those vehicles which send out their pollutant emissions along their routes (Option 5).

Table 10.3 Basic strategy options for transport planning

Option	Potential policy instruments
(1) Alter demand for transport services	Land-use planning
(2) Alter modal choices	Infrastructure provision Route restrictions Pricing
(3) Improve fuel efficiency	Mandates Pricing
(4) Alter fuel choices	Mandates Pricing
(5) Clean up exhausts	Mandates Pricing

Source: Barron and Leung 1991

It might at first appear that the appropriate response is relatively straightforward, and would involve:

- Step one: reducing the use of road-based vehicles to the minimum through a combination of Options 1 and 2 (i.e., generate fewer road-based trips and more mass transit trips).
- Step two: switching the remaining road-based vehicles to zero (local) emission options (e.g., electric battery cars and trucks).

Road and rail transport

It should be noted in this context, that significant practical considerations will limit the extent and pace at which such changes might be implemented. The question of what the minimum role for road-based public transport vehicles could be in an urban landscape such as Hong Kong is a matter of conjecture. Despite its appeal, long-term prospects for reducing transport needs through the substitution of telecommunications for travel remains speculative. While efforts continue to encourage flexible work hours as a means of reducing peak-hour traffic and moving toward an off-peak hour spread focused on mass transit, these measures also have limited potential. This is because many jobs are determined by the needs of work schedules in the increasingly important service sector and because utilisation on much of Hong Kong's transport system is already relatively high during many non-peak periods (e.g., lunchtime, weekends).

Over the longer term, the introduction of more environmentally conscious urban planning measures such as the creation of more employment opportunities near residential areas and making non-motorised travel safer and more comfortable than it is today in much of the territory should also be emphasised. However, having said this, if the past is any indicator for the future, progress in this regard in Hong Kong is likely to be piecemeal, and will take years or decades to have much positive impact.

It is clear that a reduced dependency on the current role of road-based public transport in Hong Kong will require a relatively long lead time of anything up to 15 years to allow for the full development of alternative infrastructure, and for the re-design and modification of the urban setting to support more fully other options such as considerable expansion of rail-based systems. While the role of heavy or light rail transit systems certainly needs to be expanded in Hong Kong (see Chapters 5, 8 and 12) such developments are appropriate only where population densities are relatively high (i.e., in the range of about 500 persons per hectare close to stations, where each station has a 'catchment' area of about 8 hectares at such density (CUPEM 1993)). Government continues, even today, to show considerable disinclination to provide incentives for extension of mass transit rail systems to areas of somewhat lower density as a means of reducing environmental and congestion externalities associated with road traffic.

In contrast to the relative inflexibility of rail transport, the service levels (and the average cost of such services) of bus/taxi-based public transport systems can be scaled up or down relatively quickly to match demand. Rail, with its high front-end investment, and the requirement for financial viability for both track and vehicle operations (unlike road transport systems where the track is subsidised/funded by the public sector), is invariably more inflexible.

This fact has led 'prudent' decision-makers in government to invest in rail and other fixed-route public transport services only where there appears to be a good probability that load factors will be sufficiently high to keep the costs of each individual journey within pre-specified limits of financial viability. There is thus a continued, if not increased, dependency on road transport, particularly if one takes into account the infrastructure proposals made by the Port and Airport Development Study (Lands and Works Branch 1989) notwithstanding the recommendations made by the Rail Development Strategy (Transport Branch 1993) (Dimitriou 1994). In the short and mid-term, this in turn, guarantees that transport-related air quality problems in the territory will not only continue to grow but will increase dramatically, unless more effective measures are taken to lower future road transport demand or to make road vehicles much cleaner.

In North America and Europe, much of the current focus in dealing with the amelioration of transport-related environmental problems is to alter modal choice, basically by attempting to persuade trip-makers to abandon their cars in favour of trains and buses, particularly for the journey to work. Yet in Hong Kong, car ownership remains very low at less than one car per 20 persons (i.e., about one private car for every four or five families) (Information Services Department 1997), despite per capita GDP figures of almost US$23,000 (World Bank 1997) which solidly places the territory among the world's higher income economies.

The costs of car ownership in Hong Kong are relatively high in comparison with other better endowed economies, with a 100 per cent duty on new vehicles and an annual registration fee in the range of HK$6,000 (US$775). Among public transport journeys, rail accounts for about 30 per cent of trips (Government Information Services Department 1997).

These figures are impressive compared with those of most urban centres in wealthier economies and suggest that the moves being proposed in these other places have far less scope in Hong Kong. Hence, while some additional gains could certainly be made through further changes in transport mode choice, Hong Kong already has an impressive modal mix by world standards. The real issue is whether this mix can be maintained in the face of new pressures created by current infrastructure proposals.

The relatively high cost of car ownership, along with high traffic densities and limited parking facilities, has by and large encouraged small vehicle sizes so that there is only moderate scope for improving fuel efficiency. Hong Kong's 18,100 taxis account for about 12 per cent of all public transport journeys (see Chapter 6). They are generally small, with seats for four or five passengers. Mini-buses, on the other hand, account for another 16 per cent of public transport journeys. They generally have high load factors and, it would appear, offer little room for significant expansion of the fleet (again see Chapter 6). Hence, like Options 1 and 2 cited earlier, the potential for Option 3 in Table 10.2 appears to be moderate at best.

All transport systems pollute in some form and to some degree. Heavily used electric rail systems pollute relatively little on a per person/kilometre travelled basis, but (like 'free-wheeling' electric battery cars and trucks) they add to emissions of SO_2, CO_2, particulates, and other pollutants from electric power plants. Vehicles powered by heat engines (e.g., the internal combustion petrol and diesel engines) emit most of their air pollutants directly along their routes.

In this respect, pollution from internal combustion engine vehicles is generally a more direct and serious health threat to urban inhabitants than are emissions from transport systems utilising electric power. This is true both because stationary and large pollution sources are more amenable to emission controls than are smaller and mobile sources, and because stationary sources are able to send their emissions into higher elevations where dispersal (dilution) is more effective. For the vast majority of people in Hong Kong (as in most modern urban environments), it is therefore chemical pollution (together with noise emissions) from petrol and diesel fuelled vehicles which typically present, on a day-to-day basis, the most immediate and potentially health-affecting form of environmental degradation.

As for making a significant fraction of road vehicles 'electric' or introducing some other form of zero-emission powered vehicles, it is important to underscore the fact that electric battery power systems are, despite some recent improvements still more expensive and less convenient than comparable internal combustion engine vehicles. While the day of the electric vehicle will probably come eventually, in Hong Kong this is likely to be in another one or two decades rather than earlier.

Hence, for the short- and medium-term, the internal combustion engine vehicle, with its locally polluting exhaust emissions, will continue to play a major (even growing) role in Hong Kong's passenger transport. In reality, this leaves policy-makers with only a few options if they are effectively to lessen the role of the motor vehicle in the territory. One option is to employ urban planning practices that reduce the need to travel. Another option is to encourage urban densities compatible with the needs of operating viable public transport services, especially rail and possibly other forms of fixed route systems. Both these, and other options, are further discussed in Chapter 12 of this publication.

Freight transport environmental issues

While passenger transport has traditionally occupied the centre stage of transport planning, it is important to note the major role of freight transport in Hong Kong. As shown in Figure 10.2, goods vehicles are projected to account for just over one-half the road traffic in the territory by 2001. Freight movements include the transport of goods between Hong Kong and China and internal movements within the territory. About 24,400 vehicular border crossings per day were made in 1996, 91 per cent of which were by goods vehicles (Information Services Department 1997). The cross-border container truck trade increases congestion and pollution, particularly in the New Territories and in some parts of Kowloon.

Freight movements are also a major source of road traffic throughout much of the urban area of Hong Kong. On average, each goods vehicle makes about seven trips per day with most of these being single door-to-door trips delivering food, beverages and fuels (Ng and Wang 1993). It is the door-to-door service and perishable nature of some of the goods which makes truck freight an indispensable part of the transport system. Freight movements are in some ways even more problematic than passenger transport because the freight itself cannot easily be mixed and then separated and delivered by various modes of transport. Movement by rail is appealing, but only for relatively long distances in contrast to commuter rail systems. Since most deliveries are one way, there is inevitable low utilisation on the return trip for many types of goods vehicles. Basically, the options appear to be to make use of rail where this is feasible for cross border movement, to schedule goods deliveries for off-peak periods, and to make goods vehicles themselves as clean as possible.

The first possibility would require the building of a new rail link between the container port and the Chinese border (belatedly incorporated in the Railway Development Strategy). Restrictions on the times for deliveries pose obvious inconveniences to businesses and are likely to be acceptable only where benefits in terms of road congestion are quite clear. In this regard, some form of time-of-day road pricing would be quite useful, since, if the prices were properly set, this would allow the market to determine which types of goods deliveries should be carried out during peak periods and which could inexpensively be postponed to off-peak hours.

Finally, the issue of making goods vehicles in Hong Kong cleaner is closely related to the matter of how to make road passenger transport cleaner, since virtually all goods vehicles in the territory are diesel. The next section of this chapter considers the matter of diesel vehicle emissions.

A look at diesel vehicles

Advantages and disadvantages of diesel-fuelled vehicles

Diesel vehicles, while potentially only moderately polluting, in practice, tend to be highly polluting due to requirements for relatively frequent engine tuning to achieve optimal performance. For lighter vehicles in which the petrol engine does not offer overwhelming disadvantages in terms of fuel efficiency and power (e.g., for cars and mini-vans) petrol engines with catalytic converters have far lower emissions of particulates, NO_x, and SO_2. However, clean petrol systems are more expensive to operate due to poorer fuel efficiency and hence their use (instead of diesels) results in modest increases in CO_2 emissions compared to the same-sized diesel-powered vehicle.

In the case of heavy diesel vehicles (e.g., double-decker buses and large goods vehicles), the efficiency advantages of diesel engines over petrol systems are significant. As a result, the diesel engine dominates these larger vehicles throughout the world. Hence, at least for the larger diesel vehicles, a switch to petrol is not appealing.

Within diesel technology itself, much can be done to lessen emissions in a manner less dependent on high levels of maintenance. The basic strategy in the past has been to improve the engine design, particularly the fine control of air-fuel mixtures. Other efforts have concentrated on exhaust after-treatment, particularly with regard to trapping particulates. Ultimately, a chemical catalyst system, such as that for petrol vehicles, would be highly desirable. Indeed, some progress has been made in this field, though commercial systems still seem to be some time away (Royal Commission 1991). Overall, reliable low emission diesel technology has been much slower in becoming truly off-the-shelf, than have the three-way catalytic converter emission control systems for petrol vehicles.

Fuel switching

In terms of the Options listed in Table 10.3, EPD has shown a preference for Option 4 which entails switching fuel systems (at least for lighter diesel vehicles) and Option 5 for cleaning up diesel vehicle exhaust streams.

Although the pollution implications of private motorcar use are much less of a problem in Hong Kong than in most higher income countries, it is nonetheless important to consider developments in the territory in this area. From 1 April 1991, unleaded petrol became available in Hong Kong. In 1992, more stringent emission standards were then imposed by government. This measure effectively required the use of catalytic converters (which must run on unleaded fuel) for most new petrol-driven vehicles (EPD, Environment Hong Kong 1993).

One factor which is sometimes overlooked in this regard is that while unleaded petrol itself involves some trade-offs in pollutant emissions (e.g., lead versus benzene), motorcars with catalytic converters have a clear net benefit over those without, and, as noted above, unleaded fuel is necessary for the catalytic converter to function properly. Hence, in Hong Kong, as elsewhere where unleaded petrol has been mandated, the primary goal is to allow new cars to utilise catalytic converters. The full benefits of this are only felt as the vehicle fleet gradually turns over during a five to ten year period. The point being made here is, simply, that the issue of introducing unleaded fuel should be evaluated with regard to the time at which most of the fleet is using catalytic converters, i.e., over the long- rather than short-term.

In Hong Kong, fuel-switching is most important for its indirect rather than direct impacts in so far as it helps to bring about the existence of a clean petrol alternative for smaller diesel engines. For diesel vehicles, the basic alternatives are selecting 'normal' engine designs and choosing 'state of the art' engines. Both of these design options are coupled with sub-choices of either low levels of engine maintenance or high levels of engine maintenance.

Emissions of concern from vehicles with engines using leaded petrol, unleaded petrol without catalytic converters, unleaded petrol with catalytic converters, with poorly maintained standard diesel engines, and optimally tuned technically advanced diesel engines are shown in Table 10.4. From the general information shown, petrol vehicles with catalytic converters are generally superior to those without catalytic converters in terms of overall pollutant emissions. Optimally tuned diesel engines, however, are probably comparable to petrol vehicles with catalytic converters in terms of overall emissions.

The catalytic converter has the potentially major advantage of involving low maintenance requirements for much of its normal life (e.g., 70,000-120,000 kms),

Table 10.4 Major vehicle emissions

	NO_x	RSP	Smoke	CO_2	CO	O_3	Lead	HC
Engine and fuel system	□			■	□	□	■	■
Leaded petrol w/o catalytic conversion	□			■	□	□		■
Unleaded petrol with catalytic conversion				■		□		
Normal diesel poorly tuned	■	■	■	■				■
Advanced diesel with best tuning	□	□		■				

□ *represents a moderate concern*
■ *represents a major concern*

Source: Barron and Leung 1991

so long as it is used with unleaded fuel. In contrast to the relatively low maintenance requirements of the catalytic converter and unleaded petrol system, present technology for equivalent low-emission diesel systems mean that they require relatively frequent maintenance. This consideration has potentially major implications for the implementation of a practical emissions reduction programme. Furthermore, unlike the catalytic converters where purposeful tampering (e.g., by using leaded fuel) gives relatively little benefit to the operator, the operators of potentially low emission diesels may be tempted to alter engine settings so as to 'overwork' the engine at the expense of the emission control systems. Hence, it is the uncertainty about maintenance which is perhaps diesel's greatest drawback in terms of practical emissions control efforts, as distinct from any questions regarding theoretical potentials.

In large part, therefore, the major contribution of diesel vehicles to Hong Kong's air pollution has to do with the difficulty of ensuring that all diesel vehicles are properly tuned. The Transport Department has an annual inspection programme for all public service passenger vehicles and for goods vehicles over twelve years of age. The EPD also operates a 'smoky vehicle' control programme in which spotted smoky vehicles are required to be presented for examination. Lorries, PLBs and taxis are the primary offending vehicles in terms of visible smoke emissions (EPD, Environment Hong Kong 1995). Legislation was passed in the Hong Kong Legislative Council in January 1991 to further enhance the EPD programme by including privately operated motorcar testing centres to examine smoky vehicles. Yet, as is evident on Hong Kong's roads, smoky vehicles remain very common despite the prohibitions, the efforts of trained spotters, and the inspection requirements. It is presumably the striking failure of efforts to eliminate (or even to make relatively uncommon) the sight of heavily smoking diesel vehicles on the streets of the territory, that increases the inclination of EPD to move towards relying less on maintenance and inspection as a system of emissions reduction, and place more emphasis on switching lighter vehicles to unleaded petrol with a catalytic converter system.

In preparation for its planned moves toward lighter diesel vehicles, EPD embarked on a two-step programme to make petrol vehicles much cleaner in Hong Kong. The first step was the introduction of unleaded petrol in April 1991, and the second was legislation in 1992 regarding the introduction of stringent Japanese and USA emission standards for new petrol and diesel vehicles (EPD, Environment Hong Kong 1992 and 1993). For most petrol vehicles in Hong Kong, this is likely to lead to the use of catalytic converters for newly imported petrol motorcars.

Unleaded petrol in Hong Kong costs about $HK 1.00 per litre less than leaded petrol. This is due primarily to a tax differential in favour of unleaded fuel. For the vast majority of motorists, however, there is no other penalty associated with the changeover. Some vehicles require minor tuning adjustments and a few must continue to use leaded fuel. The air quality benefits to Hong Kong from this first step alone remain modest, particularly in the short-term when most petrol vehicles do not have catalytic converters. The real gains from unleaded petrol will come when most petrol vehicles have catalytic converters. While lower lead emission is desirable in and of itself, the most important point is that lead fouls the catalytic conversion system and therefore must be avoided in order to keep converters properly functioning.

Catalytic converters chemically alter exhaust gases, turning CO, hydrocarbons (HC), and NO, into CO_2, water, and N_2. Since CO is also a greenhouse gas, and the carbon in the HC is eventually oxidised into CO or CO_2, the catalytic converter does not add meaningfully to the total carbon-based greenhouse gas emissions relative to vehicles without such converters. The catalytic converter adds approximately $HK5,000 (US$640) to the cost of the average new motorcar, a modest cost increase relative to the typical after-duty price for a new economy-class motorcar of $HK120,000 (US$15,380) (Barron and Leung 1991). The average life of a petrol vehicle in Hong Kong is about seven years. This suggests a six year period or so before 85 per cent of the motorcars in Hong Kong will either be equipped with converters or meet stringent emissions standards. While this is slower than ideally desired, it is still a relatively quick change compared with many other efforts to clean up the environment.

With a much cleaner motor vehicle fleet (and eventually, other clean fuel systems as viable alternatives), it later becomes potentially more feasible to impose much stricter emission standards on new diesel vehicles as well. If new diesel vehicles have trouble meeting the more stringent emissions standards, would-be importers of these vehicles now have the option of switching to very clean alternatives (i.e., vehicles with catalytic converters operating on unleaded petrol). This is particularly true for lighter weight diesel vehicles (most of which are goods vehicles).

As noted above, Hong Kong's new emissions standards also involve very stringent emissions control for diesel as well as for petrol vehicles. An alternative way to ensure that diesels continue to meet emissions standards over their lifetime under actual operating conditions is to require more frequent emissions testing. It is the costs and reliability of a greatly enhanced inspection programme relative to the fuel switching option which deserves careful consideration. In many respects, the best approach would involve a combination of the two.

A possible switch from diesel to petrol (or other clean fuels) and the more frequent inspection of remaining diesels have negative features, particularly in terms of costs and energy efficiency. The next section of this chapter considers these trade-offs.

Diesel/petrol trade-offs

Air quality gains, costs and penalties

The approximate expected trade-offs between air quality gains and cost and energy penalties of a switch from diesel to unleaded petrol for lighter diesel vehicles are given in Table 10.5. For petrol vehicles, the catalytic converter option has a moderate cost penalty, coupled with a small energy penalty and significant emissions benefits. It is worth noting that catalytic converter emission control technologies necessitate better engine management and combustion controls. Fuel economy improved significantly in the years following the introduction of catalytic converters on new motorcars for this and other reasons (Hillsman and Southworth 1990).

For diesel vehicles, improved engine emission requirements and frequent inspections (to ensure proper on-road maintenance) result in a moderate cost increase but without an energy penalty and with substantial emissions benefits. The option of encouraging lighter vehicles to switch from diesel to cleaner fuel systems has a moderate cost and energy penalty but the benefit of substantial emissions reductions.

As noted above, modern diesels can produce relatively low emissions. An alternative to fuel switching is to mandate use of modern diesel engine designs for new vehicles, ensure that the fuel is high quality (e.g., 0.2% or less sulphur content) and impose strict standards on emissions, with periodic inspections. However, even advanced diesels emit considerable amounts of particulates and these emissions are a

Table 10.5 Major expected impacts and trade-offs of alternative motor vehicle environmental policies

( ) denotes positive impact

	For petrol vehicles: require catalytic converter or emissions equivalent	For all diesel vehicles: require strict emissions standards and frequent inspection	For smaller diesels: encourage switch to clean fuel system (e.g., petrol with catalytic converter)
Cost & Energy Impact		**	**
Approx. cost penalty	$HK5000 for catalytic converter	$HK500/yr for cost of inspection	Approx. 10-20% higher fuel costs
Approx. energy penalty	insignificant		Approx. 10-20% increase
Net Emission Impact		*Note: runs risk of inadequate implementation*	
Lead	(90%)	NA	NA
CO_2	insignificant change	small reduction	10-20% increase
NO_x	(>90%)	(45%)	(55%)
RSP	insignificant change	(60%)	(40%)

* *Cost of unleaded & leaded net of tax is about same*

** *Given the extraordinarily high proportion of diesel vehicles usage in Hong Kong, the most important comparisons are between the two right-hand columns.*

Source: *Barron and Leung 1991*

Figure 10.3 Combined effects of the proposed control measures on motor vehicle particulate emissions in the urban areas

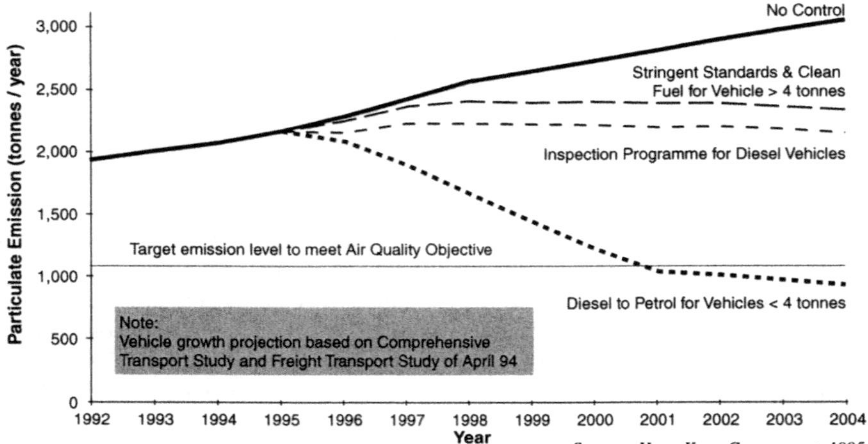

Source: Hong Kong Government, 1995

major source of public health concern, especially in congested areas. While the government will phase in even higher diesel fuel standards it is still pushing for mandatory switching of all lighter diesels to petrol (EPD, Hong Kong Environment 1995). As shown in Figure 10.3, the government argues that only the diesel to petrol switch will enable attainment of the air quality objectives (Hong Kong Government 1995).

For larger vehicles the basic option is to require better engines and fuel and impose strict and effectively enforced pollutant emission standards and testing. Such additional technology-based steps as particulate traps, might be employed. The basic policy options which the Hong Kong government has been actively considering with regard to addressing the diesel emission problem are:

- introduction of higher quality diesel fuel, strict emissions standards and inspections for all diesel vehicles; or
- couple the above option with a mandate to force certain categories of new light vehicles which are at present diesel-fuelled (i.e. mini-buses, light goods vehicles and taxis) to switch to engines using unleaded petrol with catalytic converters.

Under the first option, owners would decide whether the advantages of fuel efficiency and engine life outweigh the disadvantages of the diesel under a stricter emissions regime (i.e., greater off-road time and maintenance costs). The financial costs for the fuel-switching are likely to be in the range of roughly 15 per cent to 35 per cent higher. The improved diesel engine, higher quality fuel, and inspections approach would also raise costs, but not by as much.

The reason that EPD is arguing for the fuel-switching for lighter vehicles is that it is far more reliable and requires less on-going effort for inspections and enforcement. The concern is that the engine tuning which yields low emissions is not the same as the engine tuning which provides either greatest fuel efficiency or power. A related matter is that an inspection system for all of Hong Kong's 135,000 diesel vehicles might be difficult to implement and operate effectively. Since the majority of the existing diesel fleet in Hong Kong falls into the categories which might potentially be switched to petrol, this move would allow for easier enforcement of the diesel standards for the remaining much smaller group of larger diesel vehicles (Barron 1994).

One added dimension should be noted. Rusco and Walls (1994) argue that heavy vehicles account for most of the diesel emissions and hence should be the priority target. EPD maintains that even if heavy vehicles account for more total emissions, lighter vehicles tend to emit their pollutants in closer proximity to people and hence represent much more of a health concern. Rusco and Walls note that their data do not support EPD's contention in this regard, and that even in an area such as Mongkok (relatively typical of the congested, heavily pedestrianised commercial areas in Hong Kong), a large share of emissions are from heavier vehicles (Rusco and Walls, 1994). The concern on the part of EPD seems to be that, if the Rusco and Wall position is accepted that the larger diesels should be the priority target, there may be less opportunity to pursue emissions reductions aggressively for lighter diesels, which EPD continues to argue present the greatest health risk (Barron 1994).

The basic future options for substantially reducing road transport-related air pollution in Hong Kong are illustrated in Figure 10.4. Each of the major options is moderately costly. Yet, if Hong Kong is serious about eventually attaining relatively clean air for its inhabitants, then the diesel road-related emissions challenge must also be tackled head-on. For the imposition of frequent and regular inspection requirements on all diesel vehicles, together with moves to encourage a gradual shift

Figure 10.4 Major remaining future transport emission reduction options

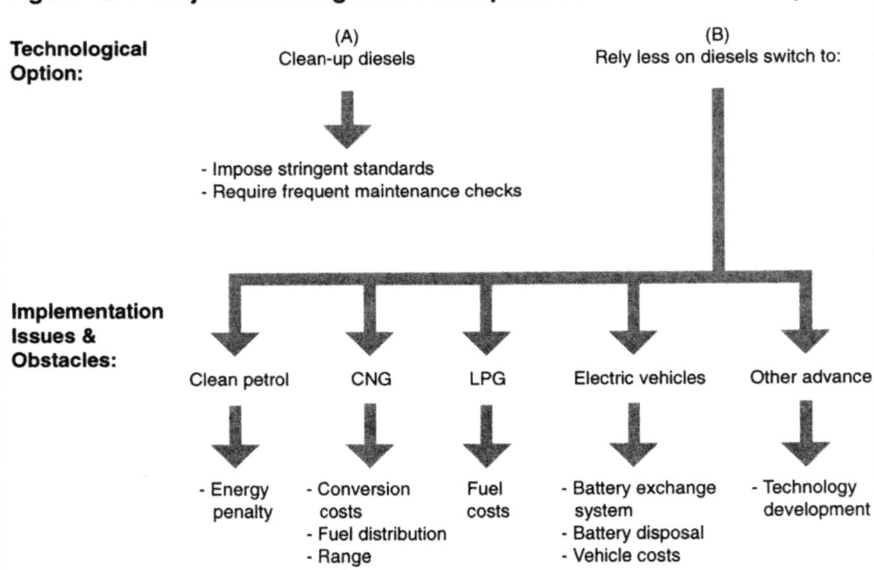

Source: Barron and Leung 1991

by lighter vehicles (e.g., mini-buses) away from diesel fuel to cleaner fuel systems (such as unleaded petrol with catalytic converters), should also be pursued. Over the longer term, however, other clean fuel options such as Compressed Natural Gas (CNG), or Liquified Petroleum Gas (LPG) may also deserve careful consideration. In addition, there may possibly be a role for electric battery vehicles in bus (and other) fleets, though probably not for private motorcars or taxis until a substantial technological breakthrough is made to improve battery performance.

Conclusions

As noted, diesel vehicle exhaust emissions are a major source of air pollution in Hong Kong, due in large part to their numerical dominance, their inadequate maintenance and inefficient operation, and to the long distances travelled by such vehicles relative to the size of the territory. In the short-term, options for environmental transport policy offer rather limited possibilities of air quality improvements through reductions in the demand for road transport services. Hence, in the short- and mid-terms the basic options relate to keeping the growth in road usage as constrained as possible and making the vehicles on the road far less polluting, particularly the newer ones as the fleets turn over.

In this context, the introduction of unleaded petrol (and its implications for the use of clean petrol road transport technologies) has created the potential for replacing the most worrying source of diesel pollution (i.e., older, smaller diesels vehicles) as these are retired, with new clean fuel systems. Such a move, while highly attractive from a pollution control perspective, is expected to take time to introduce and, like virtually any other major environmental improvement, will have important cost implications. In addition, this move should be coupled with measures to enforce higher emissions standards for remaining diesel vehicles regardless of whether some classes of lighter diesel vehicle switch to a clean petrol system or not.

The basic choice, then (as always), depends on how much the Hong Kong government is prepared to spend for specific improvements and how these costs will be imposed. The signs are that the government will move to impose much tighter emissions standards on vehicles and to expect that the costs of meeting these will ultimately be borne by the passengers and those whose goods are moved by road. More radical measures such as banning internal combustion engine vehicles in parts of central business districts or limiting far more streets or lanes to buses only, have so far not been favourably received by government. Whether such measures will receive more consideration if congestion and air quality worsen remains an open question.

For the longer term, urban and transport infrastructure planning would be improved if a more prominent role were to be given to environmental impacts assessments (EIAs) as part of project feasibility assessment and design. It is virtually a truism in planning that prevention is more effective and cost-effective than cure. This is typically true with regard to environmental impacts where better planning can take better advantage of natural conditions (e.g., wind patterns) to lessen the impacts of man-made pollutants; to employ alternative less environmentally damaging technology to accomplish the same ends, and reduce the need for transport or its consequent congestion in the first place.

For both short and longer term assessments we need to face the fact that we have been paying higher transport costs all along. However, part of this cost has been hidden because the effects of air pollution from road vehicles to health among other aspects, were inadequately understood and, in any case, did not have a dollar value attached. In an important sense, reducing the environmental effects of road transport does not raise the total costs, but rather shifts part of it from unpriced externalities to explicitly priced costs which are now becoming part of our internal calculations of the benefits and costs of moving people and goods from one place to another. As economists have long argued, such internalisation of externalities can be expected to lead to more appropriate decision-making with regard to the design of cities, transport systems and the use we make of them.

Notes

1 However, in November 1995, in the midst of several weeks of unusually high air pollution readings, the EPD issued a public warning for those suffering from asthma, announcing the worst air quality readings ever recorded in Hong Kong.

2 Alex Ng, personal communication 1993, and Energy Unit, Asian Development Bank (ADB) May 1989.

References

Asian Development Bank (1989) *Energy Indicators of Developing Member Countries of ADB*, Energy Planning Unit, May, Manila: The Asian Development Bank.

Barron, W. (1993) 'Energy Related Environmental Controls: assessing transport and industrial options in Hong Kong' *Energy Policy*, June.

——— (1994) *Market-Based Versus Direct Control Environmental Policy Instruments: conceptual and practical considerations with selected examples from Hong Kong*, the Centre of Urban Planning and Environmental Management, Hong Kong: The University of Hong Kong.

Barron, W. and P. Hills (1994) 'City Paper: Hong Kong' Programme For Asian Cooperation on Energy and the Environment (PACE-E), Sectoral Meeting of Experts and Policy Makers on Energy-Environment Strategies for Urban Transport Systems, May, The University of Hong Kong.

Barron, W. and R. Leung (1991) 'Road Transport and Air Quality in Hong Kong: a look at issues, plans and future options' in Boxall, J. (ed.), conference proceedings, *POLMET 91*, Hong Kong Institution of Engineers, Hong Kong.

CUPEM (1993) 'Supporting Urban Development Strategies for Metroplan', Student Urban Planning Workshop, Railway Development Group, The Centre of Urban Planning and Environmental Management, The University of Hong Kong (unpublished).

Dimitriou, H.T., (ed.) (1994) *Moving Away from the Motor Vehicle: the German and Hong Kong Experience*, The Centre of Urban Planning and Environmental Management, The University of Hong Kong.

Environmental Protection Department (1995) *Environment Hong Kong 1995* (also for previous years) Hong Kong:Government Printer.

Hedley, A., J. Peters, T. Lam, S. Ong, C. Wong, A. Tam, C. Betson, and J. Liu (1993) *Air Pollution and Respiratory Health in Primary School Children in Hong Kong, 1989-1992*, a report to the Environmental Protection Department, September, The Department of Community Medicine, The University of Hong Kong.

Hillsman E., and F. Southworth (1990) 'Factors That May Influence Responses of the U.S. Transportation Sector to Policies for Reducing Greenhouse Gas Emissions' *Transportation Research Record*, No. 1267, Washington D.C.

Hong Kong Government (1995) Cleaner Air: *Further Proposals to Reduce Emissions from Diesel Vehicles*; Consultation Paper, Hong Kong: Government Printer.

Information Services Department (1997) *Hong Kong 1997 Yearbook* (also for previous years) Hong Kong: Government Printer.

Lands and Works Branch (1989) *Port and Airport Development Strategy* Final Report. Hong Kong: Hong Kong Government.

MVA Asia (1993) *Travel Characteristics Survey*: Final Report, Transport Department, Hong Kong: Hong Kong Government.

Ng, Alex (1990) Chart of Detailed Energy Balance For Hong Kong, 1990 (personal communication).

Ng, S. and L.H. Wang, (1993) 'Freight Transport Management', in Wang, L.H., and A.Yeh (eds.) *Keep A City Moving: urban transport management in Hong Kong*, Tokyo: The Asian Productivity Organisation.

Royal Commission on Environmental Pollution (1991) *Emissions From Heavy Duty Diesel Vehicles*, September, London: HMO.

Rusco, F., and W. Walls (1994) *Clearing the Air: Vehicular Emissions Policy For Hong Kong*, Centre for Economic Research, The University of Hong Kong, Hong Kong: The Chinese University Press.

Transport Branch (1990) *Moving into the 21st Century - the White Paper on Transport Policy in Hong Kong*, Hong Kong: Government Printer.

——— (1993) The Railway Development Study, Highways Department and MVA Asia, Hong Kong.

Transport Department (1995) *Annual Transport Digest* (also for previous years), Hong Kong.

Wang, L.H., and A. Yeh, (eds.) (1993) *Keep A City Moving: urban transport management in Hong Kong, Tokyo:* The Asian Productivity Organisation.

Wang, L.H. (1993) 'Overview of the Transport System and Transport Management in Hong Kong' in Wang, L.H., and A. Yeh, (eds.) (1993) Tokyo: The Asian Productivity Organisation.

World Bank (1997) *World Development Report 1997* The World Bank, Oxford: Oxford University Press.

11 Institutional planning framework and effective land-use/transport planning

Colin J. Arnott

Introduction

The role of the public sector - of central and local government and of publicly owned development agencies - is pivotal to the land development process and provision of transport infrastructure and services throughout the developed and developing world. It is often (and increasingly) argued that in highly developed market economies such as Hong Kong (and the USA) the private sector is able to perform many of the development functions which, elsewhere, are the province of governments. In areas such as road and rail transport investment, power and water supply, and environmental health and management, the private sector in Hong Kong plays a very significant role.

However, even where transport infrastructure and development on land is private sector led, the government also plays an influential role in assembling, allocating and controlling land use and development (see Chapter 2). In Hong Kong, where land has been controlled through the Crown Land Lease system and has been 'created' through the process of land formation, the public sector has thus had a uniquely important role in the development of the territory.

The publication of the Territorial Development Strategy Review (TDSR) (Planning Department 1995) highlights the importance of public investment and government control in land and transport infrastructure in the territory. Whilst the Strategy has been reviewed within the traditional non-interventionary policy framework that the territory's government has always exercised, the TDSR seeks for the first time to link land use and infrastructure planning with the process of economic development and management. Thus the Review recognises that the process of economic restructuring - away from manufacturing - must be facilitated by the redevelopment of the older industrial areas for new uses as outlined in Chapter 8 (cited as Key Issue 1 in TDSR). It also recognises that the process of achieving 'job balance' requires the decentralisation of development and investment in public transport infrastructure (Key Issue 2); and that new patterns of land use and development can promote the financial viability of new rail-based passenger transport systems (Key Issue 5).

This chapter examines the increasingly significant role which the institutional framework - specifically the legal and financial mechanisms, policy environment and organisational arrangements - has exercised on land-use and transport planning in Hong Kong. A clear and successful formula has evolved for the process of land reclamation, infrastructure servicing and new development, in both the New Territories and the urban areas, during Hong Kong's period of rapid growth from the 1960s through to the 1980s and which continues today in the 1990s with the implementation of the Port and Airport Development Strategy (PADS) in the form of the core projects. The institutional arrangements through which this process has been conducted have been central to its success. However, as the urban environment and economy of the 'World City' of Hong Kong matures and confronts the problems of urban restructuring - including transport congestion, environmental degradation, urban renewal and economic restructuring - we must ask whether existing institutional arrangements will meet the needs of the new millennium.

This chapter commences by examining the evolution of the government role in land and transport infrastructure development in the territory. A distinction between the arrangements for land assembly, land-use planning and property development, on the one hand, and for transport and other infrastructure planning and implementation, on the other, is apparent. An examination of the effectiveness and adaptability of each is offered, drawing on the institutional experience of some of the recent development plans and projects in the territory.

Background to the institutional framework

Government context prior to the transfer of sovereignty

Land-use and transport planning in Hong Kong is principally the responsibility of government. Since 1841 the territory has been administered by Britain and this continued to 30 June 1997, when sovereignty was transferred and Hong Kong became a Special Administrative Region (SAR) of the People's Republic of China (PRC).

The written constitution of the British Administration gave authority to the Executive Council, the Legislative Council and the Governor's powers, which among other things, covered local legislation and the disposal of land. The government's most important and effective control over land use and planning lay in the fact that all land in Hong Kong was the property of the Crown and on this basis the government was empowered to sell or grant leasehold interests. The executive arm of the government under the Governor was headed by the Chief Secretary who was responsible for the formulation and implementation of policies and headed the Government Secretariat which was divided into policy and resource branches with the branch secretaries responsible for the direction of the government departments and agencies who implement the policies of their respective portfolios.

For an understanding of the institutional planning framework within government, the following brief outline of its evolution is helpful. From the time of its inception as a British administered trading centre, the overriding policy of the Hong

Kong government until recent times has been one of ensuring law and order and facilitating trade and other economic activity with a minimum of regulation and as little as possible of anything else. A former Chief Secretary, Sir Philip Haddon-Cave, described this as 'positive non-interventionism'. The first two chapters of Part One of this book show how little planning was undertaken before the Second World War; development was at a slow pace and the government did not need to have anything more than a Public Works Department to take care of the leasing of Crown land and the construction of highways, some land reclamation and other civil engineering works.

After the war-time Japanese occupation, the British Administration was re-established with little alteration but it was not long before the rapid changes in population growth, housing policy, the construction of new towns and the industrial and commercial development described in earlier chapters made it necessary for a revision in the structure of government to be made. With these changes, the Director of Public Works under whom town planning was inconspicuously tucked away in his Crown Land and Surveys office, found himself under the policy direction of a Secretary for the Environment who was also the policy director for transport. In the past twenty years, responsibilities for land administration, public works, planning and environmental protection has been shifted around the secretariat policy branches several times. Fairly frequent changes have also taken place in the distribution of functions among the departments that had previously been offices within the former Public Works Department.

The Town Planning Ordinance was passed in 1939 and with it a Town Planning Board was created. After the Japanese occupation ended in 1945 six years went by before the Board was re-convened in 1951. A small Town Planning Unit had been formed in the headquarters of the Public Works Department in 1947 to provide plans and surveys to aid the work of Sir Patrick Abercrombie in the preparation of his 1948 report on planning and development in Hong Kong (see Chapter 2). Among the recommendations in his report was the setting up of an office to draw up detailed development plans but five years went by before the Crown Lands and Surveys Office established a Planning Branch.

In Hong Kong, there has always been a strong 'market forces' culture in both the private sector and within the government, and for some years therefore there was a notable reluctance to provide adequate resources for planning. The major impetus to modern development planning came with the New Territories Development Department (now the Territorial Development Department (TDD)), created to implement the government's public housing policy. It had its own planning unit and meanwhile, in the government Secretariat, a Strategic Planning Unit came into being (see Chapter 3). Finally, in 1990, the planning functions were brought together in a new and separate Planning Department under the Secretary for Planning, Environment and Lands (SPEL). For the first time there was a policy branch and a department with 'Planning' in their titles.

Land-use planning that is environmentally acceptable is obviously essential but, as elsewhere, it took some time before environmental assessment and impact studies became an important part of the planning process. Advice on pollution control is provided by the Environmental Protection Department (EPD) which began life in 1981 as an agency under the Secretary of Health and Welfare. It became a full department in 1986 and is now under the policy direction of SPEL and is better placed to bring its influence to bear on land use and planning.

The institutional framework in government administration for transport followed a more orderly development but it took time for the government to accept the need to have a role beyond law enforcement and regulatory controls on the standards of service provided by the private sector. These basic responsibilities were scattered among different departments and it was not until 1965 that a Commissioner for Transport was appointed to head a Transport Office within the Government Secretariat. Various powers for licensing and the regulation of traffic were taken over from the police and the Public Works Department and the office became a separate department in 1968. Among its new responsibilities was co-ordinating the activities of the various government departments which affect traffic and transport. The Transport Department had virtually no planning capability, with the result that the early transport studies were carried out by consultants working to the Public Works Department. The Transport Department gradually became more experienced and capable and with the creation of a separate policy branch under a Secretary for Transport and a redistribution of functions between the department and a new Highways Department (previously an office in the Civil Engineering Department) the branch and the two departments formed a more cohesive framework within the government administration (see Figure 11.1).

Over the past twenty years both planning and transport have experienced a period of frequent organisational change brought about by the desire to have a structure in keeping with the changing circumstances created by the rapid economic development of Hong Kong. What further changes may transpire as a result of the political transition in 1997 is unclear but further change seems inevitable.

Policy and resources

Policy and resources for planning and transport are controlled by the branches of the Secretariat, specifically by the Chief Secretary, SPEL, the Secretary for Works and the Secretary for Transport. The co-ordinating and decision-making role of the Chief Secretary on major issues is exercised through the Land Development Policy Committee (LDPC) and for major projects separate committees are created: at the time of writing this chapter, for example, there is an Airport Development Steering Committee. In the LDPC the direction of planning and the implementation of policies are determined by the Chief Secretary with the relevant branch secretaries.

SPEL plays a key role in planning development strategies. The Secretary, through his branch and the Committees which he chairs (the Town Planning Board, the Development Progress Committee, the Port Progress Committee and the Environmental Programme Committee), can monitor and guide government policy and activities relating to land use, both strategically and locally, and to environmental protection. His branch provides policy direction to the departments for Buildings and Lands, Planning, Environmental Protection and to the Land Registry. The Secretary for

Works heads a resource branch under the Financial Secretary which is in some ways a recreation of the headquarters of the former Public Works Department. In addition to policy matters the branch acts as a manager for the allocation of resources to the six departments within the 'Works' group. It is the overall co-ordinator of the public works programme. The branch has an important consultative role in development and transport planning and this is conducted through direct inter-branch contacts

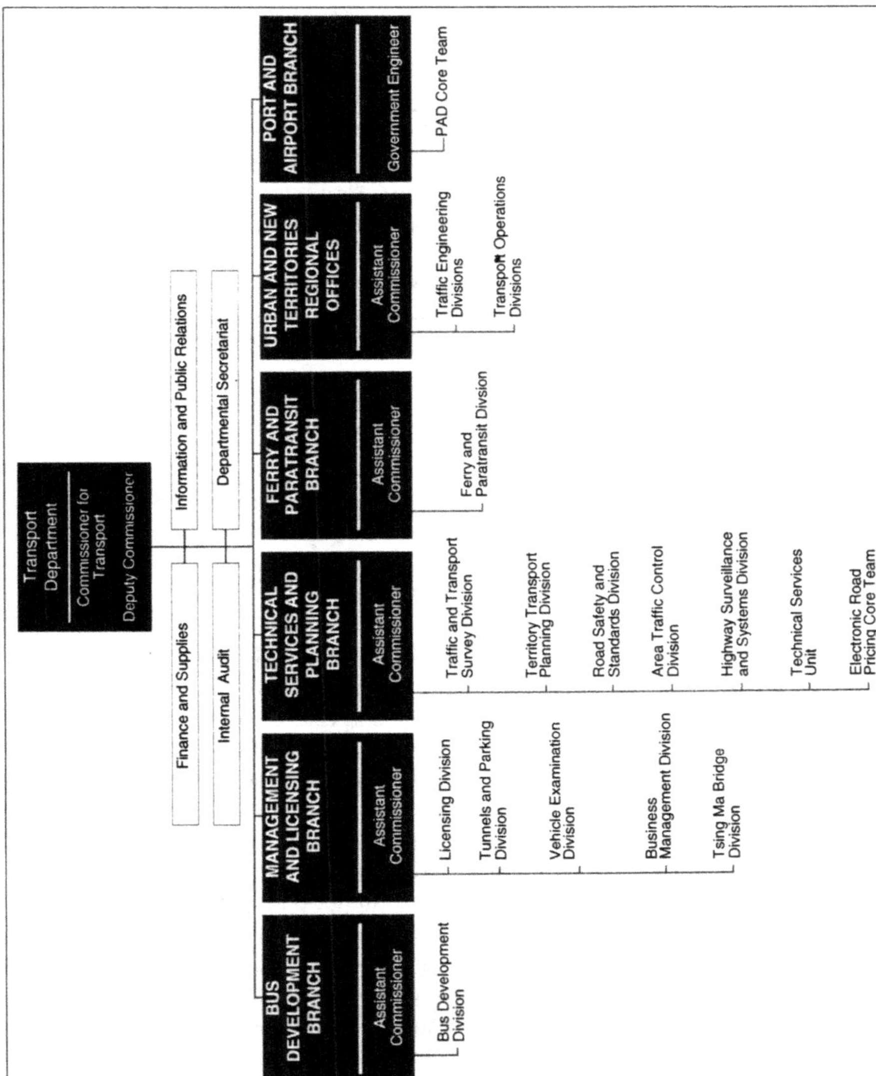

Figure 11.1 Organisational chart of Transport Department
Source: Transport Department

and cross membership in various branch committees. The Branch includes the New Airport Projects Co-ordinating Office staffed by a multi-discipline team drawn from the works departments. This office has both a planning and a works role, although, as the project progresses, the implementation stages will become more important. The office is an example of the type of temporary agency that has been successfully used for major projects; a similar arrangement was utilised for the construction of the Mass Transit Railway (MTR).

The Transport Branch is responsible for the formulation and monitoring of transport policies and strategies, and the steering of major transport studies, which currently include transport projects related to the port and new airport development. It acts as the policy branch for the Transport Department and the Highways Department, although the latter is also responsible to the Secretary for Works in respect of standards, contract procedures and coordination of the public works programme. The Secretary for Transport is the chairman of the Transport Policy Coordinating Committee which brings together representatives of the interested policy branches and departments.

The Secretary is a member of the Land Development Policy Committee, the Public Works Priority Committee, the Airport Development Steering Committee and the board of the Mass Transit Railway Corporation (MTRC). Other Transport Branch officials are members of the Town Planning Board, the boards of the road tunnel companies and public transport operators. It is the committee network which provides a convenient means of co-ordinating the transport aspects of planning and land use. Other Secretariat branches such as the Secretary for Economic Services and Secretary for the Treasury have a less direct interest in land use and transport planning, but have a significant role in the allocation of resources, resource planning, and development planning for the activities of the productive sectors such as industry, retailing, tourism and business services.

At the departmental level, most of the land-use planning which incorporates strategic transport development, is within the Planning Department. This department (as already indicated) was formed in 1990 following the creation of the Planning, Environment and Lands Branch in 1989 and it took over planning activities previously carried out by the former Lands and Works Branch, the Building and Lands Department and the New Territories Development Department. The Planning Department has, apart from departmental administration and technical administration divisions, two planning branches (see Figure 11.2); one is for territorial and sub-regional planning (the sub-regions being Metroplan, North West New Territories and South West New Territories) with units for the port and airport development strategy and transport studies. The second branch is for district planning and includes support staff for the Town Planning Board, an urban renewal section and district planning offices for the Metro district, Hong Kong, Kowloon, Tsuen Wan, New Territories, Lantau and Islands, Sha Tin and North East New Territories, Tuen Mun with Yuen Long and Sai Kung with Tseung Kwan O.

The Transport Department has been involved in a number of transport studies mostly carried out by consultants but previously had limited resources for planning work itself (see Chapters 3 and 7). However, it now has divisions for territory trans-

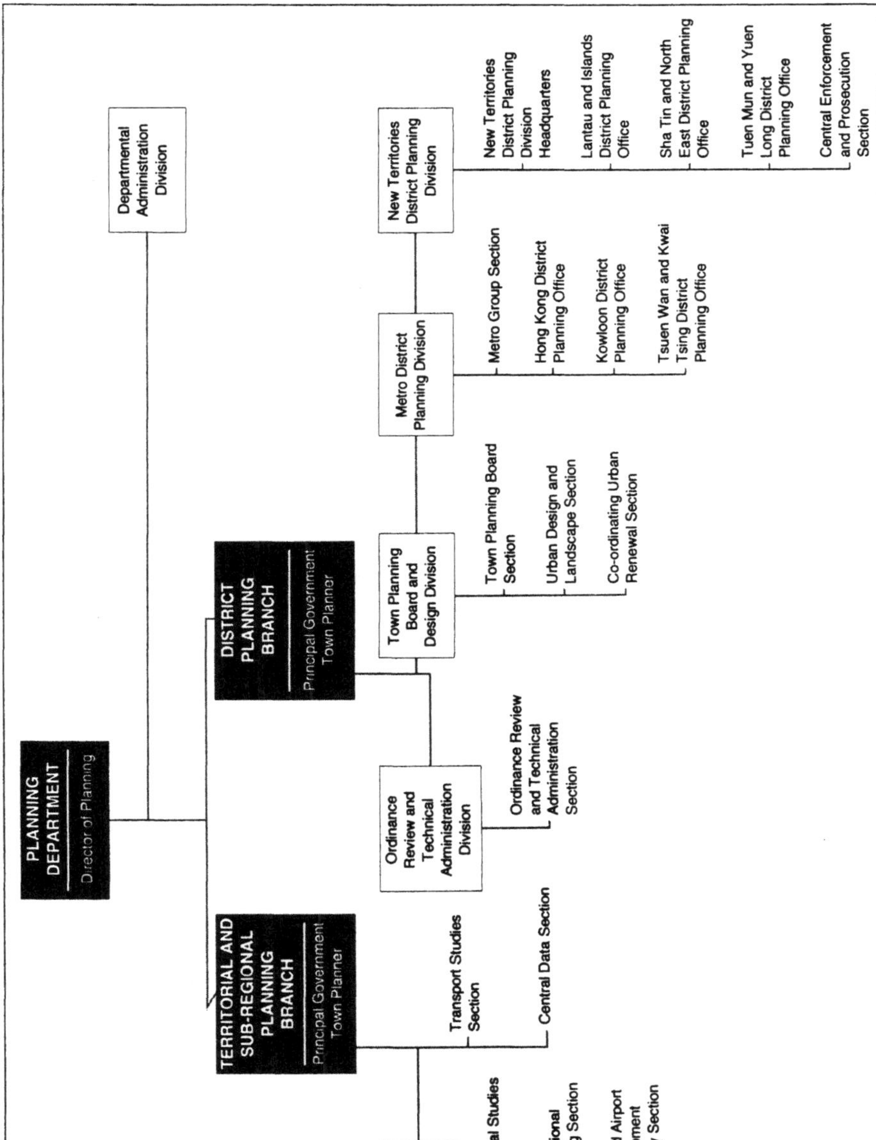

Figure 11.2 Organisational chart of Planning Department
Source: Government Information Services

port planning and the transport elements of port and airport development. Its interest and expertise is more directed towards transport operation than land use but this complements the strategic planning carried out by the Planning Department. In carrying out their work, the Transport Department and the Planning Department have contact and liaison with each other, with their respective policy branches and the departments of government which have interests and responsibilities in land-use and transport planning. In addition, there is a consultative network with other government agencies including the Housing Authority, the Land Development Corporation (LDC), the Hong Kong Airport Authority and the two railway corporations, the MTRC and the Kowloon-Canton Railway Corporation (KCRC).

Government corporations

As in the case of many other developed and developing countries, autonomous, or nearly autonomous, corporations and other government owned agencies are a feature of the Hong Kong Administration - and some of these are significant actors in the process of land-use and transport planning, and the implementation of infrastructure and urban development. The Hong Kong Housing Authority, for example, accommodates over three million people in 146 public housing estates and is continuing to build around 40,000 new housing units each year (Housing Branch 1997). The Authority's development plans, particularly in the new towns, have to be closely integrated with the government's plans in respect of population distribution, land-use development and the provision of public services, including transport. The Authority itself is independent - it is staffed by the civil service and this is conducive to smooth relations with the government departments involved in planning. On a smaller scale, the LDC, established in 1988, is involved in joint ventures with the private sector for urban redevelopment. The Corporation works in close cooperation with the Lands Department but has recently been the subject of a review of its status and operation as part of the Government's review of Urban Renewal Policy (see Planning, Environment and Lands Branch 1996).

In transport, the two government owned railway corporations, the KCRC and MTRC, have become increasingly important to the territory's development strategy as greater emphasis is placed on rail links which have been recommended in recent studies to relieve pressure on the road system. The planners of these corporations maintain close links with their government counterparts and, at a higher level, there is liaison at board level through the government members. The KCRC was formerly a government department that was, in effect, privatised without being sold. The MTRC was set up *ab initio* as a statutory corporation to build and operate the railway on commercial principles. The planning, building and operation of urban railways by the MTRC and, currently, by the KCRC has proved to be a significant force for change in the implementation of land use planning and development in Hong Kong. As discussed in more detail later in this chapter, the MTRC's remit, financial structure and development powers have combined to make the Corporation a successful property developer and planner on a significant scale in its own right (see also Chapter 5).

Political and public consultation

It is only in recent years that the public and their elected representatives have had any influence on the process of land-use and transport planning. District Advisory Boards were established in 1977. They became District Boards in 1982 with a mix of elected and appointed members, and the first directly elected members joined the Legislative Council in 1991. Public interest in planning issues was consequently sharpened.

Comments on development plans are sought from the District Boards and for the more important development plans, such as Metroplan, where the consultative process is extended to the public in general through publication and display of the reports and plans at key points in their preparation for comment (see Chapter 8). Comments are also solicited from trade associations and professional institutions. Both the District Boards and the Legislative Council have committees and panels, and these provide a forum where land-use and transport planning issues can be discussed with government officials. Some of the meetings are open to the public.

Before there were directly elected representatives, a well established system of appointing 'worthy and knowledgeable persons' to advisory bodies was already in being for a wide range of government activities. Where planning permission is sought by a developer, this is notified publicly and any objections are considered by the Town Planning Board (TPB) before their advice is given. For transport, the main body is the long established Transport Advisory Committee (TAC), although its position and influence is now less prominent with the emergence of other elected bodies. The committee considers major development projects and policy changes prior to consideration by the Executive Council.

Regional planning in the Pearl River Delta

While the territorial planning framework within Hong Kong has become better developed in recent years, the same cannot be said for regional planning in the Pearl River Delta area; that is Guangdong Province of the PRC, including the Shenzhen Special Economic Zone, the Portuguese administered territory of Macau and Hong Kong. For the most part, each administration has gone on with its own strategic planning independently.

Contacts between Hong Kong and Chinese officials at departmental levels have been established for specific joint projects, such as new border road crossings and measures to alleviate the environmental impact of development on both sides of Deep Bay, as well as day to day cooperation on such concerns as road and rail cross-border traffic, immigration and customs controls, and crime prevention. While participation by consultants and developers in planning and private sector development projects has flourished, inter-government cooperation in strategic regional planning appears to be inhibited. Prior to the Handover, contacts with the government of the PRC on many issues were routed indirectly through the Political Adviser to the Hong Kong government, who was an official of the British Foreign and Commonwealth Office. Under the Sino-British Joint Declaration, a Sino-British Joint Liaison Group and a Sino-British Land Commission were established to facilitate the transfer of sovereignty in 1997. Regional land-use and transport planning is an area where rapid change could take place once the political obstacles have been removed (see Chapter 12). There is an urgent need for a cooperative framework to be established as the rapid economic development in the region is already creating problems that affect all three territories, particularly in respect of transport infrastructure development and environmental pollution. China, as the future single sovereign power, will be the senior partner in any regional planning structure that may emerge. Notwithstanding this, the Hong Kong Special Administrative Region will be capable of making a significant contribution through its professional expertise and potential for financial investment.

Government role in land and transport planning and development

Public investment in planning and development

In many centrally planned, social market and market economies, the development process is undertaken by government or other public sector agencies. More commonly, however, development is undertaken by the private sector, and the public sector's role is confined to land assembly, management and disposal; transport infrastructure and other utility service provision; and planning and building control. In Hong Kong, the adoption of the Crown Land Lease system which separates the ultimate title of land from the right to develop, own and occupy a building on that land, has increased the importance of the process of land management and disposal. In public investment terms, however, the Hong Kong government has had its greatest impact on land development through:

- land formation and assembly which is taken to include the formation of land for development through reclamation and the assembly of existing land for development, redevelopment or renewal; and
- transport infrastructure and other service provision for new reclamation areas and for the existing urban and New Territories areas. Transport investment is the most important of these, in terms of the impact on development, including rail, road, port and airport investment but, arguably, the provision and improvement of environmental services to, for example, older industrial areas, also has a potentially important role.

It is these areas of public investment which have had a substantial impact on the budgetary and planning strategy for the territory, and which are the principal foci of this chapter.

Institutional arrangements for development

The success of public investment programmes in land development in Hong Kong, as in most territories and countries, relies heavily on the institutional arrangements for the delivery of those programmes. It is generally easier to agree 'what is to be done' than 'how it is to be done'. The accountability of public sector organisations, and for the use of public money and resources, inevitably makes it more difficult to deliver investment programmes than for the private sector. Moreover, in most countries, land and infrastructure investment is treated as a 'public good' on which a lower than market return (or no return) is sought. With a lack of market signals to guide investment, the process of public sector resource allocation is heavily reliant on the institutional arrangements for that allocation.

In Hong Kong, there is a greater emphasis on market returns on public investment (e.g., from the land sales programme, new MTR/KCR rail investments, the container terminals etc.). However, such investments are also expected to provide 'welfare' benefits, including environmental improvements, renewal of poor housing stock and so on. These potentially conflicting objectives place even more demanding obligations on the institutional arrangements for development in Hong Kong. Such arrangements can be divided into two main concerns:

- institutional mechanisms, such as legal and financial instruments which control, influence or, in many countries, give incentives and/or subsidise development;
- organisational arrangements by which public authorities/agencies are established to control, deliver, or work in partnership with others to deliver development.

In Hong Kong, it is these institutional arrangements which have been instrumental in the success (or failure) of public investment in land development. This can be illustrated from recent experiences in land development and redevelopment, as well as major transport investment projects. Some such experiences are reviewed below.

Land formation and assembly for development and renewal

Urban restructuring: reclamation, redevelopment and renewal

Based on the institutional arrangements outlined above, Hong Kong has established one of the most internationally successful systems for carrying out reclamation and new development. Government departments of the Hong Kong Administration - principally the Planning Department, the Lands Department and, in particular, the TDD - have arguably planned, formed, serviced and disposed of (rapidly and profitably), larger tracts of new land than any other city in the world.

The techniques were originally conceived and applied in the New Towns development programmes of the 1970s and 1980s (see Chapters 2 and 3), and are now being adapted for the urban area in the 1990s in order to implement the core projects of the PADS. By contrast, however, the progress of urban redevelopment and renewal, outside the central business district (CBD) and high value residential areas, has been significantly slower and on a much more limited scale than any equivalent sized city in the developed countries of the world. In this regard, a recent government 'Green Paper on Urban Renewal' (Planning, Environment and Lands Branch 1996) observed that:

Deterioration of the older built-up areas is giving increasing cause for concern, although Hong Kong is still a relatively young city when compared with others around the world. Added to the problem of the deterioration of buildings is the fact that the areas in which they are located were often developed before modern town planning practices were applied. The population has probably outgrown the original layout and design. Provision of infrastructure is often inadequate. There is a general shortage of open space and community facilities. Dilapidated buildings give rise to a poor living environment. Traffic congestion, overcrowding, acute environmental problems, a lowering of the quality of life and many other symptoms of urban decay are already common in the older areas of the city. With rising social aspirations and improving living standards, the conditions in those older areas do not meet the current expectations of the community. This points to the need for urban renewal, which will often involve the wholesale demolition, replanning and redevelopment of run down areas. In the process, planning gains, including better urban layouts, enhanced community facilities and improved housing will be achieved.

Some redevelopment, driven by the achievement of higher plot ratios or higher value uses, has been achieved in areas such as Central Kowloon, Wanchai and Causeway Bay but the renewal of the areas of highest social and environmental need (as against highest market value) has not been tackled as successfully as in the cities of the USA, Japan or Western Europe.

It might be argued that it is the very success[1] of, and the demands on, the reclamation programme which have limited the process of renewal. This conclusion emerges clearly from the work of the Development Statements for West Kowloon, South East Kowloon and Hong Kong Island West (see Planning Department 1993 and Llewelyn-Davies et al 1994). Here, plans were prepared in the period from 1991 to 1994, which were intended to operationalise the principles of the strategic approach to urban development set out in Metroplan (Planning Department 1991) (see Chapter 8). Metroplan linked the future of the older areas and the reclamations through the process of 'thinning out' of older residential and business areas and relocation of overcrowded urban activities to new development on the reclamations. The studies referred to this process as urban restructuring, but it soon became apparent that 'thinning out' potentially reduced values and made renewal less viable in older urban areas. Moreover, relocation increased the need for public housing and other 'lower value' uses on the reclamations. This dilemma, which was played out in each of the Development Statements, has come to dominate the debate about urban renewal and urban redevelopment in Hong Kong at present, although it should be stressed that very little urban renewal has in fact taken place in the territory if one utilises the definition as advocated by the editors in Chapter 8 (see note 12).

The problem was first highlighted by the Planning Department's own team which prepared the West Kowloon Development Statement (Planning Department 1993). The study was limited to the existing urban area but was invited to seek solutions to 'thinning out' on the West Kowloon reclamation. The urban renewal problems of West Kowloon are particularly acute with high densities, poor living and working environments and relatively low property values. Whereas the study was able to identify solution spaces - albeit by seeking sites on other urban reclamations, as well as West Kowloon - the financial return on nearly all redevelopment was marginal and usually negative and large areas of land on the reclamations for public or other 'subsidised' housing were required. Since the West Kowloon study was not carried out in conjunction with the planning of the reclamation, it was felt that there was limited scope to develop solutions that linked urban renewal and new development.

The South East Kowloon Development Statement (Llewelyn-Davies 1994), by contrast, was specifically asked to address an area of new reclamation together with the renewal of the adjoining older urban residential and industrial areas. This was the most comprehensive of the Statements, covering some 921 ha's of urban Kowloon including the existing airport site (245 ha's), the reclamation of Kowloon Bay around the airport (320 ha's) and the adjoining urban areas of Kowloon City, Ma Tau Kok and Hung Hom (260 ha's). The importance of this area for the successful urban restructuring of Hong Kong was highlighted in Technical Report No. 7 of the Study on the Proposed Institutional Arrangements which stated that:

> *The SEKS Study Area represents a unique development opportunity for Hong Kong. It occupies a strategic waterfront location on one of the world's great harbours and covers an area almost the size of the entire Kowloon peninsula. At the same time, the Study Area stands alongside (and in part incorporates) one of the most overcrowded urban areas of Hong Kong. It has an important function to perform in local terms, by providing areas for decanting of housing and industry, as well as for much needed open space, social and physical infrastructure. These factors, combined with those of realising its considerable development and revenue generating potential, will be prime considerations in the preparation of a Development Statement ....* (Llewelyn-Davies 1994).

Institutional arrangements for urban land assembly

The scope of work of these two Development Statements included the development of a range of institutional mechanisms - such as the linking of sites on the reclamation with the urban area, together with 'up-zoning' and plot ratio increases in industrial areas. However, the importance of 'realising the reclamation's considerable development and revenue generating potential' meant that the principles of subsidisation, or even cross-subsidisation, which are necessary to achieve restructuring, were resisted. As a consequence, the scale of urban renewal proposed in the Development Statements is more limited than required to meet social and environmental needs, and less than adequate to meet the objectives which Metroplan set itself. This failure is essentially a failure of the institutional mechanism to assemble land for urban restructuring, although there also have been accusations that it is part and parcel of real estate interests to resist the aims of Metroplan.

In practice, financial mechanisms to facilitate land assembly for urban redevelopment have been constrained by the need to maximise the marginal income from each individual development site when development leases are auctioned to the private sector. The planning system, however, has not been able to introduce mechanisms which would reflect the average long-run price of redevelopment and reclamation sites. Average long-run price levels are a better reflection of the total long-run demand for land in the urban area. The demand for urban land estimated in the TDS and in Metroplan, in part, reflects the growth of population and economic activity but it is also fueled by the process of thinning out, renewal and reclamation. If the land assembly mechanisms adopted discourage this process, then the long-run demand, and price expected, will not be met.

In practice, in virtually all 'so-called' developed nations it has been recognised that the process of land assembly and redevelopment of urban renewal areas must be subsidised. Mechanisms have proliferated and now include:

- the direct grant aiding of private developers;
- the use of public sector financial assistance to 'leverage' finance from private/public sector consortiums;
- the issuing of development or infrastructure bonds by public sector agencies providing tax breaks for private investors; and
- the assembly of sites and their clearance, reclamation and servicing by the public sector for disposal at less than reclamation costs.

Many of these mechanisms are now designed to maximise the level of private sector investment but this is achieved through the process of subsidisation. The tradition of positive non-interventionism espoused by Sir Philip Haddon-Cave has found no more faithful exponent than the land sales programme in the urban area.

The mechanisms proposed in South East Kowloon tried to overcome the constraints placed on West Kowloon without breaching the 'no-subsidisation' principle of urban redevelopment by a range of inventive devices including:

- the designation of a residential zoning, specifically for re-housing purposes, which would not, therefore, attract a full residential valuation;
- the identification of under-used Government, Institutional and Commercial (G/IC) sites, already government owned and occupied, and the linking of them to non-viable urban renewal sites[2]; and
- the maximisation of the role of the LDC and the Housing Society, which are able to carry out development on slightly more marginal terms than the private sector but also, as noted below, have limited capacity and resources.

Of course, as the Development Statement Report on proposed institutional arrangements made clear, these mechanisms do have an impact on government income through revenue foregone on deferred premiums. The G/IC site sales and the designation of residential sites for rehousing at a lower value do not directly abuse the principles of 'no-subsidisation'. This tactical approach was further developed in the Hong Kong Island West Statement by, for example, making a case for special control measures for the conservation and rehabilitation of buildings and areas with historic or architectural heritage. However, it is clear that there is still no strategic commitment to implementing and funding redevelopment on the comprehensive scale which is required.

Organisational arrangements for urban development, redevelopment and renewal

In most developed countries the public sector has also provided a framework for land assembly and financing of urban redevelopment and renewal in the form of urban development agencies (or corporations) at the central or local government level. Though most of these agencies now work in partnership with private developers, they are able to act as channels for public investment in land development. Armed with significant initial public investment capital, the most successful are able to revolve these funds through compulsory purchase, assembly of viable sites, cross-subsidisation and leveraging of private investment across large urban areas where they have the leading development role.

In Hong Kong the LDC has neither the powers nor resources to take on this role or scale of renewal and the government have been extensively criticised for this (see, for example, Cuthbert and Dimitriou 1992). By contrast, there are large scale resources available to the TDD but they have no role in the redevelopment and renewal process or in the existing urban areas. Each of the Development Statements studies has made recommendations about the vital need to strengthen or extend the roles of these and other organisations, or to create new urban development agencies at the policy or implementation level. Both approaches attempt to create an organisation with the powers and responsibilities to invest in land development and distribute resources across both existing and new reclamation areas.

The South East Kowloon study, in particular, recognised the importance of changes in organisational arrangements for urban renewal as an essential requirement to overcome existing constraints. The study highlighted the need for:

- a positive and proactive policy-making and advocacy forum for the urban restructuring process;
- dedicated financial resources for urban restructuring;
- coordination and integration of implementation i.e. of both urban renewal and reclamation/development;
- the benefits of area prioritisation - i.e., that area covering both existing urban areas and adjacent reclamation; and
- an identified executive responsibility for restructuring to rest in one organisation.

The study specifically recommended the creation of an Urban Restructuring Policy Unit within the Planning Environment and Lands Branch of the Hong Kong Administration in order to ensure that, for the first time, there was a policy and resource commitment within the government's policy branch system. With regard to implementation responsibilities, the study recommended that these could not be left to the limited remit and resources of the LDC or the Housing Society, and that urban restructuring should become the responsibility of a 'task-force' drawing on the relevant departmental skills and located in an executive department such as the TDD.

These recommendations were noted by LDPC and the ideas further developed in the subsequent Development Statements. The key message, however, from the Development Statements was the need to advocate for urban redevelopment and restructuring in the planning and land development process, and that this had to be achieved at the highest levels of government policy making and resource allocation. It is this theme which was picked up in the Green Paper on 'Urban Renewal in Hong Kong' as noted below.

Land development and transport investment

Transport strategy, policy and infrastructure provision

Following on from the above, the second significant contribution of government to the process of development in Hong Kong is the provision of transport infrastructure and other services to open up for development (and redevelopment) new reclamation areas, the New Territories and the existing urban areas. Transport investment has been the most important of these (see Table 11.1), although the development of the New Towns and the New Territories in the 1970s and 1980s also owed much to primary investment in drainage, water supply and water treatment. Moreover, in congested urban Hong Kong, considerable attention has been given to public transport (see Part Two), as well as highway investment. This is in contrast to the more sprawling metropolises of Asia and North America, where the growth of motor vehicle ownership and use have dominated the development of urban transport (see Newman 1989).

Though road congestion has increased and persists in the territory - and Hong Kong may yet become the first major city to adopt a road pricing system - the Hong Kong government can claim that it has by and large remained in control of the explosion in motor vehicle traffic and the desire for movement throughout the territory in comparison to most other cities in the region, despite the criticisms voiced elsewhere in this book. This has been achieved through its careful use of restraint policies and relatively enlightened policy of balanced road and rail development. Institutional factors have also played a significant part in this development, particularly in the implementation of transport policy.

From 1973 to 1992, the population of Hong Kong grew from 4.19 million to 5.56 million and whilst residential locations spread from the urban areas to the new towns and New Territories, employment remained concentrated in the urban areas. The demand for travel, therefore, very nearly doubled from 5.3 million trips per day in 1973 to 10.3 million trips in 1992. Notwithstanding this growth, in a densely inhabited environment such as Hong Kong, car ownership and usage has remained relatively low despite the high growth in average disposable income. The proportion of households with a car available remained static throughout 1973-1992, at 13 per cent. In consequence, the trips made by private transport (i.e., cars and taxis) declined as a proportion of total trips from 26.5 per cent in 1973 to 19.8 per cent in 1992. These developments are summarised in Table 11.1 below.

In these circumstances the major growth in travel has, therefore, been in public transport - in particular, by heavy rail (KCR) and mass transit (MTR). KCR trips have grown from 41,000 in 1973 to 544,000 in 1992. Since 1979, buses and mini

Table 11.1 Trip making characteristics 1973 and 1992

Item	1973	1992
Trips	5,320,000	10,298,000
Population	4,191,000	5,560,000
Trips per capita	1.27	1.85
Trips by private transport (i.e., motor car and taxi)	1,410,000	2,038,000
Modal split (private)	26.5%	19.8%

Source: Travel Characteristics Survey: Transport Department (1993)

buses have gained little and ferries have declined. The dramatic change in public transport use is summarised in Table 11.2.

The commitment to a public transport-led strategy, and in particular to the development of the MTR, was made in the 1970s, implemented to its present state between 1979 and 1989, and remained at the forefront of the Second Comprehensive Transport Study (CTS II) published in 1989. The future rail strategy has recently been reviewed and the commitment of the 1989 Strategy to MTR and KCR extensions, to the MTR Airport Express Link, light rail projects and now, rail links to the Pearl River Delta sub-region of the PRC, remains.

The CTS II study also first seriously considered policies for the management of

Table 11.2 Usage of transport system: number of daily boardings, 1973, 1992 and 1996

(thousand journeys)

Mode	1973 (1)	1992 (1)	1996 (2)
MTR	-	2,124	2,231
KCR	41	544	678
LRT (Tuen Mun)	-	251	342
Minibus	1,363	1,517	1,749
Bus	1,910	3,519	3,754
Ferry	581	180	214
Tram	397	364	307
Special buses (work, school)	321	947	n.a
Car	805	1,210	n.a
Taxi	675	1,124	1,293
Total	**6,093**	**11,780**	**n.a**

Note: No travel characteristics survey has been undertaken since 1992
Source: (1) MVA Asia (1993) and (2) Information Services Department (1997)

transport demand to further restrict private transport growth. Four management packages were reviewed:

- Package 1 - Car Ownership Taxes: comprising increases in annual licence fees and a first registration tax of 50 per cent.
- Package 2 - Fuel Tax: proposed doubling petrol tax.
- Package 3 - Goods Vehicle Control: proposed restricting the growth of goods vehicle licences.
- Package 4 - Area Pricing: which was the only selective management measure considered and recommended consideration of a full Electronic Road Pricing (ERP) option.

To date, only the taxation package options have been taken up and though Road Pricing remains on the agenda, Hong Kong's transport policy remains highly dependent on its public transport investments. The current focus is on the development of KCRC's West Rail to the PRC which is the first major infrastructure investment to straddle the 1997 handover (see Chapter 5). Below, we consider the institutional factors which have had such a significant impact on road investment and land development in Hong Kong, using the MTRC as the key example.

Impact of rail investment on land development

By the end of the 1980s, Meakin (1989) was able to write that:

> *After ten years of operation ... the MTR has met its primary objectives of enhancing mobility and meeting its financial targets ... (however) ... the secondary effects in city form, densities and land values are more difficult to qualify because the impact of MTR is so fundamental to all economic activity.*

Touching on these points in a recent paper for the Asian Development Bank's conference on 'Megacity' development (see Arnott 1995), it has been emphasised that, as well as providing a balanced sustainable transport solution, investment in modern mass transport systems has a significant impact on the economic development of cities. Moreover, the benefits of this relationship can be maximised for the financing of the infrastructure itself, for stimulating urban development and to meet wider economic and social development policy objectives. These concerns are also addressed by Chapters 8 and 12 of this book.

As Meakin's quotation suggests, the experience of Hong Kong's MTR is one of the world's leading examples of stimulating urban development and simultaneously meeting wider economic and social development policy objectives. It is important to note once again, however, that it is the institutional mechanisms and organisational arrangements for mass transit development in Hong Kong that made this possible, and resulted in the MTR having such an important influence on land development.

The direct impact on development by MTRC its development partners, at and around its station sites, has been substantial. In the existing urban area, the mass transit railway has led to the development at nineteen stations and depot sites of nearly 200,000m^2 of office space, over 300,000m^2 of retail space, and more than 31,000 flats. As a result, the cumulative profits to MTRC from development totalled HK$4bn by 1994 (Mass Transit Railway Corporation 1995). With the advent of the Airport Express Line (AEL) and the Lantau Line, a further 3.2 million m^2 of development will be created - this time mainly on the new reclamations. Indeed, in view of the scale, location and importance of these new developments, the MTRC might currently be regarded as Hong Kong's most significant urban developer - and certainly the catalyst for significant levels of urban land use change.

The scale of development at the existing nineteen station and depot sites are scheduled in more detail in Table 11.3. Fourteen of these developments are located

Table 11.3 MTR station and depot related developments

Hong Kong station locations	Number of residential flats	Office area (m^2)	Commercial area (m^2)	Government institutional and community area (m^2)
Telford Gardens	4,992	-	92,300	1,000
World-Wide House	-	33,300	7,100	-
Admiralty Centre	-	72,500	17,300	-
Argyle Centre	-	26,200	-	-
Luk Yeung Sun Chuen	4,000	-	15,900	9,400
New Kwai Fong Gardens	1,264	-	4,600	-
Sun Kwai Hing Gardens	600	-	17,800	-
Fairmont House	-	21,000	-	-
Kornhill	8,828	11,000	93,200	13,300
Fortress Metro Towers	757	-	5,800	-
Vicwood Plaza	-	29,600	7,300	-
Hongway Garden	412	-	2,600	-
Perfect Mount Gardens	760	-	1,200	19,000
New Jade Garden	1,488	-	11,100	5,400
Southern Garden	480	-	3,800	49,300
Heng Fa Chuen	6,504	-	26,700	6,700
Park Towers	493	-	3,700	19,000
Heng Fa Villa	56	-	-	-
Felicity Garden	732	-	-	16,200
Total	**31,366**	**193,600**	**310,400**	**139,300**

Source: Roger Tym & Partners (1992)

at Island Line stations where nearly all of the stations were built in association with office or retail property developments. The developments were undertaken as joint ventures with, or on development leases to, many of the leading private sector commercial developers in Hong Kong. MTRC developments are implemented by the government granting development rights to the Corporation at full market value land premiums. These are recouped from a group of the leading Hong Kong developers (including Hongkong Land, Sun Hung Kai, Henderson Land) appointed by the MTRC to carry out development in individual packages. These packages are then offered for open tender by the MTRC to developers, who bear all development costs. Any surplus proceeds of sales over costs will be shared by the developers and the Corporation - the latter using the proceeds 'to assist in servicing the debt for the construction of the AEL' (Mass Transit Railway Corporation 1995).

Importance of institutional arrangements for infrastructure investment

The significance of the impact of MTR on land development derives from the institutional arrangements under which the MTRC operates. These arrangements make an explicit link between infrastructure and land development (not made so explicitly in the case of highway infrastructure and land development). The Corporation, which is a public statutory body owned by the Hong Kong government, received large scale initial capital funding from government but was required to generate sufficient cash flow to service the investment and cover operating costs. The MTRC is now able to raise finance through export credits and the money market and claims to be the only underground mass transit railway in the world to cover all costs including depreciation and an operating profit margin (see Chapter 5). As a public corporation, the MTRC is not required to produce a market return to shareholders, while fares are informally regulated in consultation with government. The development of the initial system would have been paid off from revenues by 1990 had it not been for the subsequent expansion of the system which has required high levels of continuing investment.

Even for a system which claims to enjoy the highest patronage rates and fare affordability levels in the world, it is important to recognise that the growth of the MTR system has only been achieved through the ability to cross-subsidise income from rail operation and property development.

As the importance and potential of the link between transport investment and land-use development has been increasingly recognised, so a wide range of institutional mechanisms for linking the process have evolved including:

- build/operate/transfer (BOT) schemes - entitling a private transport developer and operator to development rights as well as fare box revenues for a fixed period;
- the concept of 'planning gain' - where developers benefiting from improved accessibility are required to contribute financially or indeed to build part of the transport facility concerned. This has proved more appropriate in the case of highway investment;
- land development taxation or other forms of 'betterment' - representing a levy on those benefiting from transport investment. The proceeds are then used to pay for the transport investment itself;
- direct increases in local property taxes - on those existing land and property owner benefiting from improved access and the hypothecation of these tax increases by the taxation authority to the transport agency concerned.

Notwithstanding the above mechanisms, many of which are radical not only for the region but also globally, it is essentially the broad concept of 'value capture' that represents the hidden engine of success. Through the establishment of a public transport and development agency - such as the MTRC - with access to land assembly powers to secure the infrastructure and property development rights at existing use values (so 'capturing' the development benefits to cross-subsidise transport investment within the same agency) and with the supporting mechanisms in place, Hong Kong has arrived at an extremely effective formula for developing and financing its rail expansion programme. In the medium to long run, this could prove critical to environmental protection efforts and to countering the inevitable forces of motorisation.

KCRC's Westrail expansion of their system through the urban area and the New Territories is now seeking to replicate the MTRC's approach. The potential for urban and strategic highway investment to generate such development effects is more limited. Development nodes are not created in the same way, though there are examples of linear development opportunities being created in conjunction with highway structures in urban areas. Nevertheless, it is felt by the Hong Kong Administration that a high level of innovation in the territory's transport strategy and co-ordination with land-use planning has been achieved through TDS, PADS, the urban development statements and the sub-regional strategies.

Many of these innovative approaches to transport investment and development rely on adopting institutional arrangements which may themselves be innovative. Legal and financial mechanisms, for example, which allow value capture - through taxation, compulsory land assembly or joint venture developments - imply a level of legislative development, compliance and an ability to innovate within that system which is not present in all government systems (especially in the region).

It might be argued that the benefits of infrastructure-led developments can only be secured when city government achieves not only a certain scale of transport and development demand but also a maturity and reliability in its legal and financial framework.

Even more importantly, in the case of Hong Kong, is the organisational framework and the entrepreneurial ability of those organisations to innovate to exploit the benefits of transport investment and urban development. Included here is the ability of private sector organisations to work in a complex and competitive development environment guided not only by market returns but also by the ability to deal, often long term, in partnerships with public developers, planning authorities and, of course, transport authorities. At the same time these transport, development and regulating authorities must have the manpower and technical capacity, integrity and the remit to deal with the private sector - or to take on the development role successfully themselves. It has already been suggested that Hong Kong's MTRC has become one of the territory's most influential urban development agencies, and it is this which demonstrates the sophisticated level of organisational response which has been delivered.

Conclusions

This chapter has set out the institutional mechanisms and structures for land-use and transport planning and associated urban development in Hong Kong. The mere scale of development in the last 20 years - accommodating an additional 1.5 million people, 5 million additional trips, the development of five new towns, the reclamation of more than 1,000 hectares of land and the relocation of one of the world's largest ports and airports - demonstrates that the territory's institutional systems are largely efficient and effective. However, as Hong Kong joins the ranks of the most affluent economies, so the development challenges become more complex and call for increasingly sophisticated responses.

Urban redevelopment and renewal appears to be one of the most intractable of these challenges. It has long been recognised that the failure for a territory to develop is more likely to be associated with incompetent (or underdeveloped) institutional structures than the lack of good development projects or resources. This point was made in a paper prepared for the World Bank on the constraints of institutional weaknesses (see Untraced Author 1996) in which it was noted:

> *all too often many policy makers are aware of the importance of institutions in a development strategy (but) they still do not give it the necessary priority .... And of course building and inaugurating (projects) are politically more attractive than maintaining or operating.*

Institutional arrangements for development, which impact upon organisational structures in key policy making areas and upon the allocation of financial and other resources, are at the heart of the process of government. In Hong Kong, as elsewhere, therefore, they have been at their most effective when there has been a political and/or social imperative driving government policy making; and least effective when this is absent. The early sections of this chapter showed how the present institutions grew out of the need to accommodate a rapidly growing population and economy and how this was effectively met by a narrow range of central government authorities and agencies. It is ironic to note that, in one of the leading free market economies of the world, traditional style central government (rather than federal, local, community or public/private 'partnership' agencies) has had such a significant role. Well run, well resourced and focused agencies, such as the TDD and the MTRC, are the product of this approach but so too is the failure to address the more socially and organisationally complex problems of urban renewal. In Hong Kong the key is to establish a priority at the policy, i.e., 'Branch', level of government. The above cited evaluation of institutional development programmes presented to the World Bank identified six factors explaining the success or failure of programmes:

- Exogenous factors - ranging from Acts of God to international economic shocks over which, by definition, there can be little control other than adopting flexible and responsive institutional regimes. Hong Kong, as an entrepot, world financial centre and aspiring international crossroads, is particularly sensitive to such factors.
- Leadership by outstanding individuals - is all too often critical. The best institutions should be able to rise above the stature of the individuals which manage them. The World Bank study noted: 'reliance on leadership alone is not very promising as a generalised method of achieving high level institutional performance'. Although outstanding individuals have, from time to time, had a significant role in championing the development process in Hong Kong (this is especially the case, elsewhere, in Asia), it is perhaps encouraging that most of the key development institutions have successfully evolved and survived with a changing leadership and a strong middle management. How this capacity will survive the transition period with so many senior and middle management civil servants departing the territory (see Chapter 12) is open to speculation.
- Good institutional and project planning - appears to be a necessary but not sufficient condition. That is, poorly planned organisations and projects are guaran-

teed to fail but good planning alone does not guarantee success. It is emphasised that 'good planning requires the meticulous design of a whole process, not merely of one alternative, and the knowledge of how to modify one or more alternatives in the light of experience and changed circumstances'. It has been suggested that the planning process(es) in Hong Kong has (have) been exemplary. Persistent rather than flexible but very competent. Good planning has undoubtedly been one of the hallmarks of Hong Kong's institutional development.

- The application of good management techniques - particularly in project execution, is an extension of the planning process through administration and implementation. Given the general levels of education and training in Hong Kong, its good communications, an international perspective and culture, as well as the extent of its use of international consultancy expertise, have strengthened Hong Kong's institutions. However, despite these natural advantages, it should be noted that the best results are not always achieved from them.
- Minimising the distortion in relative prices of crucial inputs or outputs - is important for distorted prices have acted as a powerful disincentive to institutional development. In many countries this takes the form of deficits being covered from the national budget leading to the mis-allocation of resources and distortion of the system of incentives inducing negligence and lower productivity. Hong Kong is particularly prone to overpricing in the short run, especially of resources. This is especially apparent in the case of land and leads to significant distortions in the urban development process in the territory.
- Commitment is a final important consideration - for achieving institutional improvement. This is required not only in the implementing agency(ies) itself (themselves) but the other relevant agencies of government, particularly those that hold the purse strings. This requires strategies to build and reinforce commitment during the planning and implementation period. In Hong Kong, experience suggests that this policy and resource commitment are most important and effective at the policy branch level of government.

The land-use and transport planning system in Hong Kong has begun to recognise and become more responsive to the importance of institutional matters. This has been brought to a head by the complexities of implementing Metroplan. In 1990, LDPC considered a paper on the Institutional Framework for the Implementation of Metroplan (LDPC 1990) which recognised that 'from virtually the commencement of the Metroplan study a predominant issue of concern has been the institutional means by which Metroplan's strategic objectives can be achieved in a timely and coordinated manner within constraints of resource availability'. The paper recognised that there were no less than twenty government departments and agencies (excluding the relevant policy branches) involved in the implementation of Metroplan. Six major issues confronted these agencies and had to be resolved if the objectives of Metroplan were to be met (see Chapter 8):

- The integration of existing and new development areas - specifically 'that the planning and development of new areas (particularly reclamation) recognises the need for the renewal and upgrading of nearby obsolete and congested residential/industrial zones. The planning of such areas must be undertaken on a coordinated and integrated basis if the city is to be improved as a whole'.

- An overall population limit had to be set for the Metropolitan area - specifically 4.2 million people.
- The gradual reduction of the scale of manufacturing industry in the metropolitan area consistent with maintaining productivity. This recognised the westward shift of port and airport activity and the outward processing movement to the PRC.
- The need to expand existing office and business centres and the development of new centres at public transport nodes in the Metropolitan area.
- The need for greater integration of land development and infrastructure provision.
- To achieve environmental upgrading in the metropolitan area to create a better place to live and work.

Since 1990, most of these issues have been directly addressed in a series of Development Statements and sectoral studies, all of which have paid particular attention to the institutional dimension. These have fallen into two groups.

Firstly, a series of 'sectoral' studies on the upgrading of obsolete industrial areas and industrial building standards (Townland Consultants Ltd 1994 and 1997), office decentralisation (Roger Tym & Partners 1997a), retail location and standards (Roger Tym & Partners 1997b) have been or are being carried out, and a major study on environmentally sustainable development (SUSDEV21) was commissioned in 1997. Broadly speaking, these sectoral studies have not encountered insurmountable institutional problems and have provided an opportunity to rewrite the relevant chapters of the Hong Kong Planning Standards and Guidelines, so introducing new institutional and financial mechanisms.

Secondly, a series of Development Statement studies have been prepared providing the next tier down of detailed area planning for Metroplan. These have tried to address the complex issues of planning for a particular population limit, integrating existing and new development areas, and land development and infrastructure provision. As noted above, the Development Statements have had considerable problems in meeting their institutional objectives particularly over the issue of urban redevelopment and renewal. In both instances, it is the political and financial commitment and the issue of organisational responsibility for a complex implementation process that cuts across departmental responsibilities (and priorities), that has proved most difficult to resolve. The work of these studies has been carried out against the background of government (branch level) committees on implementation, specifically, the Sub-Committee on Implementation of Metroplan which has brought most of the parties together to consider the institutional recommendations of each study - and a review of the role of the LDC. The latter broadened out into the 'Green Paper on Urban Renewal in Hong Kong' published in June 1996 (Planning, Environment and Lands Branch 1996) and recommended a number of improvements:

- Improving site assembly mechanisms by empowering LDC and encouraging more owner participation;
- Improving relocation and rehousing mechanisms, jointly involving the LDC, Housing Society and the Housing Authority; and

- some minor improvements to project viability by linking viable and non-viable sites, thereby introducing for the first time the possibility of cross-subsidisation;

It also recommended strengthening co-ordination among different parties and the overseeing of the implementation of the new urban renewal (redevelopment) policy, undertaken by a dedicated urban renewal (redevelopment) team in the Planning Environment and Lands Branch. The team will be multi-disciplinary and include ' ..... officers with experience in land administration and town planning'. This picks up precisely the recommendations of the Development Statements, as does a longer term proposal for the creation of a new statutory Urban Renewal Authority by upgrading the LDC into an Urban Redevelopment Agency (URA). This proposal recognises that 'the existing mechanisms cannot deliver urban renewal (redevelopment) on a large enough scale and sufficiently quickly to avoid serious deterioration in large parts of the urban area'. Finally, the Green Paper suggested some long held shibboleths, for example, the introduction of a Rehabilitation (Loan) Fund, to be administered by the URA, to assist private owners to renovate i.e., to conduct urban renewal in the more Western sense of the term.

The results of the Green Paper are encouraging and, although they have been thought to be too generalised by some commentators, it is precisely the fact that these issues are being considered at the branch level - focusing on policy and resource allocation, and organisational arrangements for urban renewal (redevelopment) - which is breaking new ground. Hong Kong's ability to take up the implementation tools and achieve things on the ground has never been in doubt. However, this is only the case when a clear policy commitment has been made. The overriding critical question is whether this commitment will survive the political transition.

Notes

1 It should be stressed that while the reclamation programme is seen as a financial success in certain quarters, it has also attracted increasing controversy and opposition as pointed out by Ng and Cook (1997).

2 G/IC sites effectively have a low 'book-value' but when linked to non-viable urban renewal schemes, can render the overall redevelopment viable, having deferred payment of premium to assist the process of rehousing and clearance.

References

Arnott, C.J. (1995) 'Transport Investment and Economic Development in Asian Megacities' Paper delivered to Asian Development Bank Regional Seminar, Asian Development Bank, Manila.

Cuthbert, A. and H.T. Dimitriou (1992) 'Redeveloping the Fifth Quarter: Practices and Prospects in Hong Kong' *Cities Journal* Vol. 9 No. 3.

Housing Branch, Government Secretariat (1997) *Homes for Hong Kong People: The Way Forward.* Long Term Housing Strategy Review Consultative Document. Hong Kong Government Printer.

Land Development Policy Committee (1990) 'Institutional Framework for the Implementation of Metroplan'. Hong Kong Government, Hong Kong.

Leeds, P.F. (1989-1994) Unpublished series of lectures on transport policy. Centre of Urban Planning and Environmental Management, The University of Hong Kong, Hong Kong.

Leeds, P.F. (1995) 'Institutional Planning Framework and Needs for Effective and Efficient Land-use and Transport Planning in Hong Kong' Unpublished paper, the University of Hong Kong.

Llewelyn-Davies et al (1994) *South East Kowloon Development Statement* Final Report for Planning Department, Hong Kong Government.

Mass Transit Railway Corporation (1995) *Annual Report 1994* Hong Kong.

Meakin R.T. (1989) *Hong Kong's Mass Transit Railway: Vital and Viable* Transport Department, Hong Kong.

MVA Asia (1993) *Travel Characteristics Survey* Report for Transport Department, Hong Kong Government.

Newman, P. (1989) *Cities and Automobile Dependence: an International Source Book* Gower and Aldershot.

Ng, M.K. and A.H.S. Cook (1997) 'Reclamation: an urban development strategy under fire' *Land Use Policy* Vol. 14, No. 1.

Planning Department (1991) *Metroplan. The Selected Strategy* Technical Report, Volume 1 Hong Kong Government.

Planning Department (1993) *West Kowloon Development Statement* Final Technical Report Volume 1 Hong Kong Government.

Planning Department (1995) *Territorial Development Strategy Review* Final Technical Report Hong Kong Government.

Planning Environment and Lands Branch (1996) *Urban Renewal in Hong Kong* Hong Kong Government.

Roger Tym and Partners (1992) *Retail Viability Study of Rail Station Development Sites* Report for Mass Transit Railway Corporation, Hong Kong.

Roger Tym and Partners et al (1997a) *Study on Office Decentralisation Strategy* Report for Planning Department, Hong Kong Government.

Roger Tym and Partners et al (1997b) *Study on Shopping Habits and Revision of the Hong Kong Planning Standards and Guidelines* Report for Planning Department, Hong Kong Government.

Townland Consultants Ltd et al (1994) *Study on Restructuring of Obsolete Industrial Areas* Final Report for Planning Department, Hong Kong Government.

Townland Consultants Ltd: et al (1997) *Study on Planning Standards for New Industrial Areas Report* (forthcoming) for Planning Department, Hong Kong Government.

Untraced Author (1996) 'Institutional Weakness: A Roadblock to Development' Unpublished paper prepared for the World Bank, Washington.

12 Transport and sustainable urban development: providing a new lead

Harry T. Dimitriou

Introduction

China, the largest developing country, has now absorbed Hong Kong, one of the world's economic powerhouses. The period leading up to this event was filled with much uncertainty. In the words of Silverman (1995):

> *So much is at stake - the world's seventh biggest stock market, 6 million lives etc...... It's natural to speculate about what's going to happen when China takes over in 1997 because no one really knows.*

Notwithstanding the uncertainty this statement justifiably conjures, if the worst scenarios are to be avoided, land-use and transport infrastructure planning efforts and environmental protection measures must continue well beyond the transition. Given the circumstances, while preparing such plans and measures, it would be more than prudent for the planners to simultaneously have an eye on what the political and economic speculators have to say about Hong Kong's future. For what is speculation one day may in this uncertain world become a reality another.

With the above in mind, this chapter first provides a brief review of various positions regarding the future of Hong Kong as a backcloth to the chapter's main concern of building a more sustainable development and transport strategy for the territory. In addressing the central theme of sustainable urban development the chapter has three objectives.

The first is to inject the concept of sustainability into the discussion about the vision building and plan evaluation process.

The second is to promote the concept of sustainability as the basis of a revised strategy for guiding the future development of Hong Kong.

The third is to contribute to the wider debate on sustainability in the context of a densely developed city experiencing strong economic growth.

Hong Kong's uncertain future: critical questions

Socio-economic and political visions

In the Hong Kong Government promotional literature distributed in the years immediately prior to 1997, the administration boasts that the territory possesses the world's busiest container port, handling more than 11 million twenty-foot-equivalent units (TEUs) a year (Information Services Department 1996). That is the equivalent of one twenty-foot -long container every three seconds! The government also pointed out that some 70 per cent of Hong Kong's marine throughput is bound for, or originates in, China. This is handled by the eight existing container terminals located at Kwai Chung, financed, built and operated by two private operators: HIT and MTL (an almost unique feature in itself) (see Figure 12.1). A ninth terminal has been planned for Tsing Yi Island, opposite Kwai Chung[1]. Meanwhile, a further three million TEUs are handled mid-stream. This same government literature also informed interested parties that over the past decade, container throughput in Hong Kong has grown at an average annual rate of 18 per cent with this growth forecast to continue well into the next century. Further estimates suggest that by the year 2011, there will be a demand for Hong Kong to handle 32 million TEUs a year. That is the equivalent of one 20-foot container every second, 24 hours a day. Looking ahead, to the territory's future, and taking account of other territorial growth trends, and the environmental impacts and traffic congestion which might ensue, a critical question must be asked. Namely, whether in the post-transition years this growth is economically and environmentally sustainable and, if so, whether there will be the management capability to service it. For the purposes of convenience, we will refer to this management issue as Critical Issue No. 1.

Figure 12.1: Kwai Chung Port in operation
Source: A.H.S. Cook and H.T. Dimitriou

The Port and Airport Development Strategy (PADS), which in part is a response to the above marine traffic growth (but which also seeks to accommodate air traffic growth and some territorial re-structuring needs as well as new residential development (see Chapters 3 and 7)), paints a future Hong Kong where in the year 2000 it will take only 25 minutes to travel by rail from central Kowloon to the new airport in Chek Lap Kok and not much more from Central to the border. Based on the anticipated successful construction of this immense infrastructure programme, the local press (the South China Morning Post) has claimed that:

> ..... *by the time the Union Jack is lowered after 155 years of British rule, Hong Kong will have been transformed from a barren rock into a world-class financial centre with a governmental system as democratic as can be expected, an efficient infrastructure marked by a modern airport and port facilities, and an affluent population that enjoys free education and heavily subsidised medical and housing benefits.*

Irrespective of the whether the reader attributes this success to the British administrative rule or the hard work of the local population and the support of the mainland[2], a more pertinent question is whether this development will grind to a halt after 30 June 1997 as an address by the Governor in October 1996 seemed to suggest. Related to this is the question whether, before the handover, Hong Kong will be presented with a clearer insight into what vision the prospective leaders of the Special Administrative Region (SAR) have for the future (Critical Issue No. 2).

The Financial Times of London in one of its leading articles on Hong Kong (see Financial Times Survey, April 1994) forecast that the years up to 1997 and immediately beyond, would severely test its status as a world-class financial centre. This article, like many on Hong Kong in the local and international press, raised a number of other questions that highlight the uncertain future of the territory. These include the issue of whether the Chinese Government has the ability to control the territory's economy, and whether business interests will look elsewhere in Asia (and even further afield) for investment opportunities. A scenario such as this would, inevitably lower investment levels in the territory and with the political uncertainties, could contribute to capital flight.[3] It should be said in response to this fear, however, that such uncertainty is not new to Hong Kong. As the Financial Times article points out, it has not deterred many of the world's leading financial institutions (especially those from the US) from flocking to Hong Kong to participate in the prospects that mainland China has to offer and to service the corporate growth needs of the city[4]. On the other hand, global investment (and investors) is notoriously mobile in this day and age, so Critical Issue No. 3 is whether this investment confidence will remain well after 1997.

Twelve months before the handover date, Time Magazine, in a feature article (see Spaeth 1996) asked whether China's takeover would 'ruin Hong Kong as it is known today, or merely alter it'. The article reported differing views on the territory's future. The general tone was pessimistic. It cited two principal schools of thought. One such school (publicly endorsed by Chinese officials) is that 'the mainland needed its 'golden goose' on which it relies for the bulk of foreign trade, management expertise and technology, and thus no harm would come to Hong Kong. Another is that China is 'incapable even of comprehending Hong Kong, in particular the personal and economic freedoms that make it thrive, and would (therefore) unwittingly grind it down'. Sharing this latter view, Martin Lee (a leading Hong Kong democrat) claimed: 'Beijing is putting a noose around the goose's neck and still expecting it to lay golden eggs'.

The same Time Magazine article also reported on another version which merged the above two schools of thought, namely, that territory will remain an 'international financial centre and essential source of capital for rapidly growing China but lose its cosmopolitan luster...'. There is in fact, much to support this scenario, for the process of China-isation is already well underway. The territory has become so closely integrated with the Chinese economy (to the mutual benefit of both economies it should be emphasised) that there is no turning back[5]. Accompanying this development, however, is Critical Issue No. 4. This is whether the severance of Hong Kong's courts of law from the British Privy Council after the transition will ultimately lead to the insistence by multinationals wishing to sign contracts in Hong Kong after 30 June 1997 that such contracts be subject to arbitration in the West.

Richard Wong, Director of the Centre for Economic Studies at the University of Hong Kong (also referred to in the above source) cites three other possible post-1997-futures.

The first is where Hong Kong remains the 'Manhattan' of the Orient, i.e., 'free and open to the world's businessmen and financiers'. The second is where, 'with heavier interference from the north, Hong Kong could degenerate into a Chinese Bombay, used mainly as a base for businessmen interested in the Chinese market'. The third possibility presented is where Hong Kong becomes the 'Miami of China, dominated by the underworld and swamped with migrants from China'.

Wong acknowledges that in all cases, the territory's prosperity will greatly depend on whether the city 'has to absorb millions of Chinese restless poor'. Notwithstanding the above possibilities, he concludes that the opportunities for Hong Kong remain good so long as the new post-1997 administration and Chinese government employ a 'very innovative and intelligent way of managing the border with flexibility and with sternness'. This raises Critical Issue No. 5, namely: what happens if the border is removed or not policed effectively.

Almost simultaneously with the above cited Time Magazine article, the Economist (June 1996) ran an editorial and feature article on the future of Hong Kong. The former made much of the argument that because Hong Kong, with its reliable legal and financial infrastructure had become a crucial conduit of money into China, it was inadvisable for the mainland to intervene economically in the territory. The feature article poured cold water on the idea, circulated by some, that Hong Kong could continue its economic pre-eminence as a 'kind of Singapore', where its new leaders would not interfere in the conduct of business and would not jeopardise the city becoming by far the wealthiest in China. The Economist dismissed this on the grounds that in the last few years Hong Kong's public have become 'much more vocal and too entrenched in their views to accept this'.

The cover story of the first edition of the Far Eastern Economic Review in the first week of July 1995 (see Silverman 1995) raised quite different concerns (but equally valid) about the future of Hong Kong. Although this well respected journal did not place too much credence on reports of the pending death of the territory, most graphically presented by the New York based Fortune Magazine in June the same year (this magazine only a year earlier designated the colony as 'the world's best city for business'), it argued that Hong Kong's biggest problem in the mid 1990s was not political uncertainty but its costs. Silverman explained that 'Chronic inflation and lagging productivity are taking a toll on growth' and cited the following as contributory factors:

- unemployment levels at a nine-year high (although the number of unemployed is in fact dwarfed by the nearly 200,000 foreigners working in Hong Kong);
- chronic inflation (near double digits throughout the 1990s);
- among the world's highest housing and office rents; and
- a devastated manufacturing sector which, since 1988, has shed 446,000 jobs.

Silverman and others have expressed concern at:

- the shortsightedness of the manufacturers (and government) at not looking sooner at more value-added manufacturing options and strategies;
- the over-dependence on the new service sector which generated 670,000 jobs from 1988 to 1994;
- the possibility that the peg between the US$ and the HK$ will be removed post-1997; and
- the dependency on real estate speculation rather than improved productivity with recent Jardine Fleming statistics showing that Hong Kong's productivity growth lags behind its Asian rivals.

This last point[6] has been picked up by the Hong Kong business community and was highlighted by a recent study commissioned by the Vision 2047 Foundation conducted by faculty members of the Harvard Business School with others (see Enright et al 1997).

This study observed that Hong Kong's competitors are improving and that some are targeting industries and activities in which Hong Kong has traditionally been strong. The study notes that:

> *Singapore is a direct competitor for Hong Kong in some areas, such as attraction of Asian regional headquarters and financial services, but in most areas of the economy, Singapore serves different markets and is not a direct competitor.......Taiwan, although faced with major political obstacles, could emerge as Hong Kong's most direct competitor. Taiwan is a potential competitor for location sensitive business and its Asia Pacific Regional Operating Centre strategy has targeted activities, such as air and sea cargo, telecommunications, media and entertainment, and multinational headquarters in which Hong Kong has been a regional leader.....Shanghai and Sydney are in many ways complementary to Hong Kong rather than directly competing.*

Enright et al claim that the territory's unique mix of firms employing what they term 'commitment strategies' and 'hustle strategies', and the interaction of these strategies, rather than the presence of any one component, has provided Hong Kong with a competitive advantage[7]. A critical issue here is whether this mix (and this edge) can be both sustained and further built upon (Critical Issue No. 6).

The conclusions of the report are positive. Its authors argue that the territory is uniquely situated to move with current global trends and respond to the growing importance of an information and knowledge-based economy, a dispersed production network and packaging and integrating abilities. By transforming its economy from one based on manual labour to an information and knowledge-based economy, and from an enclave economy to a metropolitan economy, the authors claim that Hong Kong has the potential not only to be the 'Gateway of China' but to become the 'Crossroads' between East and West, East and East, North and South, and between industrialised and developing nations, capitalist and reforming economies, and between China and everywhere else[8]. For this to happen, however, the researchers argue that the territory needs to become more cost effective. It can do this by, for example, utilising more extensively cost saving information and managerial systems[9], reducing its traffic congestion costs (estimated at US$2.3 billion per year) and keeping ahead of 'the curve in infrastructure, education and training'. In this regard, they further argue that it is essential to 'retain people with higher qualifications, advanced and specialised skills, and with the ability to think independently and make decisions' so as to accomplish this vision. Critical Issue No. 7 is whether Hong Kong will succeed in becoming an international crossroads.

By implication, the Vision 2047 Foundation study team joins Silverman and a good many other 'Hong Kong-watchers', and cautions analysts (and planners) not to underestimate Hong Kong's resilience to deal with its uncertain future. After all, as Silverman points out, 'Hong Kong has always lived dangerously'. Notwithstanding this 'macho' concluding note, all seven Critical Issues raised (see Table 12.1 for summary) point to an overriding challenge Hong Kong will inevitably have to confront. Namely, how can the territory sustain its success both in a new political context and with (no doubt) a different environmental agenda.

Technocrat visions

Despite the above cited concerns, uncertainties and speculation, Hong Kong has a territorial strategy and a set of development plans, programmes and related projects that the present government is committed to implementing (see earlier chapters, especially, Chapters 3 and 8). In so doing, the administration is locking the territory into a particular vision that is infrastructure-based and, in particular, designed to service trade, commerce, real estate and industrial interests.

While the current development strategy for Hong Kong essentially relies on 'trickle-down economics', it also depends a great deal on public sector leadership in infrastructural development and housing. Thus, the territory's reputation of being governed by free-market forces and a non-interventionist colonial administration (see Chapter 11) is only a partial truth. For the government's commitment to the

Table 12.1 Critical issues of the future confronting Hong Kong

Critical Issue No. 1: Whether in the post transition years Hong Kong's growth is economically and environmentally sustainable and, if so, whether there will be the management capability to service it.
Critical Issue No. 2: Whether, before the hand-over, Hong Kong will be presented with a clearer insight into what vision the prospective leaders of the Special Administrative Region (SAR) have for the future.
Critical Issue No. 3: Whether the investment confidence in Hong Kong will remain well after 1997.
Critical Issue No. 4: Whether the severance of Hong Kong's courts of law from the British Privy Council after the transition will ultimately lead to the insistence by multinationals wishing to sign contracts in Hong Kong after 30 June 1997 that such contracts be subject to arbitration in the West.
Critical Issue No. 5: What happens if the border is removed or not policed effectively.
Critical Issue No. 6: Whether the mix of 'commitment' and 'hustle' strategies pursued by Hong Kong that provide it with a competitive edge can be both sustained and further built upon.
Critical Issue No. 7: Whether Hong Kong will succeed in becoming an international crossroads.

execution of the Port and Airport Development Strategy (costing US$14.74 billion)[10] provided Hong Kong with more than an infrastructural foundation for the future. This initiative not only helped to sustain private sector buoyancy through uncertain times but also attracted significant levels of foreign investment, provided new employment opportunities and offered a major stimulus to domestic demand. In fact, according to Miron Mushket, chief regional economist of Lehman Brothers in Hong Kong, it kept GDP growth positive during the 1990s and 'oddly for Hong Kong, left the Government as the engine of economic growth' (Sunday Morning Post, 5 February, 1995). In this way, it can be said that the resultant infrastructure development sought to inject certainty into an environment of great uncertainty - and to a large degree succeeded! This strategy is very much in line with the Keynesian view that government can, and should, act as an agent for boosting demand in times of recession, or feared recession. Interestingly, this central government 'helping-hand approach' has also been successfully employed by other 'Tiger' economies in the region in a variety of ways as the World Bank has recently acknowledged (see World Bank 1993). The East Asian experience suggests that the argument often promoted by economic monetarists, that technocratic planners are ineffective as action-generators, is not entirely true. It is perhaps, instead, much more pertinent to situations where plans and strategies do not receive political ratification or where economies have insufficient growth. Where technocratic planners do receive support, as in the case of the New Deal in the USA (which among other things led to the foundation of the Federal Highway System) or as in the case of the European Union's current plans to use the Trans-European Transport Network to assist the integration of national economies (EC 1995), another story is told. The point missed by the monetarists is that there are different kinds of planning, and that planning need not be predominantly control-based, although all plans do require some control, together with a vision and differing levels of government enablement. It has been argued (see Dimitriou 1997) that plans - be they territorial plans, city plans or transport plans - require their own particular mix of policies and mechanisms which set the vision, the level and type of government enablement, and the public levels of control and enforcement. The nature and mix of these policies, vary according to development and cultural context, economic era confronted, available resources, affordability, the degree of certainty and economic and political confidence experienced. This aspect is returned to in the latter part of this chapter.

TDS and PADS, and to a certain degree Metroplan also, are what may be termed 'enablement strategies'[11] that incorporate a certain welfare function but are in general terms primarily a technocratic vision, ratified by 'the powers that be' to achieve specific 'stakeholder objectives'. These plans were formulated in close consultation with the 'captains of industry' and other stakeholders deemed important by government who typically share short-term goals. While one can legitimately debate the choice of vision pursued by these plans, as indeed Legislative Councillor Christine Loh and others have on a number of occasions,[12] the public at large in Hong Kong did not have the democratic channels or power to contribute to this vision. Those who were fortunate to belong to a strong constituency had the possibility of influencing the formulation and execution of certain enablement and control policies. The question of vision, however, was decided by the few.

The preceding sections have argued that Hong Kong's development strategy and plans are captive to an outdated infrastructure-based vision servicing the insatiable needs of trade and commerce, which in time, will seriously further reduce environmental qualities of the territory. With this in mind, the following pages address the need for a change in vision for the territory from one based on unrestrained economic growth to another based on sustainable development.

The concept of sustainability: the international debate

Origins and definitions of concept[13]

In 1987, with the publication of Our Common Future (WCED 1987), also known as the Brundtland Report, the concept of 'sustainable development' was placed firmly on the international agenda. While a wide variety of definitions for sustainable development exist (see Pezzey 1992), the definition in the Brundtland Report is perhaps the clearest and most widely accepted:

> *Sustainable development is development that meets the needs of future generations without compromising the ability of future generations to meet their own needs.*

The widespread international acceptance of this concept following the publication of the Report resulted in the emergence of a broad consensus for action which was formalised at the 1992 United Nations Conference on Environment and Development (UNCED) in Rio de Janeiro. This culminated in the signing of the Rio Declaration, alternatively referred to as Agenda 21. However, the promotion of sustainable development beckons the questions: what is to be sustained? for whom? for how long? (Lele 1991). It also broadens the concern for development well beyond economic output issues and resource consumption questions, and embraces the concern for capital assets, social goals as well as GNP. Daly (1990) argues that four points need to be taken into account when using the sustainability concept. These are that:

- the human scale of development should be limited so that it may be sustained by the ecosystem;
- technical progress needs to be made to increase the efficiency in the use of natural resources;
- renewable resource exploitation should be based on the sustained maximisation of profits; and
- non-renewable resources should be exploited at a rate equal to the creation of the renewable resources that replace them.

The Brundtland Report cited eight critical objectives for sustainable development:

- reviving (maintaining) growth;
- changing the quality of growth;
- meeting essential needs (for jobs, food, energy, water, sanitation, transport etc.),
- ensuring a sustainable level of population;
- conserving and enhancing the resource base;
- reorienting technology and the management of risk;
- merging environmental and economic considerations into decision-making processes; and
- adjusting to relations of the new international order of economics.

Most agencies and organisations that have pursued this line of thought and action have added a ninth, more operational objective, namely, to make development more participatory.

Whether one is referring to rural or urban areas, it is useful to breakdown the concept of sustainable development into four sub-systems: economic, environmental, social and institutional. Evidence suggests, however, that the sustainability field of concern has been much confused and burdened by the mixing of goals, which make up the vision of sustainability and the means to achieve this vision (Tengstrom et al 1995)[14].

Sustainable economic development

In economic terms, sustainable development can be defined as non-declining per capital utility (Pezzey 1992). This economic definition satisfies the criterion of inter-generational equity. The use of non-renewable natural resources in the economy poses critical issues for intra- and inter-generational equity, which have both economic and environmental consequences. From the economic perspective, the use of resources remains sustainable so long as the 'competitive rents' are invested in reproducible capital (Solow 1986). This is an elaboration of Hardwick's Rule that all rents from depleting non-renewable resources are invested in reproducible man-made capital which substitutes for resource inputs in the production process. The consumption of non-renewable resources typically depends upon imported supplies, so that economic sustainability must incorporate the idea of balanced trade and productivity sufficient to meet the potential real price for resources (Pezzey 1992).

Sustainable environmental development

While some economists (see Solow 1986) assume away the technical limits to substituting resources for depleted ones and ignore the biosphere, others (see Pearce et al 1988) more directly address environmental sustainability by emphasising the threshold effects and critical minimum stocks of resources (Pezzey 1992). Under this concept, the twin factors of human population and individual resource use must be held within the carrying capacity of the earth (IUCN/UNEP/WWF 1991). Current evidence indicates that the deteriorating environment, especially that associated with urban development and transport growth, is one of the main ways today's generation is creating uncompensated future costs (see Watson et al 1996). In essence, sustainable development is trying to get us to move away from a situation in which we are constantly drawing down environmental resources to one in which the renewal and replenishment of these resources are recognised as being essential to the long term viability of our economies (Hills 1995).

On the above basis, one may conclude that environments should be protected in such a way that environmental capacities and limits are respected over time at levels sufficient to avoid future catastrophe and at levels which give future generations the opportunity to enjoy an equal measure of environmental consumption. Incorporated within this is the concept of 'ecological sustainability', which may be defined as 'the existence of the ecological conditions necessary to support human life at a specified level of well-being through future generations' (Lele 1991).

Sustainable social development

Welfare is a central concern of sustainable development (Hills 1995). Failure to adequately involve key groups or stakeholders[15] in the design and implementation of development projects often contributes to their rejection or disownment by communities for whom they were intended, especially at the first sign of problems emerging. Similar reactions are observed in the case of institutions, both public and private, which have not been involved in the conception of such projects (see Cernea 1994). For the purposes of this chapter, coverage of institutional and social considerations will be separated, with the latter focusing on participation and equity issues as a prerequisite of sustainable social development.

To achieve inter-generational and intra-generational equity, an analysis must be made of which social groups incur the costs and which incur the benefits; which groups generate the wealth and which consume it. This analysis may eventually lead to enablement policies with incentives being offered by government to targeted groups, and/or enforcement policies with penalties being imposed on others. Such policies ensure that some kind of community balance is kept and that development efforts are 'on track' with the sustainable vision agreed.

Of particular concern regarding equity matters are 'trade-offs' made between efforts to meet the basic needs of the disadvantaged, and efforts to meet the productive needs of the city. In this regard, the Brundtland Report argues that sustainable development contains within it two key sub-concepts: the essential needs of the world's poor, to which overriding priority should be given, and the social sustainability of the society as a whole including the maintenance of its culture. The former has been the focus of numerous research exercises in development studies, particularly for rural areas, and more recently, in urban and transport studies (see Dimitriou 1992). While some work has also been undertaken in the latter, in Scandinavia (see Tengstrom et al 1995), it is a much more nebulous concept than the concept of ecological sustainability and requires more research.

Sustainable institutional development

From an institutional perspective, the concept of sustainability denotes the capability of organisations involved in the planning, management, operation and maintenance of development efforts to sustain themselves both politically and functionally. Organisations should be able to carry out their current and future tasks efficiently and effectively, and be capable of generating and renewing their manpower, financial and information resources on a continuous basis. The concept also conveys the presence of qualities of good co-ordination, transparent decision-making, clear lines of accountability and responsibility, community/employee participation, equal opportunities and access to competent.

A more mechanistic definition of sustainable institutional development is one which defines it as 'the ability of an organisation to produce outputs of sufficient value so that it acquires enough inputs to continue production at a steady or growing rate'. In these terms, a sustainable institution is 'one that has earned the adherence of a sufficient body of people so that it receives the continuing encouragement and support it needs to handle, at a minimum, a stable volume of transactions (with effect)' (untraced source).

Sustainable urban development

Given the predominant urban character of Hong Kong, the following discussion will emphasise the urban dimension and urban issues. According to Hilderbrand and Pouliquen (1993) a milestone of the international acceptance of the sustainability concept occurred in Rio de Janeiro in 1992 when the environmental debate was broadened to focus on urban priorities. This 'Brown Agenda' as it became known, sought to address air, water and waste pollution and other environmental hazards; precarious and inadequate housing; deficiencies in public transport provision; and the prevalence of poverty and inequities in many cities. A second milestone was provided by the Ford Foundation Global Review of urban research in the developing world (1993) which, finding that very little research had been undertaken on the urban environment, developed a rapid urban environmental assessment process. One may define a sustainable urban development process as the:

> ..... *means of restoring (or moving towards) a condition in which the city can again fulfill [sic] its traditional and natural tasks; a synergetic co-evolution of all the above mentioned sub-systems, leading to a non-declining level of welfare for the local population over the long run, ignoring the environmental impacts a city has outside its borders.* (Ernst 1996)

Reflecting the above agenda, the following issues were found to typically arise in the planning of land use, transport and the environment in cities of the Asia Pacific Region (after Dimitriou 1997):

- limits to economic growth and production;
- efficient resource use and depletion;
- investment viability and cost recovery;
- equity and welfare;
- affordability and poverty alleviation;
- man and machine-generated air, water and noise pollution;
- environmental heritage and protection;
- cultural heritage and protection;
- participation and consultation;
- institutional development and delivery; and
- institutional responsibility and accountability.

Each of the above may be examined from both a global basis and a local one. At the latter level, these again can be pursued on a sectoral basis (i.e., for housing, transport, utilities etc.) or on a cross-sectoral basis. They may be further investigated and responded to on a citywide, community-wide and/or on a corridor and/or street-specific, basis.

Sustainable urban development: Hong Kong initiatives

Local skepticism

Hong Kong is a latecomer to acknowledging the relevance of the sustainability concept to its development. It was not until the late 1970s that the beginnings of an environmental protection and pollution control regime were established by the Hong Kong government (Copeland 1996). This is perhaps not surprising, as the territory is a bastion of entrepreneurial success based on high returns within very short-term horizons. Given the close and long-standing working relationship between government and key economic actors in the territory, the initial encounter with the sustainability agenda was received by its leaders with skepticism, if not suspicion. This was reflected by the fact that although Hong Kong sent representatives to the Rio Summit in 1992, it did not sign the Rio Agreement.

Fortunately, there has since been a shift in attitude. This can be attributed to the activities of a number of local environmental groups (such as Friends of the Earth, Green Power, the Conservancy Association etc.). It can also be attributed to the effective dissemination of the issues associated with sustainability by the international and local media; and to the persuasive arguments made by individuals from academia, political circles and, in some cases, from industry. Perhaps the best publicised of these activities were the lobbying efforts made by Christine Loh for government to adopt a sustainable strategy in her Alternative Policy Address (1994) which advocated the pursuit of three objectives:

> *... the development of the human person to his or her full potential .....a healthy environment to live and work in .... (and an economy) to remain economically competitive in the world market*[16].

An earlier, much less publicised but perhaps intellectually more challenging contribution, came from the Faculty of Law at the University of Hong Kong in the form of a publication edited by Heilbronn (1993). This set of papers presented a whole range of legal issues affecting environmental protection and sustainability[17]. In the introduction to this publication, Heilbronn claims that 'the government in Hong Kong is not actually doing enough, despite its talk and the enormous capital reserves it maintains'.

Notwithstanding such criticisms of government, the efforts of the Planning, Environment and Lands Branch of the Hong Kong government - which published a number of locally influential reports such as A Green Challenge for the Community (1993) - has also had a positive effect. This report constituted a second review of the 1989 White Paper on Pollution. It represented the starting point of the emergence of a more conscious official position advocating sustainable development but did not provide an operational framework for the pursuit of such a strategy. Finally, the efforts of the government's Environmental Protection Department (EPD) should also be acknowledged, even though various parties have argued that given the scale of the environmental problems confronting the territory, more could have been done and at a faster pace. Having said this, on many occasions (both behind the scenes and openly) EPD became locked in conflict with other government departments and the private sector about environmental issues associated with the territory's ambitious development projects.

Perhaps the most significant of all the factors that have contributed to this shift over the last five years, has been the very apparent deteriorating quality of Hong Kong's environment reported on by many observers (for example, Chan and Hills (1993), Hodgkiss (1994), Rusco and Walls (1995), and Barron (1996)). Without doubt, there has been a very noticeable increase in air and noise pollution in Hong Kong in the 1990s as infrastructure and building construction reached pre-1997 fever levels. The territory has also experienced a noticeable rise in traffic congestion levels and a widespread decline in water quality. The adverse effects of these developments are all too apparent to the inhabitants of Hong Kong, where the deteriorating environment is a major public concern (see Dimitriou 1994a).

This deterioration of environmental quality did not go unnoticed by the international community. The 1995 Quality of Life Index presented by the United States-based magazine International Living compared 192 countries by measuring pollution, population density, deforestation and the percentage of protected land. It concluded that Hong Kong was the 'most environmentally degraded place in the world'[18]. Countering this claim, the then Director of Hong Kong's EPD pointed out in the local press that in 1994 the Hong Kong government invested more in Hong Kong's environmental infrastructure, including waste disposal plants and sewage works, than the British government in the UK[19]. He added that by 1997, the territory 'will have an environment infrastructure as good as anywhere in the world'.

New initiatives

While Metroplan emphasised the goals of enhancing the environmental quality of the territory's metropolitan area it did not address issues of urban sustainability (see Chapter 8). It was left to the Territorial Development Strategy Review (TDSR) to do this and present certain key sustainability issues and development principles which were endorsed by the Planning, Environment and Lands Branch (Planning Department 1993).

Most promising of all, in the field of land-use and transport infrastructure development, albeit somewhat belated, was the Planning Department's initiative in June 1995 to prepare a brief for a study on Environmental Sustainability and Port Development in Hong Kong (SENPORT). This was subsequently revised and renamed the Study on Sustainable Development for the 21st Century (SUSDEV21) (Planning Department 1996a). The change reflected the refocusing of the study firmly on the issues of sustainable development. TDSR (Planning Department 1996b) acknowledged the difficulties which Hong Kong faced in meeting environmental objectives in a context of rapid economic development and promotion of the hub function of the city. On the matter of sustainable development it concludes:

> *Certain issues need to be addressed on a stand-alone basis but there are many others that are interactive and bring us back to the question of what kind of urban-based style of living we would like to enjoy and can afford to provide. It is, therefore, crucial to balance the perceived economic benefits of development scenarios and the related hub functions of Hong Kong, on the one hand, against, on the other, such matters as potential resource availability, environmental objectives, the efficiency of infrastructure systems, the provision of essential land use requirements, standards of public health and the aspirations of the community with regard to the style and quality of life preferred. The urban system as a whole needs to be considered, having regard also to regional interactions. Taking account of all these broad aspects of sustainability is a highly complex issue.*

Government has nominated a sum of US$5 million (US$3.75m fees and US$1.25m equipment and contingency) for this study which commenced in 1997. One wonders whether the net is being cast too wide and too late, and whether a more targeted approach addressing the critical issues would be more appropriate.

The launching of the new study was in response to the growing concern in a number of quarters about the environmental sustainability of the territory. It followed a plea to government by the Conservancy Association to adopt a sustainable development framework for its land use strategy at the Conference on Planning Hong Kong for the 21st Century organised jointly by the Institute of Planners, Hong Kong University and the Chinese University of Hong Kong, in April 1995 (see Ng 1995)[20]. At this event, it was pointed out in a presented paper (see Hills 1995) that 'Hong Kong's success had been gained at some considerable expense to the quality of its environment' and that the territory was 'inherently unsustainable'.

In order to move toward a more sustainable development, Hills argues that there is a need to:

- integrate economic, social, political and environmental considerations within the policy-making process;
- evaluate critically the effectiveness of existing policy-making and advisory institutions;
- track changing public values concerning environmental issues to bring more clearly to the public attention the trade-offs that are associated with different strategies for development, both regionally and globally;
- identify the roles and responsibilities of other actors, particularly the private sector and NGOs;
- target specific sectors and activities; and
- explore and evaluate a wide range of policy instruments that might be used to encourage a move towards sustainable development[21].

Concurring with a view raised by the author of this chapter (Dimitriou 1994) several years earlier, and in Chapter 8, Hills underscores the point that the significance of how Hong Kong tackles its growth problems extends well beyond its boundaries. Appreciating this same position, in May 1995, the University of Hong Kong's Centre of Urban Planning and Environmental Management (CUPEM) launched a two year research programme (funded by the Hong Kong Environment and Conservation Fund) to develop indicators for measuring environmentally sustainable development in the territory. The purpose of this research is to establish 'the extent to which Hong Kong is on a sustainable path and to provide direction regarding opportunities for making Hong Kong a more sustainable place'. The key study objectives are to:

- identify a framework for achieving sustainability in Hong Kong;
- develop a methodology that will identify practical indicators for measuring and predicting where Hong Kong is on the path towards achieving sustainability; and
- apply this methodology to identify trends, project future conditions, and evaluate specific policies and mitigating measures.

In a Working Paper written for this study, Copeland (1996) raises the following issues regarding the sustainability debate for Hong Kong:

- growth versus maturity;
- valuing the environment;
- substitutions of capital and technical progress;
- human activity versus the environment;
- uncertainty and irreversibility;
- inter-generational equity;
- intra-generational equity; and
- survivability versus sustainability.

In her analysis, she points out that:

> .... *the notion of continual economic growth within the parameters of sustainable development has not proven to be a panacea for resolving the ongoing and related problems of resource scarcity, poverty and environmental degradation. The market economy, even with governmental adjustments and constraints, has been unable to place true monetary values on natural capital ... or to account for the full social and economic costs of environmental degradation.*

Copeland continues by emphasising the very significant and now widely accepted view that current measures of economic progress, such as GDP, do not incorporate these externality issues and other costs on future growth and prosperity (see Leipert 1989 and Gowdy 1994). This places Hong Kong's planners, whose government habitually cites growth statistics (especially of economic growth) as a sign of its economic virility, in an invidious position in any search it may undertake for a sustainable strategy.

Any study of Hong Kong's options for sustainable development must confine itself to some limits of geographic enquiry. While at one level this can be determined by the current boundaries of the territory, as the Special Administrative Region (SAR) of China, in practice, this confinement is somewhat unreal. For as an entrepot and world financial centre, and a major 'gateway to mainland China' in particular, Hong Kong is intrinsically linked to, and dependent upon, economic and other activities in the global market far outside its own terrain.

This inevitably involves it in unsustainable practices both outside its territory as well as within. Confronting this problem, Copeland argues the need for Hong Kong to 'think globally and act locally' which she claims, entails among other things, assessing its import/export activity for implications of environmental damage. This immediately raises the above cited issue of survivability (of Hong Kong) versus its sustainability.

While acknowledging the validity of this broader line of analysis, the remaining pages of this chapter focus more on how Hong Kong's land use, transport infrastructure and services, and environment, can be planned, managed and maintained in a more sustainable manner in the interim, before a sustainable development strategy is fully in place. How this can be best related to its regional context of the Pearl River Delta is another central issue. Once a sustainable development strategy has been formulated, it will need continual re-appraisal in the context of the territory's involvement in both regional and global activities.

Given Hong Kong's international status as a world class city, and its frequent use as a role model for other major cities in the region, it is incumbent upon those who have been closely associated with its success and with planning for its future to look toward a more sustainable strategy for its development. Progress in this regard will have the added benefit of tackling head-on some of the critical issues raised earlier in this chapter regarding the territory's uncertain future. This is only possible with a vision that goes well beyond present attempts at meeting the insatiable appetite for additional growth. Such a vision needs to move the territory more rapidly into a high-tech, information and financial-based economy, to create a congestion-free transit nerve-centre, with a greatly enhanced quality of life and environment that befits a world-class city of the next millennium. It is imperative that Hong Kong take up the challenge as: 'sustainable development is a concept that few environmental policy-makers can afford to ignore at the present time. It is a concept that is widely advocated but which few countries or cities have yet managed to translate into practice' (Hills 1995).

Toward a new lead: the role of transport

General

The starting point of this part of the chapter is the issue of a new vision for Hong Kong, and the simultaneous call for the movement away from a dependency on unsustainable growth. Its purpose is to offer some further leads with which to appraise the current hierarchy of strategies, plans and programmes in the territory as a basis for suggesting more environmentally sensitive land-use/transport planning goals of sustainable development. The comments offered and points made are by no means exhaustive. They have drawn from a number of sources. These include: the Public Law 102-240 of the US Government 1991; the World Bank's recent Transport Policy Paper (1996); the work of Maddison et al (1996); and past research conducted by the author (Dimitriou 1992, 1994a, 1995 and 1997). The findings of each are briefly summarised below.

US developments: ISTEA

At a time when the US government had a Republican President, and the praises of monetarist thinking were being transformed from Thatcherite economics and Reaganomics to a seemingly respectable basis for land-use and transport investments (with their inevitable environmental impacts), the USA saw the introduction of one of the most revolutionary and radical pieces of environmental and transportation legislation ever. For in December 1991, the Inter-modal Surface Transportation

Efficiency Act (ISTEA) became public law. This legislation broadly addressed the needs of the development of a national inter-modal surface transportation system and authorised a significant increase in funds. These funds were assigned 'for the construction of highways, for highway safety programs, and for mass transit programs, and for other purposes' it stated that it would do so not only on the basis of economically efficient criteria but also on grounds of 'environmentally sound' and 'energy efficient' criteria.

This was a visionary piece of legislation that sought to shift significantly the current land-use, transport and environmental paradigm for the US toward one with the following qualities (after Tomazinis 1993):

- greater efficiency, aesthetics and equity;
- more community, amenity, safety and health;
- further opportunities, flexibility and freedom; and
- greater sustainability (both locally and globally).

The above translated into:

- significant increases in funding for transport and related concerns (totalling US$155.3 billion nation-wide[22]);
- the greater flexibility in the use of funds;
- the redefinition of the national highway system;
- a renewed federal government interest in public transit (assigning a total of US$31.5 billion to transit out of a total US$155.3 billion for the transport sector as a whole);
- an emphasis on inter-modalism (and thereby integration);
- a greater involvement in transport planning issues by the Metropolitan Planning Organisations (MPOs), introducing a fifteen factor planning process with a much greater concern for environmental issues;
- the funding of traffic congestion mitigation and Clean Air Act measures (to a level of US$6 billion); and
- a revised federal funding share level of major transport schemes (up to 90 per cent in the case of interstate projects and 80 per cent for others).

This initiative also saw financial support being given to efforts to reduce motorised movement, such as car and van pool schemes, park and ride schemes, cycleway and pedestrianisation schemes and certain rail projects. The act made available significant additional funds for highway and transit research, and assisted in the set-up of a number of national centres of transport research. Particular attention was given to urban traffic congestion mitigation and air quality improvement in cities, both in terms of government aid and research funding. Interestingly, the legislation also communicated a more environmentally caring administration willing to offer a greater lead in the field of land use, transport and the environment than has been the case for sometime. This was not only displayed by the new higher funding levels offered but by changes in the names of certain government agencies, such as the previous Urban Mass Transit Authority, which was renamed by the act as the Federal Transit Authority (FTA).

There are a number of issues of relevance to Hong Kong from this US experience. Three particularly come to mind. Firstly, and perhaps most significant of all, is the acknowledgment that even in the most powerful capitalist country in the world, government has a very important strategic role to play in the provision of transport infrastructure, and in the way it is managed. Second, is the realisation that a bold alternative vision is necessary and firm proactive action required, when the limits to growth have been exceeded and the current situation is found unsustainable. Third, is the appreciation that, for all the efficiencies that market forces generate through competitiveness, inter-modal and inter-agency co-ordination is essential. Although the Hong Kong government appreciates this position, it seems unable to convince the private sector operators of the public bus transport industry of the merits of co-ordination (see Chapter 6).

The World Bank's new position

According to the World Bank, there is no doubt that transport is central to territorial development, be it urban or rural, regional or national. It performs best when it does not 'crowd out' other productive investments, but rather, serves them effectively and efficiently (World Bank 1996). This position coincides with present and past perspectives of the Hong Kong government, particularly given the territory's entrepot functions, which is why the territory has to date so heavily invested in transport infrastructure.

The main challenge ahead for transport cited in the World Bank's recently published Transport Policy Paper (1996) includes:

- inadequate asset maintenance;
- the need to improve the efficiency of service supply;
- increasing the access to, and affordability of, transport;
- adapting to the globalisation of production and trade;
- responding positively to growing aspirations for transport quality as incomes increase; and
- reconciling the increased availability of the private automobile and road freight movement with the maintenance of a tolerable environment.

The same source adds: 'Short term palliatives will not suffice. The simultaneous achievement of long term economic, environmental and social sustainability is the key to meeting these challenges'. The development agency advocates a five pronged new approach to transport investments, namely:

- The pursuance of economic and financial sustainability which involves making the best use of available resources and maintaining physical assets by, firstly, creating (and then maintaining) a competitive market based transport sector and, secondly, introducing economically efficient charges.
- The pursuance of environmental and ecological sustainability which entails the reduction of air, noise and visual pollution. This is to be achieved by firstly focusing on the mechanisms through which environmental and ecological impacts emerge (taking into account the values that society places on them) and the

range of potential remedial actions open, and secondly, by finding ways of integrating the less critical but widely dispersed environmental concerns within economic incentive structures.

- The pursuance of social and distributional sustainability which affects the wealth creation effects of transport and poverty alleviation, firstly, by designing general transport policies targeted on the poor and disadvantaged, and secondly, by managing problems created by the unwanted social effects of economic reforms in the transport sector.
- A changed function of government in transport which reduces its role as supplier or regulator of the sector but accentuates its roles as facilitator of market orientated transport supply and trustee of the interests of environmental and social sustainability. This is to be achieved firstly, by reforming institutions concerned with the development of the transport and putting them at 'arms length' from direct political influence involving non-governmental local community organisations in the planning, supply and management of local services, and secondly, by developing government capabilities to perform functions necessary for the market to work.
- A changed role of the World Bank group in the implementation of investment in physical facilities in the transport sector with greater emphasis on facilitating institutional reform. This is to be achieved by focusing on instruments that better accommodate increased private sector financing, and secondly, by encouraging the developments of partnerships, within countries, with governments, with the development community and with other actors in aid and international capital markets to create a common framework within which to evaluate transport sector investments and better tackle the global motorisation process.

The above statements and position stake out a new agenda for the World Bank's lending policy for the transport sector. However, there is much evidence to suggest that as an international development agency, it is somewhat unsure of how best to implement this strategy. In the past, the transport sector has typically accounted for about 15 per cent of the World Bank's portfolio and generated an estimated rate of return on completed projects of 22 per cent, compared with an average of 15 per cent for all Bank projects (World Bank 1996). The critical question here for the Bank is whether the same rates of return are achievable with the new path it has charted out for itself, and whether such measures are indeed sufficiently meaningful to measure progress made toward the spectrum of sustainable development goals.

For Hong Kong, the critical question is what aspects of this new agenda could/should influence the decision-makers and planners of the territory when formulating a sustainable approach to the development of its future land use, transport and environment. This question is especially pertinent given the World Bank's conclusions from its experience with developing economies and the fact that Hong Kong has now become part of China, the largest of such economies. It is clear that some aspects of the private sector's greater participation in transport as advocated by the Bank, are more prevalent in Hong Kong than elsewhere. Equally apparent, however, is the fact that Hong Kong has too long overlooked the equity and environmental priorities raised by the Bank and other international development agencies such as the UNDP and UNCHS. This is a situation in urgent need of attention and the World Bank's new position is a good starting point.

The Blueprint Five position for UK

The volume in the 'Blueprint' Series entitled The True Costs of Road Transport followed four previous publications in the same series. The first was the Blueprint for a Green Economy in 1989 edited by Pearce et al, which looked at the meaning of 'sustainable development' in the UK context. The remaining three dealt with:

- the application of environmental economics to global environmental issues (Blueprint 2, edited by Pearce et al 1991);
- how sustainable development could be measured in the UK context and what issues of concern existed (Blueprint 3, edited by Pearce et al 1993); and
- how the key to solving major environmental problems lies in economic incentives (Blueprint 4, Pearce 1995).

Blueprint 5 is described by Pearce as a departure from the previous publications in that it is the first volume to address a specific sector. The importance of the transport sector, he claims, lies not only in the high capital costs incurred by it but by its high social costs. In the UK, for example, the social cost of transport has been estimated to lie between GBP 22.9bn - GBP 25.7bn (Maddison et al 1996). The same source suggests that 'there were many journeys (in the UK) for which the private benefits were outweighed by the wider costs to society'. A more controversial claim by the same authors is that 'pollution from road transport in UK kills more people than die in traffic accidents'. These are very worrying accusations not only for the UK but for places such as Hong Kong which is planning to accommodate a quantum leap in traffic growth in the next decades.

The heart of the argument(s) presented by Blueprint 5 is that the external costs of road transport, especially urban road transport, are too often greater than the economic efficiencies it generates, and that a better understanding of external costs is a prerequisite of sustainable land-use/transport planning efforts. External costs of particular concern include:

- enhanced greenhouse effects and climate change;
- the economic cost of air pollution;
- the social costs of noise pollution;
- congestion costs and road damage;
- the external costs of accidents; and
- the aggregate externality charges and taxation of road transport.

According to Maddison et al (1996), 'Having a sustainable road transport system means making each road user pay at least the full marginal cost of his or her journey'. At present, it is common knowledge that users of road networks pay only a fraction of these costs. Such costs are, furthermore, not properly allocated between different road users and different journeys, and have in the UK and elsewhere subsequently grown out of control, particularly in urban areas. Blueprint 5's response to this is not to argue the need to reduce mobility or limit the private ownership of motor vehicles but rather to argue that externality costs such as pollution should be taxed or licensed. Thus considerable dependence is placed on market-based instruments in the correction of current transport planning practices toward a more sustainable future. Notwithstanding this, the authors acknowledge that such instruments can only be applied in situations in which the externality costs (polluting activity, for example) can be monitored. Unfortunately, too often this is not possible. Maddison et al, therefore argue that:

> ...... *the best long-term strategy for a government is to enhance its capability to monitor the activities of all road users more fully in preparation for charging them for their environmental impacts. In the shorter term, the appropriate policies involve adaptation of the existing fiscal system rather than command and control policies relating to vehicle construction and use.*

Two aspects about the conclusions of Blueprint 5 are significant for Hong Kong. First, if the fuller estimates of the true costs of the kind presented by this publication in respect of road transport were applied to the territory's forecasts for all transport modes (including freight transport modes, on land, sea and in the air), one would confront a set of costs that would shock even the most pessimistic of environmentalists. The cost estimates would, furthermore, be politically unacceptable - which is most probably why they have never been fully calculated by government and industry alike, nor are likely to be in the future.

Second, the seemingly wholesale translation of the externality issues associated with transport by Maddison et al is to suggest that economics (market mechanisms in particular) is capable of adequately handling these issues - a position with which the author does not concur. The dependence upon economics, and marginal cost theory in particular, as the basis for developing transport and environmental policy responses are up to a point - and a political point at that - in line with some of the thinking on road pricing of the Hong Kong government (and others) as a means for tackling the growing traffic congestion produced by the motor car (see Hau 1988, 1992 and 1995, Transport Branch 1994, and Barma 1995). This is not, however, akin to the rationale used, or proposed for the territory's rail investments (see Chapters 4 and 5).

The selective use of marginal cost economics by the Hong Kong Administration as a basis of again attempting to introduce electronic road pricing (ERP) into Hong Kong[23] has even attracted the support of some liberals in Hong Kong who are attracted to the argument that the revenues generated by the scheme would help finance (subsidise?) improved and more extensive public transport in the territory, and assist in addressing the transport needs of the disadvantaged. A technical consultancy on a future ERP scheme was initiated in 1995, and commissioned in 1997.

In good company with others, the author has not supported road pricing as the basis for controlling urban traffic growth and as a major plank for urban transport policy-making. The author considers that such a scheme would further enhance the mobility of the rich (whose opportunities are already greatly enhanced by their privileged access to other transport modes and to new developments in telecommunications technology). It would do so without guaranteeing any benefits for the public at large, and the less privileged in particular (see Chan 1994 and Lau 1996). Another objection cited by Werner Rothengutter of Karlsruhe University in a recent infrastructure conference in Rotterdam (EMIP 1996) is that road pricing was not intended to be a transport policy measure to resolve all aspects of the urban transport problem i.e., traffic congestion, environmental issues and related equity problems. Rather it should be a part (albeit an important part) of a package of measures which ensures road users pay a realistic price for the road space used and traffic congestion contributed to.

The author of this chapter concurs very much with Rothengutter's view and is wary of the (politically convenient) selective use of marginal costing exercises in the form or road pricing as a panacea for tackling traffic problems in fast growing cities. This makes for inconsistent policy-making and planning. The dependence, furthermore, on the promises of the future use of revenues from ERP to finance public transport services and other needed infrastructure, is highly dangerous, and therefore, undesirable. For not only may the promises never materialise but by implementing ERP the government is, in effect, further extending its financial dependency on the motor car driving elite as a revenue generator. In so doing, this gives this community greater political power. A much preferred approach is to:

- first put in place land-use/transport policy and environmental protection measures on the basis of a particular agreed vision;
- provide supporting land-use and transport policy and planning enablement and enforcement measures;
- pre-invest in the public transport schemes, pedestrian networks and other priority projects needed to enhance accessibility; and then
- use road pricing as a means of supporting and complementing these already decided and funded priorities, simultaneously seeking to give priority to 'productive' rather than 'unproductive' motorised trips.

The 'Developmental Approach'

In a book written by the author in 1992, entitled Urban Transport Planning: A Developmental Approach, an approach to land-use/transport planning was advocated in which transport is seen as the 'spearhead' to urban development but, nevertheless, an agent of development that is subservient to broader development (including environmental) goals. Much of the fundamental thinking underlying this approach is grounded in a critique of the conventional wisdom of urban transport planning provided earlier in the same book. While the approach has been somewhat modified in subsequent publications and research, including one publication that addressed transport issues of Hong Kong (see Dimitriou 1994a) and unpublished research in support of the development of a generic sustainable urban transport strategy for medium-sized cities in the Asia Pacific Region (see Dimitriou 1997), it tackles a number of the concerns of sustainable development cited earlier and some raised by ISTEA.

What essentially differentiates the 'developmental approach' to urban transport planning from more traditional methodologies is that the former relies more on developments planning for urban development as opposed to planning for transport. Emphasising this point in the 1992 publication, the author states:

> *It has been suggested ..... that the conventional wisdom bias in traditional methodologies with their concern for transport systems efficiency above all else, exists because those most intimately involved in such approaches are well equipped with tools and techniques to design and plan 'operationally efficient' networks, whereas the equivalent expertise in the planning and management of more (sustainably) 'development effective' transport systems is much less advanced.*

The elements of this advocated approach, illustrated in some depth in the case study of several cities in Indonesia and shown in Figure 12.2 (see Dimitriou 1995), include:

- The adoption of the principles of urban development planning within the goals of transport planning which looks to:
  - the maintenance and creation of the growth of real resources per capita;
  - the decrease of poverty and deprivation;
  - the promotion of increased accessibility to, and responsiveness of, public sector administration; and
  - the promotion of adaptable urban physical and spatial arrangements accommodating identified needs, and cultural and social identity.
- The pursuance of the concept and principles of integrated urban infrastructure programming which seeks to:
  - stimulate increased generation of local revenues to especially cover the maintenance and operation costs of existing transport and related infrastructure;
  - expand low-cost infrastructure investments and maintenance in a planned and coordinated manner;
  - decentralise planning and implementation responsibilities and provide a commensurate enhancement of institutional capacity;
  - strengthen provincial and municipal capabilities and encourage greater local participation; and
  - incorporate goals and measures of sustained economic productivity and basic needs fulfillment (including, in both cases, those associated with the informal sector).
- The matching of appropriate transport technologies to settlement hierarchies, structure and development contexts, emphasising the greater role of non-motorised movement and public transport, based on criteria of:
  - transport technology operational efficiency;
  - sensitivity to settlement and community size (looking at density, urban form, topography, community structure and population size); and
  - sensitivity to available resources and social, cultural, political and development constraints and preferences.

- The use of a sketch-plan methodology which bring all the above elements together into and relies on:
  - scenario-building, plan robustness and flexibility,
  - cost-savings effort,
  - a set of methodological guidelines and handbooks based on selective empirical research, and
  - a systematic (computer-aided) set of interdependent steps of analysis and decision-making.

Figure 12.2 Developmental approach to urban transport planning

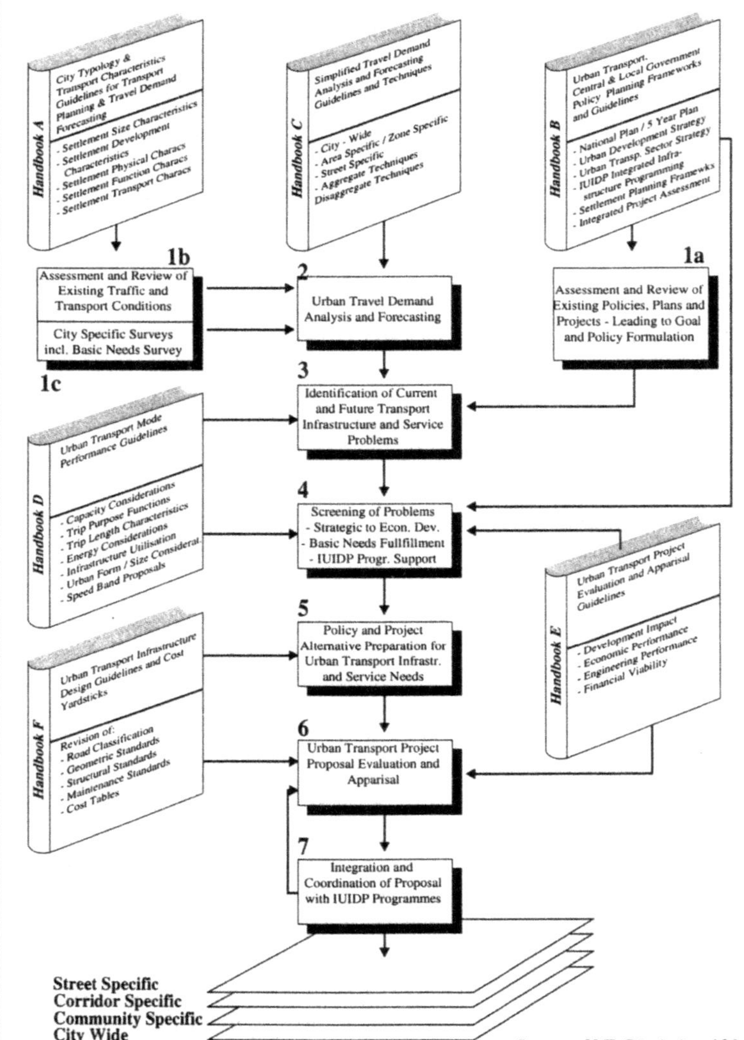

Source: H.T. Dimitriou 1996

Because much has been discovered and written about the shortcomings of past urban transport planning practices since the initial formulation by the author of the 'developmental approach' to urban transport planning, particularly regarding environmental implications (both locally and globally), it is readily acknowledged that it inadequately covers many of the environmental and participation concerns required of sustainable development. It is also appreciated that the 'developmental approach' inadequately recognised the important role of the fast growing involvement of the private sector in urban transport.

On a related note, one would also have to accept, particularly in a broader Asian context, that the approach perhaps conveys too great a reliance on a 'presumed capability' by the public sector to enforce the guidelines and actions advocated. It also conveys too little about the use of enablement measures to encourage private sector involvement. Notwithstanding these limitations, the author stands by the kind of vision and strategic directions the approach offers, and the principle that transport should be subservient to broader development goals and, simultaneously, used as a 'spearhead' for urban development.

A comparison of the 'developmental approach' to urban transport planning with Hong Kong's approach highlights a number of differences. The most notable of these is that while the current strategic plans for the territory (i.e., TDS, Metroplan, CTS I & II[24] etc.) make strong use of transport infrastructure as a agents of change, they do so in a manner whereby land-use planning and environmental protection efforts are almost totally sub-servient to goals of transport systems operations efficiency.

A second notable difference is that the entire land-use/transport development strategy for the territory relies on the same kind of outdated conventional wisdom. This has been discredited by the author and many others (see Dimitriou 1992 and Yanaguaya 1993), and criticised by the local press (South China Post 1989) at the time of the announcement of CTS II with the following comment:

> *The Second Comprehensive Transport Study is ineptly titled. It provides only limited responses to the worsening traffic situation in Hong Kong. What the study discloses in statistical terms, it lacks in ideas and vision, which are essential if the territory is to untangle its traffic snarl-up .... it relies too much on old formulae that have been found wanting in the past.*

Although the 'developmental approach' advocates the adoption of urban development planning principles within the goals of transport planning, the experience in Hong Kong is that planning focuses much more on the needs of short-term unsustained growth than on sustained long-term growth. This is inherently in conflict with the principles of urban development planning. Adding to the problems of a short-term focus are the higher costs of operating out of Hong Kong which are threatening the productivity of the territory, as pointed out earlier by Silverman (1995). The Hong Kong government has also concentrated a great deal more on responding to private sector business interests than to those of the local community and the public at large. In addition, there has been a general reluctance to acknowledge the existence of the transport needs of the disadvantaged in general, and the poor in particular who are reported to represent six per cent of the territory's population (Lau 1996). Finally, there has been an almost complete disregard for the territory's heritage in infrastructure developments.

There is, however, evidence in Hong Kong of both government successes and failures with regard to the advocated pursuit of the concept and principles of integrated urban infrastructure planning and programming. The implementation of PADS as an integrated set of core projects has been admirable, whereas, government attempts to encourage bus companies to integrate their services fully with those of rail have clearly failed. The transport infrastructure 'lowerarchy' (including pedestrian facilities and small urban streets) have been greatly under-emphasised if not neglected in Hong Kong, while the more expensive highway and rail investments are more aggressively pushed.

Scope also exists for raising, and recouping, higher levels of public revenues from real estate developments, particularly along rail transport corridors which as yet have not been sufficiently exploited.[25] Although recent developments in this field (see Chapter 11) suggest that this aspect of urban and transport development offers a great deal of promise to Hong Kong (and the region at large), there are increasing worries about the government's ability to retain its capacity to adequately perform the institutional functions and responsibilities required.

This threatens any idea of decentralisation of responsibilities, and the ability of the public sector to respond to increased levels of public participation as advocated by the 'developmental approach' to urban transport planning. The concern for institutional capacity was raised by an article in the Hong Kong Window Magazine, (Fitzpatrick 1995) which stated:

> *A shortage of talent in senior civil service ranks could force the government to recruit more professionals from the private sector to fill gaps in the administration. Faced with an exodus of senior officers, the government is set to introduce unprecedented open competition for promotion to senior civil service posts, according to sources. While senior officials say they are trying to put in place an administration to straddle 1997, the constant changes and departure of officers in key posts have fueled fears that the civil service is plunging into an era of instability.*

Despite the acceptance by government planners of the need for a hierarchy of transport systems that relates strongly to land use and urban development, and the need for an integrative planning framework within which to plan and manage the assigned roles for such modes, these are often altered or prevented from materialising by the influences of market forces. Perhaps Metroplan has gone furthest in this direction by acknowledging the strategic importance of MTR stations as nodal points for higher urban density developments within the existing metropolitan area. However, the failure to plan systematically for easy pedestrian and public transport access to such nodes jeopardises their success. There is room here for more innovative thinking at the urban design level.

Moving toward a stakeholder approach

Building on many aspects of the above reviewed positions, the author has more recently advocated an approach that addresses in a more significant way issues of participation and consultation, environmental sustainability and the transport needs of industry, commerce and households (particularly the poor). This was in the context of preparing a generic sustainable urban transport strategy for Asia and the Pacific Region to help medium-sized cities better tackle the pending forces of motorisation (see Dimitriou 1997). Given that the work was incomplete at the time of writing this chapter it is not possible to provide a summary here of the recommendations of this study. What can be said, however, is that considerable emphasis has been given to enhancing stakeholder involvement in urban transport policy-making and planning based on the principles provided by the World Bank's Sourcebook on Participation (1994) (see Figure 12.3). Key among these principles is the premise that the more informed the affected parties (the 'stakeholders') are about (transport) problems and opportunities that confront them, and the more say they have in the development of responses to them, the more sustained the nature of support the decision-takers are likely to receive in their pursuit of a strategy and its related proposals.

A more general observation that can be made about a 'stakeholder' approach to transport planning is that there is a need to clearly distinguish between public participation and public consultation. The former implies a more active involvement by non-government organisations (NGOs) in the decision-making process with perhaps even an assignment of responsibilities. The latter denotes a more passive involvement, where government typically feels obliged to search out views from different sections of the community, but does not convey any obligation to involve such outside parties in the final formulation and/or execution of decision-making. Another important observation is that the cultural and political contexts, together with historical backgrounds, present very different starting points for participation and consultation exercises to take root. As a result, the most prudent approach is often to commence from a basis of what is currently practised and if possible, build on the process from thereon (World Bank 1994).

Relating the above to Hong Kong's situation, it is clear that there are aspects of these guidelines that would greatly assist the territory in the development of a more sustainable strategy. Equally, however, it should be noted that there are aspects of Hong Kong's experience that potentially offer a great deal to the operation of such a strategy. An attempt will be made below to summarise very briefly where Hong Kong might benefit from these stakeholder guidelines and where the territory's experience can contribute to the operation of these proposals.

In helping the territory achieve a more environmentally and economically sustainable land-use/transport strategy, it is important that it commits itself to the principles of sustainable development as a basis for the vision of the future. Hong Kong's current vision, which is committed to economic growth at almost any cost is increasingly seen as being potentially self-defeating in the medium to long run. A decision to employ sustainability as the basis for Hong Kong's development and the formulation of a new vision based on the sustainability concept, will inevitably lead to changed priorities. The role of transport, for example, is likely to be subjected to much greater environmental constraints, while efforts at wealth creation will be accompanied by much greater concern for the needs of the disadvantaged and other equity considerations. Such a change of strategy will prove beneficial, not only for the territory itself, but for the Pearl River Delta and the Asia Pacific Region, given the extent of Hong Kong's influence.

To assist in future decision-making about changes, the Hong Kong government needs to work through a number of development scenarios, starting with a projec-

Figure 12.3 Principal steps of a 'Stakeholder Approach'

Introduction of Stakeholder Participation and Consultation

- This requires a great deal of preparatory work
  - It is never too early or late to commence.
- Government permission must be obtained at the outset
  - It often takes a great deal of time and diplomacy to obtain.
- Workshops for selected stakeholders can provide a very effective way of bringing to light the transport problems
  - These can provide the basis for broadening their audience as the persons in power realise the value of earning public support.
- Face-to-face interaction of key parties
  - This is essential to the stakeholder if one is to convince key parties of the benefits of involving local people, as well as the business fraternity.

Identifying and Involving Stakeholders

- The potential 'winners' and 'losers' of change need to be identified
  - An important issue to take into account here is whether the 'winners' and 'losers' are also winners and losers in other sectors.
- Knowing the 'stakeholder enablers of change' is critical to innovative project development
  - Such parties typically include those that take the most risk (and often the benefits that can go with this risk). Some are knowledgeable about their status, others are not, and need to be informed of their potential influence.
- The aid of sponsors and communicators is often necessary to ensure that stakeholder groups which are not normally included in public sector decision-making have a voice
  - This may entail the use of exclusive participatory and consultative exercises and certain 'leveling techniques' so as to avoid the influential totally overshadowing the views of the less vocal and influential.

Participatory and Consultative Techniques

- Aims and objectives need to be set at different levels and presented
  - This needs to be done from both expert and stakeholder stances and a dialogue be allowed to ensue with a view to developing a composite position.
- The appropriate participatory and consultative techniques need to be selected
  - These are required to generate constructive collaboration among stakeholders who may not be used to working together. To foster this, one can again rely on workshops and/or field survey techniques.
- Injecting a common purpose and interrupting conventional and/or habitual thinking
  - This is essential both for experts and non-experts alike, to make room for new ideas and innovative approaches.
- Developing a 'problem tree' and a mirror image 'objective tree'
  - This can prove an ideal way of using the feedback gained from the stakeholder consultative and participatory techniques to construct a profile of a joint stakeholder planned response to the identified problems.
- Critical to taking the stakeholders through the steps of public-sector decision-making about choice(s) of action is the availability of information
  - The availability and accessibility of information enable informed trade-offs to be made more easily and permit the positions and perceptions of others to be better understood.

tion of current trends (and project impacts), and realistically appraise their short-term and long-term implications for the various goals of sustainable development. The spectrum of these scenarios needs to be much broader and bolder than those previously examined by earlier territorial development strategic studies, and include the possibilities of:

- more rapid motorisation than has hitherto been experienced, as traffic movement across the border is made easier;
- reduced institutional capabilities to direct, support and enforce government decisions;
- more widespread and acute pollution generated both by growth phenomena and declining enforcement capabilities; and
- a weakened frontier, whereby, (illegal) population movements across the border from the mainland to Hong Kong, particularly of the poor from rural regions in China, present major problems of housing and crime.

These impacts need to be set clearly against the anticipated benefits and costs of pursuing a sustainable vision of development, and its implementation mechanisms.

Once the scenario appraisal has taken place, further elaboration of the sustainable development vision needs to be undertaken with stakeholder representatives from both the private and public sectors, and not just the captains of industry and commerce (as in the past). In this regard, Hong Kong would greatly benefit from applying the kind of guidelines outlined in Figure 12.4, the outcomes of which should lead to more meaningful consultation, participation and feedback.

It will also contribute to the acceptance/modification and even rejection of the proposed vision and the adoption of a new agenda for action. A new vision and agenda would inevitably lead to the different prioritisation of strategy elements. It would require the adoption of a different planning framework for choosing among policy options; re-appraising plans, programmes and projects already in the pipeline; and recommending modification to those schemes already under construction or completed.

Acknowledging that different and changing development situations, and political contexts, often call for a change in the direction, mix and balance of strategy support mechanisms, the post-1997 era in Hong Kong will warrant a re-examination of the mechanisms employed to date, not withstanding their immense success in many instances. This is because the mechanisms may require 'tuning-in' with the new development strategy along the following lines:

- Normative Mechanisms - These are expected to substantially reflect the new direction being pursued, and are more likely than not to require substantive re-orientation and redefinition, even the introduction of new laws and regulations to reflect the principles and elements of a sustainable strategy.
- Enablement Mechanisms - These, which in the case of their support to the private sector, have been so innovative, will need to extend their coverage to address more than the concern for economic growth and the interests of the captains of industry and commerce. They will need to include environmental matters as well as the interests and concerns of communities and NGOs.

Figure 12.4 Coverage by Hong Kong's transport strategy and plans of internationally cited approches to sustainable transport development

Key:
Coverage by Hong Kong's transport strategy and plans
- Limited
- Varied
- Extensive

US Developments: ISTEA transport-related concerns

- Greater efficiency of public transport service delivery
- Aesthetics considerations of provision & impacts
- Equity implications of service & infrastructure provision
- Community impacts of service and infrastructure provision
- Amenity features provided by systems
- Safety aspects
- Flexibility & freedom of service provision & operation
- Health impacts of systems, especially regarding vehicle emissions
- Contribution to local environmental sustainability
- Contribution to global environmental sustainability

World Bank New Transport Policy Position

- Enhanced asset management of systems & services
- Greater efficiency of public transport service delivery
- Increased provision of access
- Increased affordability of services
- Adaptation of transport industry to globalisation of production
- Adaptation of transport industry to globalisation of trade
- Positive response to rising incomes and growing aspirations
- Provision of improved quality of service
- Reconciling increase in motor car ownership/use with maintenance of tolerable environment
- Reconciling increase in road freight traffic with maintenance of tolerable environment
- Pursuance of environmental & ecological sustainability, incl. air, noise & visual intrusion concerns
- Pursuance of social and distributional equity and poverty alleviation
- Change of role of government from supplier or regulator to 'facilitator'

UK Developments: Blueprint 5 Position

- Enhanced greenhouse effects & climatic change as a result of increased motor vehicle emissions
- Economic costs of air pollution
- Social costs of noise pollution.
- Congestion costs
- Road Damage
- External costs of road traffic accidents
- Aggregate externality charges
- Taxation on road transport. use and provision

Developmental Approach to Urban Transport Planning

- Integration of goals of urban development planning with those of transport planning
- Maintenance & creation of sustained economic growth
- Poverty alleviation, deprivation & basic needs fulfillment
- Adaptation of transport to urban spatial arrangements
- Adaptation of transport to cultural & social identity
- Pursuance of integrated urban infrastructure programming & planning
- Stimulate increased generation of local revenues
- Expansion of low cost (including non-motorised) infrastructure
- Decentralisation of planning and implementation responsibilities
- Sustained economic productivity
- Matching appropriate transport technology to settlement size & structure
- Enhancing public transport operations efficiency
- Sensitivity to community development & size
- Sensitivity to available resources
- Use of scenario building
- Use of innovative low cost forecasting techniques

- Control Mechanisms - These will need to be modified, especially in areas of traffic management, public transport operations, environmental protection, concerns about equity and affordability. However, institutional capacity constraints may come into play during the transition period if the anticipated exit of senior and middle management staff in government materialises, which may pose a new set of problems.
- Resourcing Mechanisms - This is yet another area in which Hong Kong has excelled, particularly in the provision of infrastructure finance, and where it has much to offer in making sustainable transport concepts operative. The use of funds generated by transport induced land appreciation as a means of cost recovery hold particular promise. Using the private sector (and possibly including communities) in staff resourcing also looks promising, while local developments in information technology could also greatly assist here - both in project implementation and in dissemination to the stakeholders about plans, projects and programmes.

The significant difference in approach between a sustainable development strategy and that currently pursued for the territory relates to the criteria used to appraise the transportation proposals and projects, and their impacts. The overriding importance of economic growth and the focus on short-term high rates of return, together with the failure to internalise many of the external costs, mean that the current proposals will not perform well against the new sustainability evaluation criteria. It is here that most research is required (see Copeland 1996) and where much international experience and research can be beneficial to Hong Kong. Once this very important aspect has been addressed, and the current strategy has been re-directed to a sustainability course by following the preceding steps, there is no doubt in the author's mind that the world leadership of Hong Kong's project management and fast-track implementation capabilities will take on new international significance, not only for the region but world-wide. Failure to achieve this turn-around will, at best, leave the territory with a reputation of merely being very efficient at implementing and administering a mis-directed and unsustainable vision for the future. Re-appraisal is therefore necessary.

Conclusions

Any attempt at concluding the preceding discussion and observations will be partial by virtue of the very nature of the chapter's coverage. Acknowledging this, two sets of conclusions are offered - admittedly both are speculative. The first set is largely self-explanatory and derived from a comparison of Hong Kong's current technocratic cum entrepreneurial approach to transport developments and their land-use, environmental and other impacts with the concerns and aims of the new positions/approaches in the quest for sustainable urban and transport development (see Figure 12.4). The second set of conclusions, relates back to the critical issues confronting the territory that were raised at the outset of this chapter.

From Figure 12.4 it may be noted that Hong Kong's coverage of the sustainability agenda is very partial with the exception of certain notable instances, such as concerns for:

- efficiency in public transport service delivery;
- enhanced asset management of systems and services;
- adaptation of transport industry to globalisation of production;
- adaptation of transport industry to globalisation of trade;
- positive response to rising incomes and growing aspirations;
- provision of improved quality of service;
- energy saving;
- enhancement of investment and cost recovery options;
- increased generation of local revenues; and
- institutional development and delivery.

On the other side of the coin, the territory's coverage of some of the more frequently cited concerns regarding transport operations efficiency (particularly by the World Bank) potentially offers some very useful contributions to the developers of more sustainable approaches to urban transport policy-making and planning.

Regarding the critical issues of the territory's future, the following conclusions are offered:

- Hong Kong's current growth is economically and environmentally unsustainable, including the management and institutional capabilities to service its present vision. A more regional strategy needs to be pursued building on Hong Kong's past successes but re-orientated towards sustainable development ends.
- Hong Kong has not been be presented with a clear insight into what vision the prospective leaders of the Special Administrative Region (SAR) have for the future. Such a vision is urgently needed and should clearly differ from that pursued in the past and be based on a broader stakeholder basis.
- The investment confidence in Hong Kong will remain well after 1997 so long as the territory continues to function as a financial, trading and transportation centre for the region and a gateway to China for the world. This relies not only on infrastructure developments, and their efficient operation and management (including telecommunication infrastructure), but also on the efficient operation and management of the enablement, control and resourcing support mechanism; a broader stakeholder basis for decision-making; and an institutional and manpower expertise capable of servicing the cosmopolitan role enjoyed in the past.
- The severance of Hong Kong's courts of law from the British Privy Council after the transition will ultimately not lead to the insistence by multinationals that contracts signed in Hong Kong after 30 June 1997 be subject to arbitration in the West. This is argued on the basis that other countries in the region operate profitably without such a legal framework but, nevertheless, the territory could act as a regional leader in providing an internationally recognised legal framework that goes beyond its British heritage. This could/should be developed with the assistance of relevant internationally recognised bodies concerned with international trade, banking and intellectual property rights.

- If the border with the Special Administrative Region is removed or not policed effectively, the territory's development will automatically become unsustainable, whatever strategy is pursued at the territorial level. The pressures resulting from the likely population influx, and affecting land use, housing, freedom of movement and/or crime, will inhibit its sustained performance as an effective international centre.
- The mix of the current 'commitment' and 'hustle' strategies pursued by Hong Kong, that up until now provided it with a competitive edge, cannot be effectively sustained. As the preceding discussion and conclusions suggest, a much more sophisticated and complex set of support mechanisms and a new direction for development is needed. The experiences to date can, however, be built upon, but not repeated per se.
- The question of whether Hong Kong will succeed in becoming an international crossroads is dependent on much more than its entrepreneurial skills, infrastructure provision, and the colonial heritage of its laws. The territory requires all of these, plus adequate managerial and technical expertise (supported by a world-class education and research system) capable of dealing with the cultures and languages both regionally and globally on a sustainable basis.

The Chinese saying that all moments of crises offer the best opportunities comes to mind as a useful basis for the closing remarks for this chapter. With economically successful cities in the region, such as Singapore and Sydney, placing a very high value on environmental quality and sustainable development, any attempt by Hong Kong at this point of its political, economic and environmental transition, to turn its back on the sustainability debate (as Bangkok has done) and to continue on a 'business as usual' basis will be at its own peril. Such action will cement the territory into an all too soon outdated, heavily infrastructured and more polluted city and will represent a mammoth 'missed opportunity' for reappraising Hong Kong's transport and urban development vision for the future.

Notes:

1 The green light given by the Chinese Government for this US$1.2 billion project in January 1996 was subject to a certain amount of delay. Clearance for the project to go ahead was seen at the time to mark a turning point in Hong Kong's relations with Beijing.

2 This same newspaper source points out that 'through hard times and lean years, China has never failed to provide Hong Kong with much needed water and food since 1978 which is the underpin to the territory's metamorphosis into a financial centre'.

3 More recent evidence suggests that the predicted real gross domestic product of Hong Kong would rise by 5.5 per cent in 1997 compared with 4.7 per cent in 1996. The basis of this additional growth originates from capital inflows and liquidity and not capital flight (official statement made by the Financial Secre-

tary on 27/2/1997).

4 Salmon Brothers put Hong Kong on equal footing with New York, London and Tokyo as one of its regional business centres. Over 60 per cent of foreign investment in China is challenged through Hong Kong, and foreign investors produce two thirds of China's exports (Economist 1996).

5 Hong Kong manufacturers employ more workers on the mainland than in Hong Kong itself. Whereas the territory's manufacturing sector accounted for 24 per cent of GDP in the 1970s, today it has transformed itself into a service economy whose largest client is China, and manufacturing now accounts for about 9 per cent of Hong Kong's GDP.

6 Concerns for this issue in more general terms have been well articulated by Edwin Mills of Northwestern University at a World Bank conference in 1991 (see Mills 1991).

7 'Commitment strategies' are based on large-scale investments in industries with relatively stable cash flows such as transportation, infrastructure, property and utilities. 'Hustle strategies' rely on flexibility, rapid response capabilities, competitive pricing in light industry (such as toy and garment manufacturing) and trading (Enright et al 1997).

8 To a degree, Hong Kong has already taken on this role. It has become the *de facto* capital of the Overseas Chinese business community, which has not only played such an influential role in Southeast Asia but also in East Asia, and parts of the developed world.

9 There are signs that this process is underway with a reported 32 per cent increase in imports of labour-saving office equipment to the territory in 1994 alone (Silverman 1995).

10 These are 1991 estimates cited in *Beyond 2000: A Source Book for Major Projects* published by the Major Projects Association and Templeton College, Oxford, May 1994. Costs have since escalated.

11 An enablement strategy may be defined as a strategy that seeks to capture market forces and other potential resources to the benefit of the vision pursued (Dimitriou 1997).

12 Christine Loh advocated a number of policies and visions that differed considerably from those pursued and implemented by the Hong Kong government, on matters concerning issues such as sustainable development, traffic congestion, urban reclamation, environmental protection, public participation, social welfare, migrant problems and relations with China. These views are perhaps best summarised in *Hong Kong: An Alternative Policy Address* (Loh 1994).

13 This part of the text is drawn from research conducted by the author as a consultant to the World Bank (with the assistance of John Ernst, now of IIEC, Washington D.C.) for the UNDP and the Economic and Social Commission for the Asia Pacific (ESCAP) (see Dimitriou 1997).

14 To introduce the uninitiated reader to the subject and help investigate the literature in this field, a short bibliography of key international publications is incorporated within the references at the end of this chapter.

15 A stakeholder may be defined as any party that is affected by a particular issue

and thus has an interest in its development and change. There are 'enabler stakeholders' i.e., those who are most influential in the decision-making that affects change. These typically include those who fund/finance change and take the risks (and the benefits that go with this) as well as the less influential (even voiceless) stakeholders who possess little if any influence to affect change, but who nevertheless, are affected by it (Dimitriou 1997).

16 Christine Loh was fortunate in being assisted by a young and energetic political advisor cum assistant - Evans Ward - who with Loh's encouragement marshalled intellectual support and ideas from a wide spectrum of sources to add credence to the press releases and speeches made in the Legislative Council.

17 The fields covered by this publication included the Law of Nuisance; the principle of making polluters pay; environmental legislation, litigation and science; corporate liability; ideas from environmental economics, the applicability of international environmental law and a discussion about the right to pollute as a tradable commodity.

18 Government apologists for Hong Kong's environmental plight claimed that the survey's use of population density as one of the main factors in measuring environmental quality of life weighed heavily against the territory.

19 It should be said that this in itself is not a claim to be proud of given the UK government's recent treatment of its environmental agenda.

20 It should not be deduced from this, however, that the sequence of events that followed regarding the launching of the new study had any possible correlation.

21 Foremost among these instruments, it was argued, would be market-based measures for environmental policy initiatives.

22 The breakdown of this figure is: US$120.8 billion for highways, US$31.5 billion for transit and US$3 billion for other purposes (US DOT 1991).

23 This is the second attempt by the Hong Kong government to introduce road pricing to the territory (see Transpotech 1985). In 1985 a briefly introduced ERP scheme failed politically, mainly because of the use of vehicle-mounted transponders that enabled the authorities to track citizens and because of the fact that charges were only applied to private cars (Hau 1995).

24 While updates have since been undertaken of CTS II findings, a full review, CTSIII, commenced in 1997.

25 The consultants, contractors, bankers and investors involved in the proposed rail developments in the territory, especially the link to the new airport, are many of the same parties involved in infrastructure projects elsewhere in the region. This has encouraged a great deal of technology-transfer and has had an impact on the decision-making of other governments in the region, not only in the field of construction but in aspects of infrastructure funding and cost recovery.

References

Barma, H. (1995) *Motion Debate on Traffic Congestion*, Record of Secretary for Transport's Speech in the Legislative Council, Hong Kong, February.

Barron, W. (1996) 'Energy for Sustainable Development: Substitution, Efficiency and Technological Change', *Invited Presentation to the Hong Kong General Chamber of Commerce*, Sustainable Development Seminar, October.

Cernea, M.M. (1994) 'The Sociologist's Approach to Sustainable Development' in *Making Development Sustainable: From Concept to Action* by Serageldin, I. and A. Steers (ed.), Environmentally Sustainable Occasional Paper Series No. 2, The World Bank, Washington D.C.

Chan, C. and P. Hills (eds.) (1993) *Limited Gains: Grassroots Mobilization and the Environment in Hong Kong*, CUPEM, The University of Hong Kong, Hong Kong.

Chan, M. (1994) 'The Poverty Phenomenon in Hong Kong', Unpublished Student Paper, CUPEM, University of Hong Kong.

Copeland, A.M. (1996) 'Defining Sustainability - A Conceptual Framework for Hong Kong', *Sustainability Indicators Project: Working Paper No. 1*, CUPEM, The University of Hong Kong. *Ecological Economics*, Vol. 2.

Daly, H.E. (1990) 'Toward some operational principles of sustainable development',

Dimitriou, H.T. (1992) *Urban Transport Planning: A Developmental Approach*, London: Routledge.

———— (ed.) (1994a) *Moving Away from the Motor Vehicle*, CUPEM, The University of Hong Kong, Hong Kong.

———— (1994b) 'Bending Under the Strain: Operating and Implementing an Integrated Transport System in Hong Kong', *The Wheel Extended: A Toyota Quarterly Review*, Vol. No. 88, July.

———— (1995) *A Developmental Approach to Urban Transport Planning: An Indonesian Illustration*, Aldershot: Avebury Press

———— (1997) *Toward a Sustainable Urban Transport Strategy for Medium-sized Cities in Asia and the Pacific Region*, Unpublished Draft Consultant Report to World Bank for UNDP and ESCAP, Bangkok.

Economist (1996) 'Can Hong Kong Stay Free?', *Economist*, 22 June, London.

Enright, M., E. Scott, and D. Dodwell (1997) *The Hong Kong Advantage* Hong Kong: Oxford University Press.

EC (1995) *Trans-European Networks*, Group of Personal Representatives of Heads of State or Government, Office for the Official Publications of European Communities, Luxembourg.

Ernst, J. (1996) Extract from unpublished notes prepared for *Toward a Sustainable Urban Transport Strategy for Medium-sized Cities in the Asia Pacific Region* by H. T. Dimitriou, Consultant Report to World Bank for UNDP and ESCAP, Bangkok.

Financial Times (1994) 'Hong Kong: A Test of Nerves', Financial Times Survey, *Financial Times*, London.

Fitzpatrick, E. (1995) 'Mandarins in Disarray: Constant shuffling threatens morale and efficiency', *Window Magazine*, Hong Kong, 22 September.

Ford Foundation (1993) *Urban Research in the Developing World*, Reports from the Final Meetings, Social Research Centre, American University, Cairo.

Gowdy, J.M. (1994) 'Natural capital and the growth economy', *Sustainable Development*, Vol. 2, Part 1, Spring.

Gunn, L. (1966) *The Role of Government in the Allocation of Resources to Science*, OECD, Paris.

Hau, T.D. (1988) 'Road Pricing: A Viable Alternative?', Unpublished Discussion Paper for *Meeting Point Hong Kong*, Department of Economics and CUSUP, University of Hong Kong, November.

———— (1992) 'Congestion Charging Mechanisms for Roads: An Evaluation of Current Practice, *World Bank Policy Research Working Paper* Series WPS No. 1071, Washington D.C.

———— (1995) 'Public Policy on Private Cars in Hong Kong: The Case for Demand Management', *The Wheel Extended: A Toyota Quarterly Review*, No. 91, May.

Heilbronn, G.N. (1993) 'Environmental Law in Hong Kong: Problems and Prospects', *Papers presented at a series of three Twilight Seminars*, Faculty of Law, University of Hong Kong.

Hilderbrand, M. and L.Y. Pouliquen, (1993) 'Foreword' in *Rapid Urban Environmental Assessment: Lessons from Cities in the Developing World* by J. Leitmann, UNDP/UNCHS/World Bank Urban Management Programme, World Bank, Washington D.C.

Hills, P. (1995) 'Is Hong Kong's Development Sustainable?', Paper Presented to *Conference on Planning Hong Kong for the 21st Century*, Organised by the Hong Kong Institute of Planners, CUPEM, The University of Hong Kong, and Department of Architecture, Chinese University of Hong Kong, April.

Hodgkiss J. (1994), 'Hong Kong's Sewage: A Worsening Scenario of Pollution', *Interflow*, The University of Hong Kong Magazine, July.

Information Services Department (1996) 'The World's Busiest Container Port', Hong Kong: Government Printer.

IUCN/UNEP/WWF (1991) *Caring for the Earth: A Strategy for Sustainable Living*, Gland.

Lau, J. (1996) 'Evaluating the Performance of Public Transport Operations in Hong Kong from the Point of View of the Disadvantaged', Unpublished M. Phil. Thesis, CUPEM, University of Hong Kong.

Leipert, C. (1989) 'Social Costs of the Economic Process and National Accounts: The Example of Defense Expenditures', *The Journal of Interdisciplinary Economics*, Vol. 3.

Lele, S.M. (1991) 'Sustainable Development: A Critical Review', *World Development*, Vol. 19(6).

Loh, C. (1994) Hong Kong: An Alternative Policy Address, October 3rd., Hong Kong.

Maddison, D. et al (1996) *Blueprint 5: The True Costs of Road Transport*, London: Earthscan Publications.

Major Projects Association (1994) *Beyond 2000: A Source Book for Major Projects*, Major Projects Association and Templeton College, Oxford.

Mills, E.S. (1992) 'Urban Efficiency, Productivity, and Economic Development', Proceedings of the World Bank Annual Conference on Development Economics in 1991, The World Bank, Washington D.C.

Ng, G. (1995) 'Conservation and Environment', Paper Presented to *Conference on Planning Hong Kong for the 21st Century*, organised by the Hong Kong Institute of Planners, CUPEM University of Hong Kong, and Department of Architecture, Chinese University of Hong Kong, Hong Kong, April.

Pearce, D., E. Barbier and W. Markandya (1988) *Sustainable Development and Cost Benefit Analysis*, LEEC Paper 88-03, London Environmental Economics Centre, University College, London.

——————— (1989) *Blueprint for a Green Economy: A Report for the UK Department of the Environment*, London: Earthscan Publications.

——————— (1990) *Sustainable Development: Economics and Environment in the Third World*, London: Earthscan Publications.

Pearce, D. (ed.) (1991) *Blueprint 2: Greening of the World Economy*, London: Earthscan Publications.

——————— (1995) *Blueprint 4: Capturing Global Environmental Value*, London: Earthscan Publications.

Pezzey, J. (1992) *Sustainable Development Concepts: An Economic Analysis*, The World Bank, Washington D.C.

Planning, Environment and Lands Branch (1993) *The Hong Kong Environment: A Green Challenge for the Community*, Government Secretariat, Hong Kong, Hong Kong: Government Printer.

Planning Department (1993) *Territorial Strategy Review Development Options*. Public Consultation Digest. Hong Kong: Government Printer.

——————— (1996a) 'Brief for the Study on Sustainable Development for the 21st Century'. Planning, Environment and Lands Branch September (unpublished).

——————— (1996b) *The Final Technical Report on the Territorial Development Strategy Review*. Part 3 Section A 'A Recommended Long-term Strategy'. Hong Kong: Government Printer.

Rusco F.W. and W.D. Walls (1995) *Clearing the Air: Vehicular Emissions Policy for Hong Kong*, The Hong Kong Centre for Economic Research, Hong Kong: The Chinese University Press.

Silverman, G. (1995) 'The Price of Success', Cover Story on Hong Kong, *Far Eastern Economic Review*, July 6th, Hong Kong.

Solow, R.M. (1986) 'On the Intergenerational Allocation of Natural Resources', *Scandinavian Journal of Economics*, Vol. 8. No. 1.

Spaeth, A. (1996) 'One Year and Counting', *Time International*, 1 July, New York.

Tengstrom, E., E. Gajewska and M. Thynell (1995) *Sustainable Mobility in Europe and the Role of the Automobile: A Critical Enquiry*, Gothenburg: University of Gothenburg.

Thijs de La Court (1990) *Beyond Brundtland: Green Development in the 1990s*, New York: New Horizon Press.

Tomazinis, A. (1993) 'Planning for Sustainable City Transport Systems: Key Features of INSTEA'91'. Unpublished paper presented at the CPD Course on *Planning for Sustainable City Transport Systems: the International Experience*, conducted by CUPEM in association with the Hong Kong branch of CIT and HKIP at The University of Hong Kong, February.

Transport Branch (1994) *Traffic Flow or Gridlock: The Choices We Face*, Hong Kong Government, Hong Kong: Government Printer.

Transpotech (1985) *Electronic Road Pricing Pilot Scheme:* The Main Report, Report to the Hong Kong Government, Hong Kong.

Turner, R.K. (1991) 'Environment, Economics and Ethics' in *Blueprint 2: Greening the World Economy* edited by Pearce, D., London: Earthscan Publications.

UNCHS (1990) *People, Environment and Development*, Habitat, Nairobi.

US DOT (1991) *Intermodal Surface Transportation Efficiency Act of 1991: Public Law 102-240*, Washington D.C., December.

Watson, R.T., M.C. Zinyowerea, R.H. Moss and Dokken (ed.) (1996) *Climate Change 1995: Impacts, Adaptations and Mitigation of Climate Change: Scientific-Technical Analysis*, Published for the Intergovernmental Panel on Climate Change, Cambridge: Cambridge University Press.

World Bank (1993) *The East Asian Miracle: Economic Growth and Public Policy*, World Bank Policy Research Report, Published for the World Bank, New York: Oxford University Press.

——————— (1994) *The World Bank Sourcebook on Participation*, Washington D.C.

——————— (1996) *Sustainable Transport: Priorities for Policy Reform*, Environmentally Sustainable Development, Transportation, Water and Urban Development Department, Washington D.C.

WCED - World Commission on Environment and Development (1987) *Our Common Future* (The Brundtland Report), Oxford: Oxford University Press.

Yanaguaya, W. (1993) 'The Use of UK Transport Models in Developing Countries', *Traffic Engineering and Control*, Vol. 34. No. 10.

Conclusions

Harry T. Dimitriou and Alison H.S. Cook

While the opening chapters of this book highlight the dramatic events of Hong Kong's evolution since 1841 that led it to become the economic powerhouse it is today, the closing chapters suggest that the time has come in the territory's history to carefully reconsider its future. The basic options it confronts are between accepting further unsustainable growth or employing a development strategy that strives for a more balanced and sustainable development. By choosing the latter, Hong Kong will again take the lead in the region. It will do so after many decades of pursuing an economic growth model of development based on a 'compact city' form and well-endowed provision of public transport, aided by an enablement strategy for entrepreneurial activity offered by government. The territory's planners and government can move ahead toward a more sustainable vision that builds on this past success by genuinely incorporating the concerns for environmental sustainability and the need to vastly broaden stakeholder involvement in planning.

'Business as Usual' Versus Sustainable Development: The nature of the impacts and issues associated with a 'business as usual' growth-based model indicates that the territory does not, in fact, have a real choice if it is to survive in the medium to long run. This is because of the incredible demands this growth has placed on the planners, administrators and inhabitants of Hong Kong alike. Such demands indicate that while the administration of Hong Kong with its technocrats and its captains of industry have excelled in certain areas in the past, this approach can no longer cope with the environmental and traffic problems it spawns. Quite apart from this, other more specific observations made by individual chapters warrant a reiteration. These are summarised below.

Integration of Land Use and Transport Planning: Chapter 1 observes that an examination of the brief history of public transport provision in the territory suggests that it is only in recent times that serious steps have been taken to integrate land use planning and transport. The chapter concludes that while a market-led response to public transport provision has serviced Hong Kong well up until now, government has 'been less successful in the related, and politically more difficult area of sustaining an effective policy to deal with increasing traffic congestion ....'

Development Opportunities Offered by Shenzhen: The scenario of much closer land-use integration of Hong Kong and China, notably between the territory and the Shenzhen special economic zone, is presented by Chapter 2 as a distinct possibility. The materialisation of this scenario into reality could lead to the reduction of development pressures in the territory by the release of land across the border for Hong Kong investment, and more journey-to-work and freight movement taking place to these areas, away from already heavily congested areas.

Limits to Technocratic Planning: The need to place the appropriate degree of faith and skepticism on transport modelling in strategic land-use/transport planning exercises for the territory, and an acknowledgment that the evaluation components of such exercises are 'quantitative and qualitative' and entail 'high levels of professional judgement' is the combined concluding message of Chapter 3. The open acknowledgement of the limits of the technocratic planning process that has characterised land use/transport planning in the territory up until now points to a future where environmental planning and a broader stakeholder involvement in the planning process will be more pivotal to decision-making.

Improving the Quality of Life: Chapter 4 observes that relative to most other cities in the region and elsewhere, Hong Kong's land use/transport interface 'would appear to be in harmony' because of the impressive performance of its public transport system. Notwithstanding this (and the innumerable planning studies that have been undertaken), the chapter concludes that the territory could substantially further improve its position in terms of the quality of life 'if new investment were to be regarded more proactively..... rather than when existing systems reach saturation'.

Role of Railways: An observation of recent developments in Hong Kong that flies in the face of current world trends in the privatisation of railways is 'that the government is the principal beneficiary of urban rail development'. This is pointed out by Chapter 5, and is so, not only because it in effect owns two railway corporations that are successfully and profitably operated but also because these same corporations own a great deal of valuable property. What this observation conveys to the outside world is that even railways operating in the highest urban densities and most affluent contexts need to rely on additional revenues outside the railway transport sector for their full commercial viability. By retaining the railway corporations within the sphere of government influence, the Hong Kong Administration, maintains control over a very important potential means of wealth generation which through the government ownership of land is increasingly seen as a key to planning the territory's future.

Intervention, Integration and Participation: The paradox of public sector involvement in directing land use and transport infrastructure developments in Hong Kong highlighted in Chapter 6 is that the territory is considered one of the least controlled economies in the world. This chapter calls for 'an integrated approach (to urban public transport planning), encompassing the contributions of land use, transport and environmental planners'. It also advocates that 'short term personal interests should be compromised in favour of long term community interests, particularly in areas of environmental concern' which can only happen with a more fully representative government and increasing public participation.

Conflicting Objectives: Noting the land-use/transport model used to assist in the Territorial Development Strategy for Hong Kong is far from ideal, Chapter 7 concludes that 'the problems associated with the various land-use/transport model forecasts ....did not lie essentially with the territorial transport model itself, but in how it has been used, and the validity of the future that it had presumed capable of forecasting'. This has implications that go far beyond the boundaries of the territory. The chapter concludes that any land use strategy must address the 'legitimate objectives of the transport sector as set out by government' and claims that such objectives in Hong Kong have in reality contradicted what represents the main urban development issues cannot be blamed on the modelling exercise.

Prisoner of Past Planning Visions: Among the most important conclusions drawn from Chapter 8 is that 'although Metroplan was publicised as having a strong vision of its own, it relied (too) heavily on preceding studies for this ...', especially for its transport developments. While this has been seen as a 'lost opportunity' to push aside past planning errors, the attempt at utilising urban rail to mould future metropolitan development and the positive steps to open-up the planning process are presented by the chapter as major milestones in Hong Kong's planning history. However, an overriding disappointment is that Metroplan's implementation has not been supported or directed by an effective implementation agency that could cut across departmental interests and jealousies.

Professional Change of Focus: Traffic engineers and transport planners in Hong Kong have been obliged to change their focus - from one in the 1960s and 1970s that principally looked to satisfying the mobility demands of the territory at as low a cost as possible, to another in the 1980s and 1990s of addressing the transport needs generated by increasing affluence and the rapid growth of trade. The changed focus now obliges these professionals to work more closely with urban planners and environmental planners, and the ordinances and guidelines they adhere to. With this in mind, Chapter 9 concludes there is a delicate balance between 'laissez-faire' and 'over-regulation', and that arriving at a workable balance for urban transport, in concert with land-use planning and environmental protection, 'will be the principal challenge for the future'.

Raising Emissions Standards: The introduction of unleaded petrol in the territory and the gradual replacement of older, smaller diesel-fuel vehicles by new vehicles with clean fuel systems will take time to take effect. As a result, Chapter 10 recommends that higher emissions standards for the remaining diesel vehicles should be introduced regardless of 'whether some classes of lighter vehicle switch to a clean petrol system or not'. On the premise that 'prevention is more effective and cost-effective than a cure', this chapter argues for a far more prominent and serious role for Environmental Impact Assessment procedures as part of transport project feasibility and design exercises. Failure to do this, the chapter concludes, is to 'continue with business as usual' behind a smokescreen that suggests (rather than confirms) that something new is being done.

Institutional Mechanisms and Commitment: Notwithstanding the innovative institutional mechanisms introduced in the territory to help finance/fund new infrastructure projects, particularly railways, there remain major obstacles within the Hong Kong institutional planning framework for effective integrated land-use/transport planning. According to Chapter 11, among the most important of these are the questions of political and financial commitment, and the concerns about organisational accountability in the implementation of plans and programmes. The chapter concludes that unless inter-sectoral matters can be dealt with effectively, and unless political commitments take place across departmental boundaries, the institutional delivery of whatever is planned will always be in question, especially during a stage of political transition of the kind currently being experienced.

Acceptance of Limits to Growth: Despite the impressive successes of Hong Kong, the need to accept limits to growth and to acknowledge the benefits of a more balanced sustainable development are the key overriding issues which emerge from the analysis presented in Chapter 12. These are problematic issues that are closely related to the territory's vulnerability to international changes in economic fortunes - both regionally and worldwide. They also relate to the uncertainties of its transition. For up until now, the very survivability of Hong Kong has been closely tied to its ability to efficiently service the growth of the region, in particular, the Chinese mainland. Any failure to do this will lead to a loss of its competitive edge and economic stability, and ultimately it is believed, will contribute to the decline of the Hong Kong. Yet, to continue on the present planned trajectory will 'cement the territory into an all too soon outdated, heavily infrastructured and (inevitably) more polluted city', which makes the search for an alternative sustainable vision imperative.

Finally, much more attention needs to be paid to the nexus between employment and residential development, with the enhancement of environmental quality and railway development going hand-in-hand with this concern. The employment/residential relationship will help underpin the future sustainable land-use/transport system of the territory. This is not a repackaged argument for self-containment but a case for determining the location of feasible (and most likely) employment patterns based on public transport and freight transport accessibility functions consistent with environmental considerations and constraints. The acknowledged, belated, strategic role of rail transport as a basis for building on the existing pattern of the hub function of the central business district in Hong Kong (utilising rail junctions as higher intensity nodes) warrants much more investment in both public transport and freight transport than hitherto has been acceptable. Complementing this, land-use planning should seek to minimise new residential areas away from employment and new railway nodes. Meanwhile, enablement mechanisms that facilitate the use of appreciated land values (resulting from these transport infrastructure investments) could help finance a sustainable transport system and urban development strategy. This would act as an antidote to the mounting motorisation forces in Hong Kong and help the territory once again become the envy of other world-class cities as a result of the enhanced quality of life it could offer.

Subject Index

For Product Safety Concerns and Information please contact our EU representative GPSR@taylorandfrancis.com Taylor & Francis Verlag GmbH, Kaufingerstraße 24, 80331 München, Germany

Batch number: 10725921

Printed by Printforce, the Netherlands